Astrophysics and Space Science Proceedings
Volume 37

For further volumes:
http://www.springer.com/series/7395

Andy Adamson • John Davies • Ian Robson
Editors

Thirty Years of Astronomical Discovery with UKIRT

The Scientific Achievement of the United Kingdom InfraRed Telescope

Editors
Andy Adamson
Gemini Observatory
Southern Operations Center
La Serena, Chile

John Davies
Royal Observatory
Astronomy Technology Centre
Edinburgh, UK

Ian Robson
UK ATC
Royal Observatory Edinburgh
Edinburgh, UK

ISSN 1570-6591 ISSN 1570-6605 (electronic)
ISBN 978-94-007-7431-5 ISBN 978-94-007-7432-2 (eBook)
DOI 10.1007/978-94-007-7432-2
Springer Dordrecht Heidelberg New York London

Library of Congress Control Number: 2013955020

© Springer Science+Business Media Dordrecht 2013
This work is subject to copyright. All rights are reserved by the Publisher, whether the whole or part of the material is concerned, specifically the rights of translation, reprinting, reuse of illustrations, recitation, broadcasting, reproduction on microfilms or in any other physical way, and transmission or information storage and retrieval, electronic adaptation, computer software, or by similar or dissimilar methodology now known or hereafter developed. Exempted from this legal reservation are brief excerpts in connection with reviews or scholarly analysis or material supplied specifically for the purpose of being entered and executed on a computer system, for exclusive use by the purchaser of the work. Duplication of this publication or parts thereof is permitted only under the provisions of the Copyright Law of the Publisher's location, in its current version, and permission for use must always be obtained from Springer. Permissions for use may be obtained through RightsLink at the Copyright Clearance Center. Violations are liable to prosecution under the respective Copyright Law.
The use of general descriptive names, registered names, trademarks, service marks, etc. in this publication does not imply, even in the absence of a specific statement, that such names are exempt from the relevant protective laws and regulations and therefore free for general use.
While the advice and information in this book are believed to be true and accurate at the date of publication, neither the authors nor the editors nor the publisher can accept any legal responsibility for any errors or omissions that may be made. The publisher makes no warranty, express or implied, with respect to the material contained herein.

Printed on acid-free paper

Springer is part of Springer Science+Business Media (www.springer.com)

UKIRT at 30: Foreword

Welcome to this volume of highlights from the celebration of UKIRT's 30th anniversary, held at the Royal Observatory Edinburgh in September 2009. This volume is filled with wonderful accounts of the many technical innovations and newly-blazed trails over the past three decades, together with comprehensive reports of the excellent past, present and potential future science. UKIRT has been a truly remarkable success story.

Despite these many successes, UKIRT's operational funding has been under almost continual threat. At the time of this meeting, the observatory's strategy was to complete the UKIDSS survey programme in 2012 and then to move into planet hunting with a new instrument, the UKIRT Planet Finder. The proposal for this instrument, ably led by Hugh Jones and supported by the UKATC, was, at the time of this meeting, under assessment by STFC. Regrettably, but entirely predictably given the difficult economic constraints facing the organisation, the proposal was not approved. The search for habitable-zone Earth-mass planets around nearby M dwarfs will instead be carried out by other telescopes.

This decision, in combination with other financial pressures, led eventually but inexorably to the biggest single change in UKIRT's long history: in late 2010, UKIRT adopted a "minimalist" operating mode. In this new mode, the observatory's mission is to complete the UKIDSS programme as expeditiously as possible, and UK time is now committed almost entirely to this one programme. There is no more open time, no TAG, and no visiting observers: the telescope is operated remotely from the JAC building in Hilo. It is an enormous credit to the JAC's technical staff that this watershed change was implemented in just 8 months, from the date of approval by the UKIRT Board to the first night of remote operations.

UKIRT has thus evolved from a general-purpose infrared observatory, offering the astronomers of the world a suite of instruments with a range of unique capabilities to one with a single operational instrument and a very narrow, dedicated focus in infrared survey astronomy. This transformation has been enormously successful and UKIRT continues to deliver world-class science data to its users, continuing its long tradition of technical innovation and imaginative solutions.

UKIRT has been a major success story for British astronomy over its 30-year lifetime. On reflection, there are many reasons for this: the excellent infrared site on Mauna Kea; the superb mirror, allowing sub-arcsecond imaging which was not remotely imaginable when the telescope was built; an aggressive development programme leading to a succession of ambitious, world-leading instruments with unique capabilities; and innovative operations, including a suite of software tools for observation execution and data reduction, which make UKIRT second to none for its observers. But the most important reason for UKIRT's success over the years has been the technical excellence and singular dedication of its staff, some of whom are represented in this volume and some of whom are shown in the accompanying photograph. Without them, none of the science described in this volume would have been possible. I am honoured to be the Director of this remarkable observatory.

Hilo, HI, USA
January 2012

Professor Gary Davis
Director Joint Astronomy Centre

Some of the staff of the Joint Astronomy Centre (2009)

Contents

Part I The Development of UKIRT

1. **The UKIRT Success Story** 3
   Richard Ellis

2. **UKIRT – The Project and the Early Years** 11
   Terry Lee

3. **The IRCAM Revolution** 29
   Ian S. McLean

4. **Continuum Submillimetre Astronomy from UKIRT** 39
   Ian Robson

5. **CGS4: A Breakthrough Instrument** 53
   Phil Puxley

6. **The UKIRT Upgrades Programme** 63
   Tim Hawarden

7. **Operational Innovations** 75
   Andrew J. Adamson

8. **The UKIRT Wide-Field Camera** 87
   Mark Casali

9. **Polarimetry at UKIRT** 97
   J.H. Hough

10. **UKIRT in the Mid-Infrared** 113
    P.F. Roche

Part II UKIRT Science

11 **Thirty Years of Star Formation at UKIRT** 129
    Chris Davis

12 **Comets and Asteroids from UKIRT** 143
    John K. Davies

13 **Spectroscopic Tomography of a Wind-Collision Region** 151
    Peredur Williams, Watson Varricatt, and Andy Adamson

14 **Highlights of Infrared Spectroscopy of the Interstellar Medium at UKIRT** 159
    Thomas R. Geballe

15 **UKIRT and the Brown Dwarfs: From Speculation to Classification** . 173
    Sandy K. Leggett

16 **White Dwarfs in UKIDSS** 185
    P.R. Steele, M.R. Burleigh, J. Farihi, B. Gänsicke,
    R.F. Jameson, P.D. Dobbie, and M.A. Barstow

17 **Discovery of Variables in WFCAM and VISTA Data** 193
    Nicholas Cross, Nigel Hambly, Ross Collins,
    Eckhard Sutorius, Mike Read, and Rob Blake

18 **Near Infrared Extinction at the Galactic Centre** 201
    Andrew J. Gosling, Reba M. Bandyopadhyay,
    and Katherine M. Blundell

19 **Observations of PAHs and Nanodiamonds with UKIRT** 207
    Peter J. Sarre

20 **Nearby Galaxies with UKIRT: Uncovering Star Formation, Structure and Stellar Masses** 213
    Phil James

21 **WFCAM Surveys of Local Group Galaxies** 229
    Mike J. Irwin

22 **HiZELS: The High Redshift Emission Line Survey with UKIRT** 235
    Philip Best, Ian Smail, David Sobral, Jim Geach, Tim Garn,
    Rob Ivison, Jaron Kurk, Gavin Dalton, Michele Cirasuolo,
    and Mark Casali

23 **The HiZELS/UKIRT Large Survey for Bright Lyα Emitters at z ∼ 9** 251
    David Sobral, Philip Best, Jim Geach, Ian Smail, Jaron Kurk,
    Michele Cirasuolo, Mark Casali, Rob Ivison, Kristen Coppin,
    and Gavin Dalton

24 Observations of Gamma-Ray Bursts at UKIRT 259
Nial Tanvir

Part III   UKIDSS and the Future

**25 The UKIRT Infrared Deep Sky Survey (UKIDSS):
Origins and Highlights** 271
Andy Lawrence

**26 A Billion Stars: The Near-IR View of the Galaxy
with the UKIDSS Galactic Plane Survey** 279
P.W. Lucas, D. Samuel, A. Adamson, R. Bandyopadhyay,
C. Davis, J. Drew, D. Froebrich, M. Gallaway,
A. Gosling, R. de Grijs, M.G. Hoare, A. Longmore,
T. Maccarone, V. McBride, A. Schroeder, M. Smith, J. Stead,
and M.A. Thompson

27 The UKIDSS Galactic Clusters Survey 291
Sarah Casewell and Nigel Hambly

28 The UKIDSS Deep eXtra-Galactic Survey 299
A.M. Swinbank

29 UKIDSS UDS Progress and Science Highlights 309
W.G. Hartley, O. Almaini, S. Foucaud,
and UKIDSS UDS Working Group

**30 Exploring Massive Galaxy Evolution with the UKIDSS
Ultra-Deep Survey** 323
R.J. McLure, M. Cirasuolo, J.S. Dunlop, O. Almaini,
and S. Foucaud

31 The UKIRT Planet Finder 329
Hugh R.A. Jones, John Barnes, Ian Bryson, Andy Adamson,
David Henry, David Montgomery, Derek Ives, Ian Egan,
David Lunney, Phil Rees, John Rayner, Larry Ramsey,
Bill Vacca, Chris Tinney, and Mike Liu

Part I
The Development of UKIRT

Chapter 1
The UKIRT Success Story

Richard Ellis

Abstract This is a personal overview of the great success of the UKIRT facility; a tribute to those who designed, built and operated it and helped put UK astronomy at the forefront of world infrared observations. I will illustrate this success with a small selection of science highlights.

Introduction

It's a pleasure to come to Edinburgh to celebrate 30 years of UKIRT operations! All of us feel a close and personal affinity to this remarkable telescope. As with all observatories, it's a combination of excellent support staff and innovative instrumentation that provides the basis for scientific success and it's a privilege to summarize, albeit briefly, the key aspects and discoveries that continue to make UKIRT a world-beating facility.

The UKIRT story really is a remarkable one. For a modest financial investment, the Observatory has played a pivotal role in infrared astronomy for 30 years. It has an unrivalled reputation for technical innovation, cost-effective operations and reliability. It hosts a dedicated staff who willingly take on new responsibilities as required, and work endlessly to achieve the best, often in a hostile environment. And above all, UKIRT has delivered many scientific 'firsts' and built the astronomical careers of many.

In this written version of my Edinburgh talk, I have inevitably had to skip over some of the more amusing aspects of UKIRT's history, but I hope this brief summary conveys the spirit of my talk and the admiration I hold for this 4-m telescope.

R. Ellis (✉)
Department of Astronomy, Caltech, Pasadena 91125, USA
e-mail: rse@astro.caltech.edu

Early History

The engineering concept for an infrared flux collector goes back to the late 1960s and the influential figure of Jim Ring at Imperial College London. The Science Research Council funded the facility to the tune of £2.5M in 1975; in 2009 currency this still only amounts to £10M – the cost of a modest spectrograph for an 8-m telescope! Construction was rapidly completed during 1976–1978 with first light on July 31 1978 and a formal dedication ceremony on October 10 1978.

Others present will remember those early events better than I, but I do have the benefit of a remarkable filing system from my time on SRC/SERC/PPARC committees and can offer some insight into the early discussions.

The case for UKIRT as originally envisaged is succinctly captured at a Royal Astronomical Society meeting held on May 9th 1975 (soon after its funding approval) where Jim Ring states the case for a flux collector and justifies little attention to image quality the requirement for which is only 2 arcsec:

> the consequent saving in cost over an optical telescope is significant..(and) can be put into a larger diameter rather than accurate figuring and support systems (Observatory **163**, p. 95, 1975).

But in February 1977, the minutes of the Astronomy II Committee of the ASR Board of SRC show that the Director of ROE (Vince Reddish) reports:

> now that experience has been gained in figuring the blank, it is apparent with comparatively little extra work, 1 arcsec images can be obtained.

The additional cost for this dramatic improvement in image quality was estimated by David S. Brown (Grubb Parsons) at only £12K! It's actually typical of Grubb Parsons and the late Dr Brown that the company should be so cooperative in maximizing the effectiveness of the telescope.

I have also located the first UKIRT Newsletter (Fig. 1.1, left). (Notice the methodical Ellis stamp and prominent 'tick' that indicates I carefully digested its contents in 1979) Peredur Williams was the editor and promises a UKIRT Users' Manual 'soon' (actually it appeared in 1981!). There was no integrating TV, the control computer had a majestic 28K RAM and the Newsletter describes how the observatory earthquake protection system was usefully 'commissioned' with a real Richter 5.1 event. In this historic document one sees a good example of the pioneering, even swashbuckling, spirit that was to become characteristic of UKIRT!

Technical Innovation

Before attempting to review some of UKIRT's science highlights, let's consider the remarkable innovations that have become characteristic of this observatory during the past 30 years.

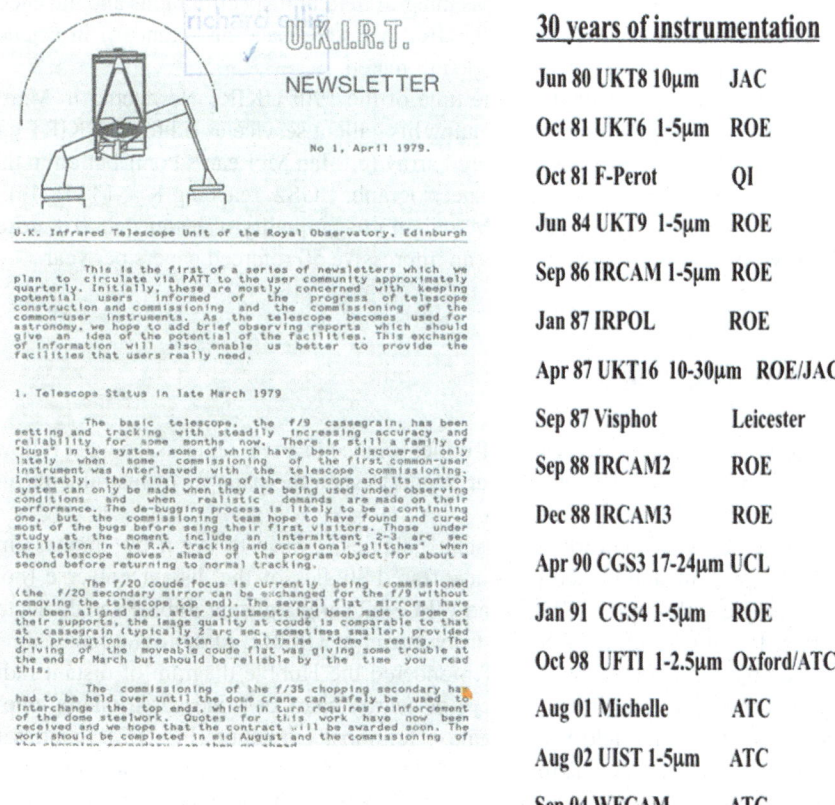

Fig. 1.1 (*Left*) UKIRT begins with the first Newsletter. (*Right*) 30 years of innovative instrumentation – a new capability every 18 months!

From the outset the telescope set a new standard with its lightweight thin primary, and later the tip-tilt secondary and cooled primary mirror. However, UKIRT has also been in the vanguard in pioneering new operational concepts. It was amongst the first to routinely offer remote operations (from Edinburgh) and later service observing and flexible scheduling. The current model whereby selected PIs observe on behalf of a group of programs is unique and very popular with the community.

An unchanging telescope is a stagnant facility and UKIRT has also pioneered a continuous set of new instruments and upgrades[1] (Fig. 1.1b). Infrared capabilities

---

[1] Soon after I returned to Pasadena from the enjoyable meeting at Edinburgh, I learned of the sudden passing of Tim Hawarden who had been present at the meeting and did so much for the UKIRT upgrades program. Here is not the place for a tribute but this is a sad loss for Edinburgh and the global infrared community.

undreamt of during the 1970s such as integral field unit spectrographs and mosaiced arrays not only first appeared on UKIRT but set a pace and standard in science discovery that other facilities struggled to match.

Ten years after first light, by the time of the 19th UKIRT Newsletter in March 1989 (with Peredur Williams continuing his gallant service as Editor!), UKIRT had the world's first common-user infrared array (c.f. Ian McLean's contribution in this volume), a world-beating infrared spectrograph, CGS2, reaching $K = 14$ (1σ 1s), a control computer with 16 Mb RAM and an image quality of better than 0.9 arcsec FWHM for 90 %. Productivity was an impressive 50 refereed papers per year.

Science Highlights

It's hard to summarize 30 years of frontier science in a few remarks, so what follows is very much a personal selection arranged to span the panoply of instruments and science territories.

As someone who works on distant galaxies, I have to begin by emphasizing that the present dominance of near-infrared studies of the distant universe (now commonplace through NICMOS and WFC3/IR campaigns on Hubble and dedicated survey facilities such as VISTA), truly began at UKIRT. Despite the handicap of single-object photometers, UKIRT pioneered the Hubble diagram of distant radio galaxies and even undertook surveys of nearby galaxies – measured one by one – to construct the first field K-band galaxy luminosity function (Fig. 1.2). Later, with early infrared arrays, came the study of distant clusters of galaxies and their lensed arcs and faint galaxy counts. Quite simply, UKIRT moved into a league of its own. Most recently, the key role of UKIRT in pushing the frontiers was demonstrated in the discovery of the most distant $z = 8.26$ gamma ray burst (Tanvir et al. 2009).

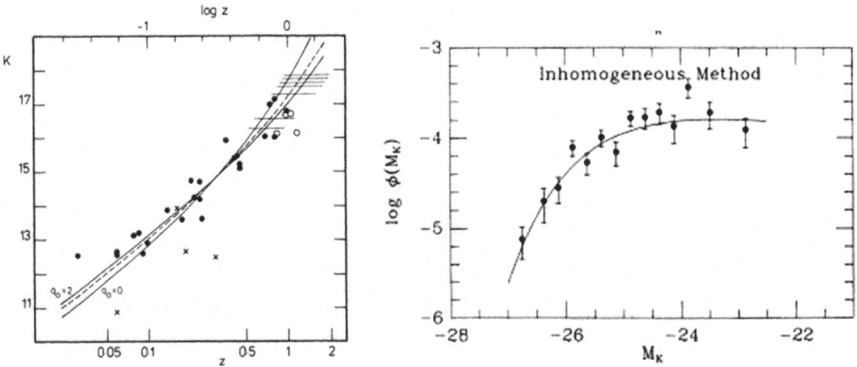

Fig. 1.2 (*Left*) The K-band Hubble diagram of distant radio galaxies from the study by Lilly and Longair (1984); (*right*) the first field galaxy luminosity function at near-infrared wavelengths from the study of Mobasher et al. (1993)

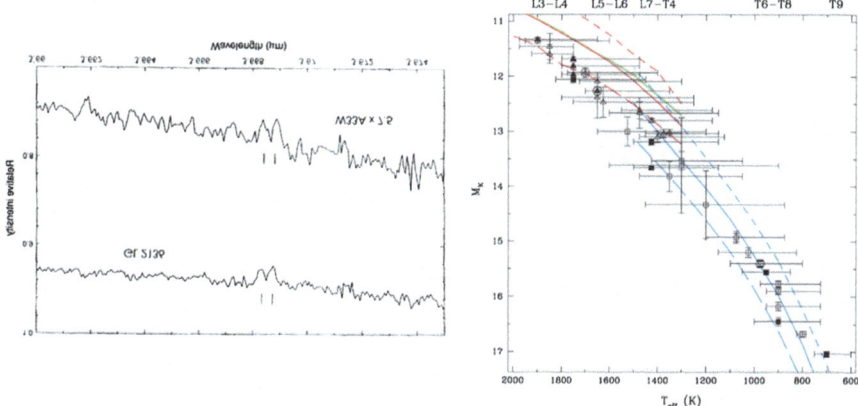

Fig. 1.3 (*Left*) Detection of the ortho-para doublet of H_3^+ (marked) in the spectrum of two YSOs from the study of Geballe and Oka (1996). (*Right*) Luminosity-effective temperature relation for SDSS-selected L and T-dwarfs from Golimowski et al. (2004)

Cooled grating spectrographs, CGS2 and later CGS4, championed wider territories in astrophysics. Highlights include the detection of the molecular H_3^+ ortho-para doublet seen in absorption in young stellar objects by Geballe & Oka, a challenging observation that confirms the key role of this molecule in the production of complex molecules and ion-neutral interstellar chemistry, and systematic observations of L and T-dwarfs by Leggett, Geballe, Golimowski and collaborators, which led to new classification methods and tests of model atmospheres in the substellar regime (Fig. 1.3).

For a telescope originally conceived without regard to image quality, I have chosen some examples that demonstrate how UIST and UFTI, instruments designed to match the excellent seeing on Mauna Kea, pushed the frontiers. Willott et al. (2003) used UIST to locate and measure the black hole mass in a $z = 6.41$ QSO; this is a pioneering observation exploiting UKIRT's unique access to the redshifted Mg II emission line. Swinbank et al. (2005) used the integral field unit of UIST to map the resolved dynamics of a $z = 2.385$ sub-millimeter galaxy – consider the angular scale in Fig. 1.4 in the context of Jim Ring's remarks in 1975! Lucas and Roche (2000) exploited UFTI to chart young brown dwarfs and planets in the Trapezium region, thereby constraining the initial mass function down to an unprecedented 20 Jupiter masses.

Finally, to the present day and WFCAM. As others will discuss this remarkable survey capability in some detail, I thought I would steer clear of UKIDSS and just touch on two surveys that have exploited narrow-band imaging with this mosaiced array. HiZELS is an impressive set of surveys that exploits a narrow-band capability to either search for Hα imaging at $z \sim 1$ or, more speculatively, Lyα imaging at very high redshift. In the former case, Sobral et al. (2009) have surveyed the COSMOS

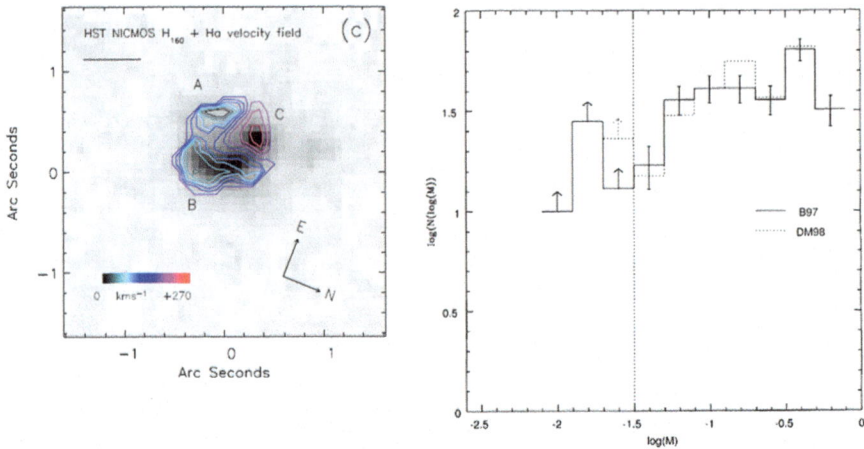

Fig. 1.4 (*Left*) Resolved velocity field in Hα for a z = 2.385 sub-mm galaxy from UIST IFU observations overlaid on a HST NICMOS H-band image demonstrating the importance of galactic outflows in early star-forming systems. Note the remarkable angular resolution (Swinbank et al. 2005). (*Right*) Sub-stellar initial mass function (in solar units) from the Trapezium study of Lucas and Roche (2000)

and UDF fields to undertake a measurement of the star-formation density at z ∼ 1. Together with similar measures at other redshifts, a self-consistent star formation history has been deduced for the first time using the same tracer.

Closer to home, C. Davis and colleagues (Davis et al. 2009) have mapped a phenomenal 8 deg^2 in Orion in the 2.122 μm emission band of molecular hydrogen, locating over 100 objects and associating them with protostellar objects. The level of detail in these maps is truly astonishing (Fig. 1.5).

Studies of the productivity of UKIRT have underlined its global role in astronomy. Trimble et al. (2005) and Trimble and Ceja (2007) examined the output and citation rate of the world's 4-m telescopes and place it alongside CFHT, NTT and WHT – optical telescopes that serve a more diverse community. Chris Benn (unpublished – see http://www.ing.iac.es/~crb/cit/9903_prelim.html) examined the fraction of highly-cited (top 2 %) of world literature over 1999–2003 and reached similar conclusions. UKIRT ranks equal to CFHT, WHT and AAT and outpaces all other dedicated infrared telescopes (e.g. IRTF) by a significant margin.

Summary and Future

So why is UKIRT so successful?

- It is blessed with being on a good site; Mauna Kea offers excellent seeing and a stable thermal environment.
- The joint operations with JCMT via the JAC is remarkably cost-effective.

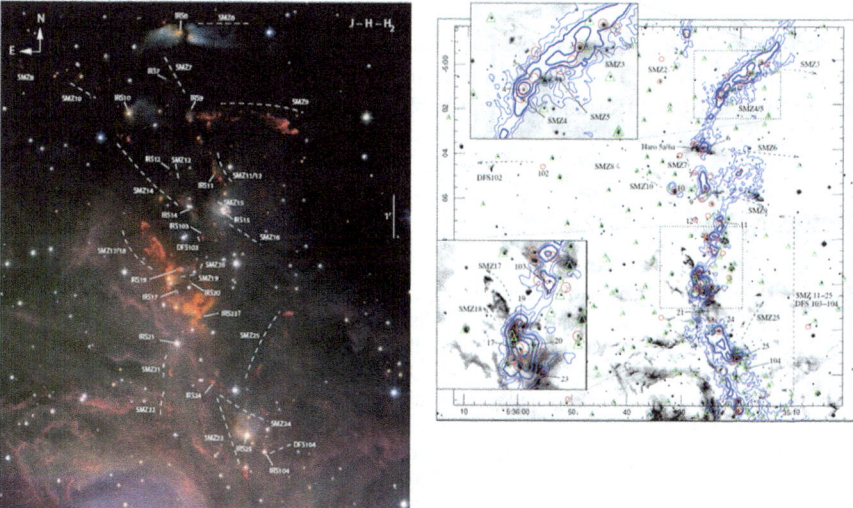

Fig. 1.5 The remarkably detailed mapping of Orion in molecular hydrogen from the study of Davis et al. (2009). (*Left*) Color composite of the region around M43 based on J, K and 2.122 μm imaging; H_2 flows are labeled. (*Right*) 2.122 μm image with protostars marked (with *inset* zooms at twice the scale). Contours refer to 1.2 mm MAMBO data

- It has maintained a flexible suite of front-rank instruments.
- It has championed creative operational models including remote observing and flexible scheduling, noting the importance of training young enthusiastic observers who then become its supporters.
- It has provided powerful online data processing capabilities.
- It has shown again and again that it can reinvent itself, e.g. moving into high resolution imaging (UFTI) and now panoramic surveying (WFCAM/UKIDSS).

The future of UKIRT should therefore be bright! In 2005 I chaired an international review of UKIRT with a charge to make recommendations to PPARC on its "role and international context during 2005–2015 and options for development of the facility". We took this charge very seriously but perhaps as an omen of what has since passed we received no financial guidelines from PPARC. It's hard to see that our carefully-argued case for a bright future for UKIRT had any impact on STFC. As I write this summary of my September 2009 talk, the outcome of the Science Prioritisation Exercise 2010–2015 has just been published by STFC and UKIRT's future in this period looks decidedly unclear.

UKIRT has been a great success not only for UK science but also in promoting the growth of infrared astronomy worldwide. It is one of the best examples of a telescope that has adapted to new technologies, far surpassing the modest goals envisaged in the 1970s. It is sad that STFC should, apparently, not appreciate both the cost-effective operational model that has been a highlight of UKIRT over the past few years as well as the strategic importance of maintaining a UK presence on

Mauna Kea amongst a set of international ground-based observatories that remain, collectively, the most productive scientifically. It will be sad indeed if the UK detaches itself from the Mauna Kea community, particularly given the glorious history blazed by UKIRT and its remarkable staff over the past 30 years.

References

Davis, C.J., et al.: A census of molecular hydrogen outflows and their sources along the Orion A molecular ridge. Characteristics and overall distribution. Astron. Astrophys. **496**, 153 (2009)

Geballe, T., Oka, T.: Detection of H_3^+ in interstellar space. Nature **384**, 334 (1996)

Golimowski, D., et al.: L' and M' Photometry of Ultracool Dwarfs. Astron. J. **127**, 3516 (2004)

Lilly, S.J., Longair, M.S.: Stellar populations in distant radio galaxies. Mon. Not. R. Asron. Soc. **211**, 833 (1984)

Lucas, P.W., Roche, P.F.: A population of very young brown dwarfs and free-floating planets in Orion. Mon. Not. R. Asron. Soc. **314**, 858 (2000)

Mobasher, B., et al.: A Complete Galaxy Redshift Survey – Part Five – Infrared Luminosity Functions for Field Galaxies. Mon. Not. R. Asron. Soc. **263**, 560 (1993)

Sobral, D., et al.: HiZELS: a high-redshift survey of Hα emitters – II. The nature of star-forming galaxies at z = 0.84. Mon. Not. R. Asron. Soc. **398**, 75 (2009)

Swinbank, A.M., et al.: Optical and near-infrared integral field spectroscopy of the SCUBA galaxy N2 850.4. Mon. Not. R. Asron. Soc. **395**, 401 (2005)

Tanvir, N., et al.: A γ-ray burst at a redshift of z~8.2. Nature **461**, 1254 (2009)

Trimble, V., Ceja, J.A.: Productivity and impact of astronomical facilities: A statistical study of publications and citations. Astron. Nachr. **328**, 983 (2007)

Trimble, V., et al.: Productivity and Impact of Optical Telescopes. Publ. Astron. Soc. Pac. **117**, 111 (2005)

Willott, C., et al.: A 3×109 Msolar Black Hole in the Quasar SDSS J1148+5251 at z = 6.41. Astrophys. J. **587**, L15 (2003)

Chapter 2
UKIRT – The Project and the Early Years

Terry Lee

Abstract This is a personal overview of the early years of UKIRT, from its inception to the first observing years, showing how many of the decisions taken have turned out to be crucial in giving UKIRT such a high scientific profile. The theme of a common-user facility in terms of the user-experience was novel for the time but set the scene for the next generation of 8 m telescopes many years later. I also pay tribute to the many people who have made UKIRT such a tremendous success.

Introduction

Thirty five years ago I was on a bus in Geneva. I heard a strong baritone voice a few rows behind say "So the UK is going to build an infrared telescope on MK. No one in the UK has done much IR except Aitken and Jones". Well Eric Becklin, who was to become a dear friend, was right, especially if you exclude those Brits who had spent time at Caltech such as Mike Penston – who he was addressing at the time. Nonetheless, many of us had recognized the potential of that part of the electromagnetic spectrum. Indeed, the proposal for a major UK IR facility was first made in 1968. At that time astronomers in the USA were reporting observations using lead sulphide (PbS) detectors (e.g.: stellar photometry by Harold Johnson, a 2 μm sky survey by Leighton and Neugebauer), while for work at longer wavelengths Frank Low was using the gallium doped-germanium bolometer, his invention (sadly, Frank died this summer).

T. Lee (✉)
Formerly UKIRT, Royal Observatory, Edinburgh, UK
e-mail: terrylee@club-internet.fr

Jim Ring, Professor at ICST, was interested both in infrared astronomy and telescope design. In particular, he wanted to explore the limits of passive designs for seeing-limited telescopes as the primary diameter increases. [At that time one arcsec was assumed to be the seeing-limit from the ground.] He also sought to challenge the cost-diameter relation for their construction: the cost of building a telescope of traditional design increased as the third power of the primary diameter or faster. By replacing the primary mirror of 6–1 diameter-to-thickness ratio with a thinner one supported by a sufficient number of air-filled pads, the mass of mirror plus cell would be significantly reduced. In turn, the mass of steel and concrete needed to support the optics would be less. Constructing a telescope with this type of design would not pose a risk to observations that do not need seeing-limited images. At the time the argument was that because the internal noise in the PbS detectors was greater than the thermal noise from the sky and telescope for apertures of several arcsec, photometric measurements did not require good image quality.

As a first step the Science and Research Council (SRC) funded a 1.5 m precursor, 'the IR flux collector' that was installed at Izaña on the Island of Tenerife in 1972. This instrument was mainly used by UK and Spanish astronomers, many of us learning how to set up instruments and observe in the infrared for the first time. Researchers who had gained experience in the USA were also able to use it to continue their work. It also indicated that the design concepts for a thin-mirror telescope were sound.

In 1973 Jim Ring and Gordon Carpenter, the senior engineer at ROE, made a proposal for a fully functional, major IR flux collector; either a 3.8 m instrument to be sited alongside the 1.5 m at Izaña or a 3 m instrument on Mauna Kea, a much higher and drier site in Hawaii. In the event the Astronomy Committee led by its chairman, Professor Walter Stibbs, forwarded a proposal for the 3.8 m version on Mauna Kea! This was approved in June 1974. Gordon Carpenter was appointed as Project Manager and the '3.8 m Flux Collector Steering Committee was formed, drawing members from the active infrared groups in the UK. The first task was to study the draft specification and recommend changes, especially in the light of recent developments.

There are significant differences between the design of telescopes used for infrared observations and optical telescopes, the break-point being around 2 μm, longward of which thermal emission from the sky and the telescope dominates the background radiation. For photographic work in the optical stray-light baffles were generally placed around the primary and secondary mirrors and inside the central hole of the primary. Effectively these have a thermal emissivity close to unity. However, for the infrared, any structure in the light path must be minimised. Furthermore, sky emission and gradients must be subtracted from the object and for the single detector instruments of the time, chopping and nodding techniques were used. The subtraction of gradients is more accurate the closer in time that the measurements in the two beams are made. In addition, nodding placed requirements on the dynamic performance of telescope.

Key Elements of the Design

- Lightweight primary supported by 80 pneumatic pistons
- Fast primary ($f/2.5$) to keep tube short
- Clean structure thermally
- Light structure, no central box thermal and cost considerations
- Large diameter gears quick movement
- Position control loop closed in computer
- No control system independent of computer
- Keyboard input supplemented by a few buttons

Specification:

- Primary diameter 3.8 m
- $f/9$ Cassegrain, $f/20$ coudé
- Primary image quality 98 % EED 2.4″
- Short nod time (2s)
- Tracking (5 arcsec per hour)
- Pointing 30 arcsec circle rms
- Dome building to be as small as possible (to contain costs and maximize slit to volume ratio to work against dome seeing)

The principal amendments to the spec as recommended by the steering committee were the following: (1) increase the maximum payload on the instrument rotator from 100 to 200 kg. [At that time a complete photometer might weigh 25 kg and the possibility of mounting multiple instruments and indeed some increase in size was anticipated (though the like of CGS4 was clearly not)]; (ii) a modest increase in the dome size would enable a chopping secondary to be accommodated; (3) the possibility of including an option for improving image quality to 90 % in 1 arcsec should be explored.

The proposed optical and mechanical design was based on such primary mirror blanks as were immediately available – Owens-Illinois Cer-Vit. Where practicable British suppliers and contractors were chosen for building and telescope, and Grubb-Parsons of Newcastle on Tyne for grinding and polishing the optics (having recently done a fine job with the AAT mirrors). The contract contained a provision for continuing to figure the primary mirror beyond the initial specification if testing and evaluation at that stage indicated that it was possible to do so and a price could be agreed. Hadfields of Sheffield was the contractor for the telescope structure and drives. The telescope columns carrying the north and south bearings are tied together, their bases rest on concrete piers via steel thrust-races which allow movement between the piers and telescope structure. Normally horizontal movement is constrained by pins between the steel bases and the piers; these shear to protect the instrument during significant earthquakes. The estimate for the mirror mass was 7 tons as opposed to 15 for a conventional one and the rotating mass was estimated to be 60 tons as opposed to about 250 tons.

A novel feature of the project was to give the tasks of slewing and tracking to a digital computer rather than a set of hard-wired electronics. The user interface was a keyboard and text display, while the RA and Dec axis encoders and sidereal and UT clocks were interfaced via CAMAC, the standard interface for high energy physics at the time. A DEC PDP11 was chosen as the real-time computer running RT11 in 28K words of 16 bit memory. By today's standards this seems very pedestrian, but remember Moore's Law: 30 years worth is a factor of more than a million! Design and programming of the real-time software was done at ICST. A computer was needed to calculate corrections due to atmospheric refraction and uncompensated mechanical movement. Giving it the additional task of closing the position loop saved on the cost of building the digital and analogue circuits and panels. In practice it provided the bonus of an easy upgrade path both to interface and control aspects.

Early Project Developments

The project as originally funded was for the telescope alone, which fell short of what was required for observational astronomy. Clearly there were a number of ways in which the efficiency and effectiveness of the facility could be assured.

1. The addition of a chopping secondary, which experience in the U.S. showed can yield a given photometric observation in a much shorter time than with a focal plane chopper.
2. A detector test and development programme. High purity InSb detectors had become available that were much more sensitive than earlier versions and the standard PbS device of the time. Indeed, developments led to detector noise becoming smaller and smaller through the late 1970s and it had become vital to understand how to get the best out of detectors and to match the optical and cryogenic performances of instruments to suit their needs.
3. The provision of common-user instruments to allow and indeed encourage investigators without infrared equipment or experience to make infrared observations.
4. The setting up of a computer system to control instruments and to log and reduce data in real-time. This enables the astronomer to assess the content and quality of data and to modify the programme accordingly; it also lessens fatigue and the tendency to make mistakes at altitude.
5. The use of intensified TV acquisition and guiding facilities to make target identification easier, to reduce the total duration of an observation (and hence increase efficiency), and to take the guide telescope out of the astrometric loop.
6. The construction of a central Telescope Simulation Facility where investigators can bring their instruments to ensure compatibility with the telescope and test elements of performance (Fig. 2.1).

The content of a possible development programme was discussed at a number of meetings of the Steering Committee leading to a formal proposal by the Royal

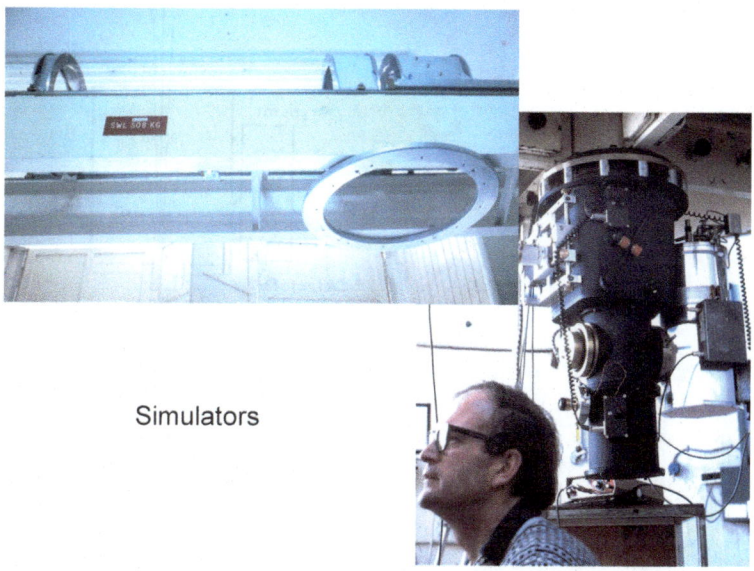

Fig. 2.1 Picture of simulation rigs – forerunners of two more generations now at ROE

Observatory Edinburgh to the Science Research Council. During this time I visited several of the labs in the US and the groups in the UK interested in active involvement. In the summer of 1976 a working group met to define details of the first phase of common-user instruments.

The Common-User Instrumentation Plan

- Focal plane choppers for $f/9$ (1 ROE 1 ICST).
- Photometers for $f/9$ Cassegrain: two cryostats using InSb detectors with bandpass filters and CVFs covering 1–5.6 μm.
- Photometers for $f/35$ Cass: two cryostats with bandpass filters and CVFs covering 1–5.6 μm
- Cryostat with bolometer detector and filters covering 3.8–40 μm
- Cryostat with bolometer or doped Si detector with bandpass filters covering 8–40 μm, and a CVF from 8 to 14 μm
- Polarimeter insert (in collaboration with Hatfield Polytechnic)
- Fast photometer
- Visible photometer (UBVRI)
- Cooled grating spectrometer

One of the crucial decisions to be made was the configuration of the instruments at the Cassegrain focus. Infrared detectors must be cooled to their optimum working

Fig. 2.2 The side-looking configuration

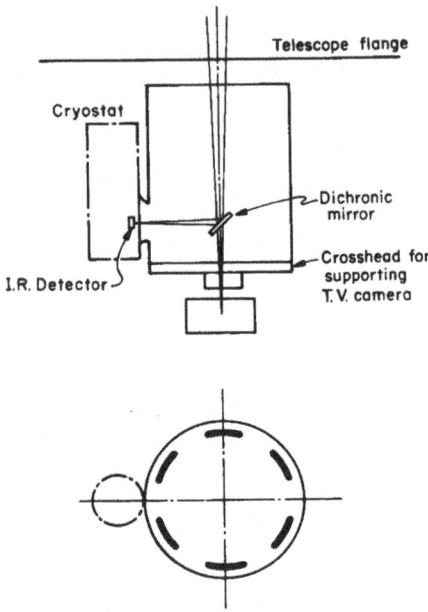

temperatures, which is 77K or lower for near-IR and 4K or lower for longer wavelengths. Cryogenic engineering constraints meant that normal practice was to build the cryostat so that the work surface is the bottom face of a cryogen tank. This configuration is optimal for cryogenic performance and also works well for mounting the detector and optical components with the light beam entering radially, i.e. a bent Cassegrain configuration requiring a tertiary mirror outside the cryostat. This warm mirror is not best suited for thermal infrared work. Some US astronomers were using upward-looking cryostats that were optically efficient (doing away with the reflecting mirror). However, these were more complicated to build, had relatively short hold-times and needed to be demounted from the telescope to fill. The side-looking configuration had the particular advantage that if the tertiary mirror is also a dichroic, the visible light passing through it can be used for acquisition and guiding. This was an important consideration for a new telescope of new design where acquisition, tracking and offsetting properties were not predictable. Furthermore, if this tertiary is mounted on a hollow bearing concentric with the telescope axis then several instruments can be mounted on the telescope at any one time providing backup in the case of problems, or choice. Longer hold-times allow refill operations to become a routine daytime task. So the choice was made. We bought six cryostats from Oxford instruments to a jointly developed design. These featured a long hold-time, flats on the outer casing for mounting motors and windows along with fittings for pumping on the inner vessel (see Fig. 2.2).

InSb detectors were sourced from SBRC or Cincinnati Electronics in the USA. Interference filters were purchased from OCLI Europe in Scotland. We specified

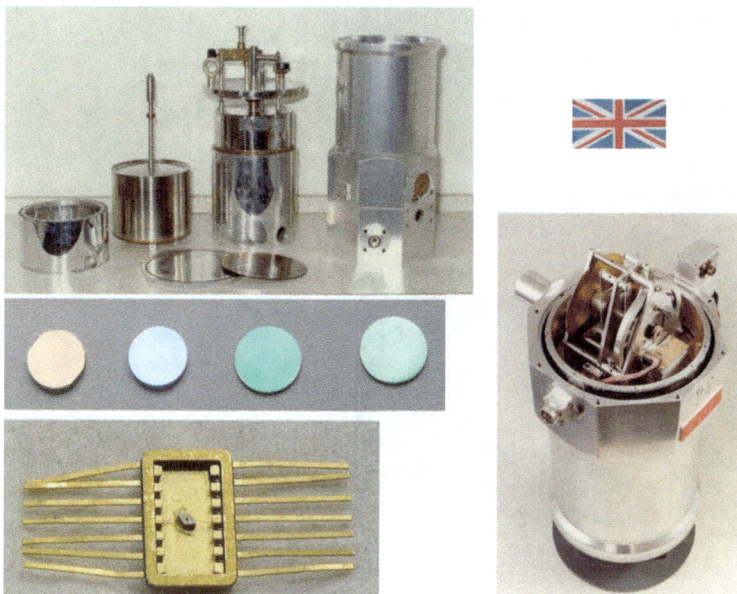

Fig. 2.3 Photometer cryostat with components

filters to match the atmospheric windows (rather than hunting for closest available from surplus catalogs). To limit cost to the project we ordered 50 sets and sold about half to IR groups around the world (Fig. 2.3).

Telescope Realisation

The first landmark in the manufacture of the telescope occurred when David Brown of Grubb Parsons informed us that the specification for the primary figure had been achieved. Furthermore, we were told that it could be improved to a quality of 95 % encircled energy in 1 arcsec for a payment of £20,000. When the proposal was put to the steering committee the decision to accept was made in record time. This subsequently turned out to be a first-class value-for-money decision in the scientific productivity of UKIRT.

The structure and optics were shipped to Hawaii at the end of 1977, and the telescope was assembled in the dome, which had been completed on Mauna Kea early in 1978. Interestingly it was as part of the shipping process that the '3.8 m Flux Collector' became known as UKIRT, analogous to the recently arrived IRTF and CFHT. This gave us an obvious identity and a place on the map locally in Hawaii.

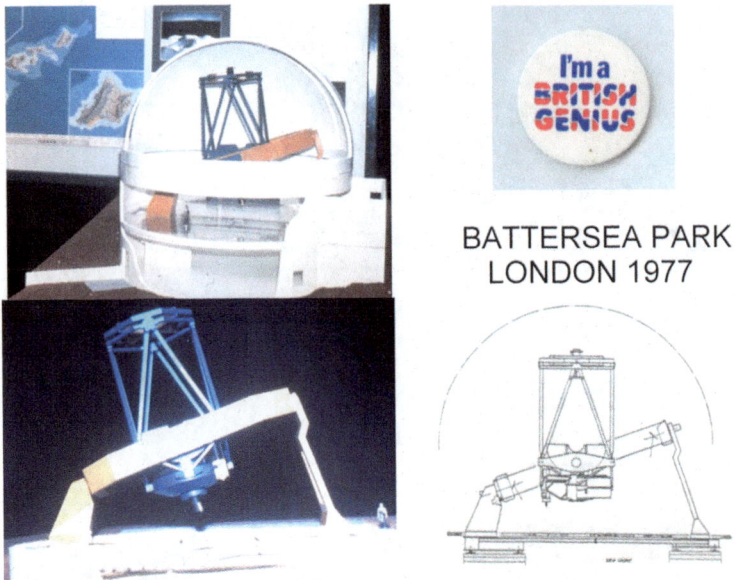

Fig. 2.4 A model demonstrating the design of the telescope was part of the British Genius Exhibition of 1977 in Battersea Park in celebration of the Queen's Silver Jubilee

In 1977 as the project was well underway, two contrasting events happened in close unison: Her Majesty the Queen visited the Silver Jubilee exhibition on May 27th (see Fig. 2.4) and tragically, Gordon Carpenter died on June 1. I was in Sheffield at Hadfields at the time. Following Gordon's death, management of the project was taken on by Colin Humphries. Colin was already part of the project as manager of the optics.

First Light

First light was obtained at 03:23 am on July 31st 1977. Alpha Peg images appeared circular and about 2 arcsec in size. Testing various aspects of the telescope in its $f/9$ configuration continued through 1978. The optical image quality was investigated using a Hartman screen and plate photography, knife-edge tests and examination of out-of-focus images. Two effects were shown to deteriorate the basic image quality much of the time: dome seeing due to warm air in the building, and oscillation in RA (of up to 5 arcsec) due to the nature of the position encoder.

The good news was that the mirrors and supports behaved within specification. Later, in early 1979, when there was a TV camera mounted and differential air temperatures were low, from time to time images appeared to be smaller than 1 arcsec in diameter. Jim Ring's telescope concept was demonstrated. A telescope with excellent image quality had been built for a modest cost.

Other results from these shakedown tests were:

- Pointing within spec (30 arcsec rms).
- Tracking was strictly within spec but the oscillation was not acceptable for observing.
- Nod performance was poor, typically about 5 s for small nods.
- Guide telescope flexed too much with respect to the main structure.
- Crosshead for the autoguider had drives errors.
- Dome drive was marginal even in moderate wind.
- *Dome crane was not safe to lift top end.*

This last was to cause a major delay as strengthening work had to be carried out through the summer of 1979. After strengthening of the dome the dome drive continued to be unsatisfactory and it was some years after the replacement of the track, bogies and drive that reliable automatic dome following of the telescope was achieved.

By the end of 1978, the main testing was completed; mounting of the $f/35$ was rescheduled until after dome work completion. Hadfields were to return to do this and try to improve dynamic performance of the telescope. The telescope became available to the operations team to install wiring and piping through the structure to the Cassegrain area to enable the instruments to be installed. The tasks were then to improve pointing and tracking and to commission the infrared photometers.

Early in January 1979 Dave Beattie prepared and I mounted UKT1 on the telescope. At dinner Eric Becklin said "I hear you guys are going to get first IR, can I come and watch?". We agreed. At the telescope the conversation went something like this:

Eric: "What size aperture do you have"
Dave: "5 arcsec",
Eric: "Do you have a bigger one"
Dave: "Yes"
Eric "Are you going to use it?"
Lee and Beattie in unison: "No"

Back in the control room we set the telescope on the target, centred it on the TV and immediately we had signal. Even Eric was a little impressed. PATT-allocated operation started with the last quarter of 1979. The ICST group took the first slot (Fig. 2.5).

Fig. 2.5 Terry Lee with the first photometer on UKIRT

Dedication

On the 10th October 1979 the Duke of Gloucester opened the facility, now officially called UKIRT, with a ceremony in the dome. Among the dignitaries were Vince Reddish (director ROE), Geoff Allen, (Chairman of SRC), other telescope directors, Harry Atkinson (SRC director of science), University of Hawaii Chancellor, those involved in constructing the facility and many others. Sadly, not Jim Ring who, perhaps wisely, had a fear of flying (Fig. 2.6).

This was followed by an evening dinner in Hilo. Round about 3 am myself, Gareth Wynn-Williams and others finally persuaded a relaxed Harry Atkinson to divulge who was to succeed Vincent Reddish as Director ROE. The answer was of course Malcolm Longair. For me this was an unknown quantity and the changeover was nearly a year away. I had my own Observatory to run and a year's worth of users to serve. Since it was a low-cost telescope we were expected by certain quarters to operate it in proportion cost – but users would expect good service!

It's worth recalling that in those days, long distance communication was a completely different picture from today. There was no internet; written communication was by letter or TELEX (110 b/s). It was not possible to direct dial an international phone call from Hawaii. Summit phones were via a low capacity microwave link able to support voice only. These realities, combined with the time difference between Hawaii and the UK, meant that volume and frequency of communication were not great. We worked rather independently. We were only a dozen or so from the UK at the time and added staff recruited locally.

Fig. 2.6 Vincent Reddish (Astronomer royal for Scotland) and Duke of Gloucester. *Top*: Geoff Allen (Chairman SERC), Min Lee, Des Hickinson (of Hadfields). *Bottom*: Colin Humphries, Terry Lee, Duke of Gloucester

How We Observed

Before we could take on scheduled observing there were issues that had to be addressed. There was no secondary chopper yet, therefore the only choice was to work at $f/9$. For acquisition, no intensified TV camera was yet available, and the combination of the lower sensitivity of the silicon target camera and the restricted field-of-view of the $f/9$ photometer was inadequate for finding guide stars. Guide telescope issues meant that pursuing its use was not fruitful.

The solution was to convert the $f/35$ gold dustbin to $f/9$ use by mounting a focal plane chopper on the end of an arm fixed in one of the six ports. The TV was mounted on the X-Y stage at $f/9$. This gave an offset field of + or-13 arcmin which was very useable. However image motion while tracking was still up to 5 arcsec due to the fine code error of the encoder. This was too big for most projects. I remembered that I had lent a constant voltage source to the telescope project team for factory tests of the telescope drives. It was therefore possible to inject small signals to the RA and Dec amplifiers to drive the telescope in fine motions. After acquisition the computer controlled tracking could be switched off and the analog box switched into the circuit. Inside the analog box was essentially a battery with a few resistors, a couple of potentiometers and a connector for an Atari joystick.

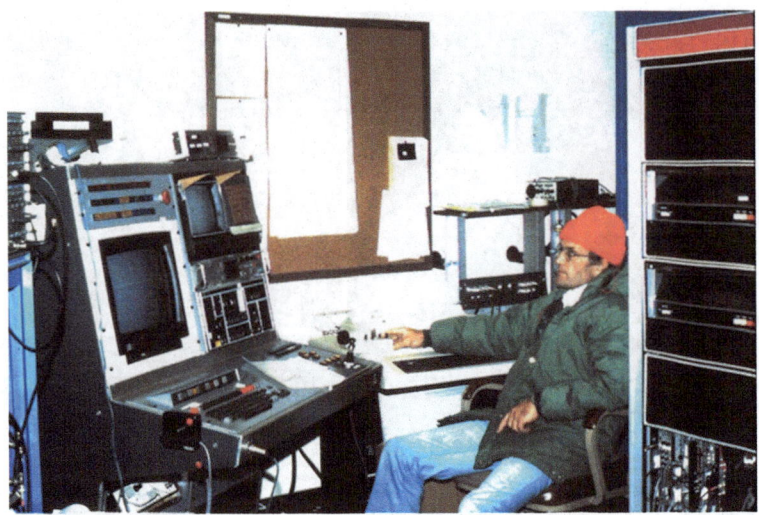

Fig. 2.7 David Beattie at the original UKIRT controls with the PDP11

The size of the observing team would depend on the scientific programme but might need:

- One person to operate telescope and guider
- One to operate instrument
- One to review data and check the dome position
- One to compute guide star and guider offsets
- At times one quality control, backup, make coffee.

Over time and with very skilful software engineering of the limited computing capacity more functions were put into the computer. Later, an autoguider was fashioned by fixing a quadrant photodiode to the TV monitor and feeding signals into an analog box. The $f/35$ secondary and InSb photometer cryostats (designed by John Harris) arrived in 1980 (Figs. 2.7 and 2.8).

The Longair Effect

Malcolm Longair arrived on the scene in Hawaii even before he was in post. It became clear that we would be seeing him in Hawaii both as Director and telescope user; and in time that he expected us to become a world beating facility. The deal would be that we had to get the results and he would find the resources. Failure was not on the agenda – action was. He moved to regenerate and re-energize our connection to Edinburgh, directly to himself as Director and to the instrument work there. As a first action item the next instrument to be built would be CGS1: a 1–5 μm spectrometer with Richard Wade as project scientist. Tim Hawarden was appointed

Fig. 2.8 The f/35 instrument cluster

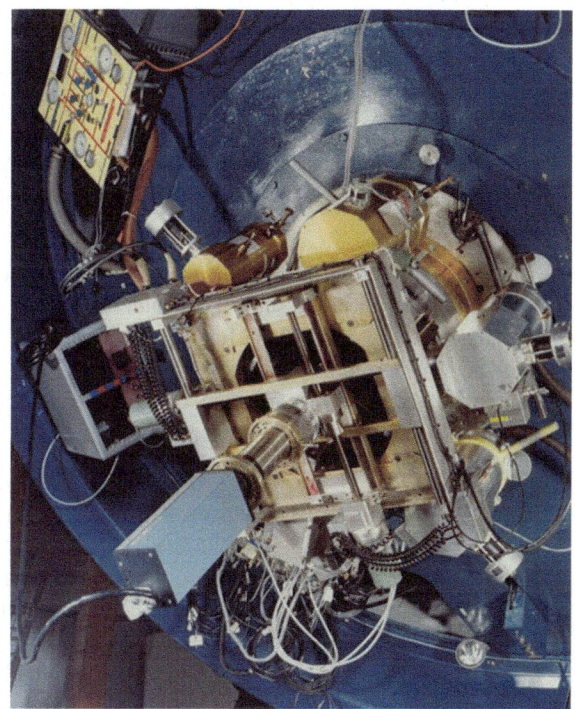

as scientific liaison generally and to the UKIRT user committee in particular. A decade later Tim thought out and carried through an upgrades programme to yield images of superb quality (see contribution by T Hawarden).

Users

The Agreement with the UH allowed them 15 % of telescope time as site rent. The rest was allocated by PATT including Dutch time of about 5 %, part of the agreement on IRAS. Users fell into broad categories. There were those who brought their own instruments, experienced observers who used common-user instruments and observers new to the infrared. In the very early years we did not have telescope operators since user interfaces and instrument systems were not stable enough. So the staff astronomer allocated to each observing team acted as telescope operator and where appropriate instrument operator and sometimes tutor. Prominent among those who brought their own instruments were far infrared observers and heterodyne spectroscopists. Most of the groups in existence worldwide using these techniques came to take advantage of the large aperture of UKIRT and the atmosphere of Mauna Kea. Observers new to the IR came from the UK, Netherlands and Japan. Experienced observers came from UK, Hawaii, US, Japan and Australia.

Examples of Observing Programmes

In general, observing programmes were designed to benefit from the greater sensitivity of the large aperture and the drier, more transparent and more stable atmosphere of Mauna Kea. This enabled more effective mapping, faster time resolution photometry, and more secure polarimetry and spectroscopy. Early examples included:

- Mapping star-formation regions to search for embedded objects
- Photometry of Variable stars at 1-s intervals
- mm/submm photometry and line searches, maps of higher molecular transitions
- Diffraction limited mapping at 12, 20, 30 μm
- Spectroscopy of IR masers, planetary nebulae
- Galactic centre, mapping, spectroscopy
- Detection of Dust features in other galaxies

Extragalactic observations became a reality. At high redshifts the peak of starlight is redshifted into the 2.2 μm window. For example Simon Lilly and Malcolm Longair came for several heroic observing runs plugging away for hours at a time to get a K magnitude detection of high redshift galaxies. Infrared photometry was shown to be a major tool for cosmology.

Mauna Kea had had a mixed reputation as an observing site. The effects of altitude on people and the difficult logistics relative to US mainland sites were cited as impediments to good science. Observations from UKIRT, the IRTF and CFHT in their first years of operation quickly dispelled these myths. Each of these large telescopes demonstrated the unparalleled quality of the mountain. Data published from UKIRT observations began to establish the telescope as a force and the UK as a major player in infrared astronomy.

Evolution

Although UKIRT was a low budget telescope, the expectation of some users was a performance and service on a par with other large telescope facilities. This was quite a challenge.

In parallel with observing support there was a continual programme of improvements. Pointing accuracy increased to 4 arcsec rms. Nod-time was reduced by running control feedback through computer-enabled dynamic tuning. Other shortcomings were diagnosed and solutions found or components replaced (e.g. 24 bit encoders in place of the original 20 bit). For guiding, intensified cameras and their postprocessors were utilised.

The user interface was improved with the incorporation of online catalogues and routines for calculating and commanding guider offsets. Communications between instrument and the telescope computer were enhanced and a 'list-processing' system for standard observing sequences was introduced. Phone capacity at the summit greatly increased so we could network from the Hilo Base to the summit and also back to Edinburgh. This opened up the possibility of remote observing, which Malcolm Stewart first demonstrated. Later a remote observing room was set up in Edinburgh using a dedicated phone line. Subsequently this was extended within the UK via the networks. UKIRT set the stage for today's practices at large observatories.

Spectroscopy

CGS1 arrived in 1982, adding a new dimension to our capabilities. Even with just one detector it enabled significant observations to be made. It was soon put to good use in mapping shocked molecular hydrogen in the Galactic Centre region. Molecular hydrogen is what had brought me to astronomy. After the galactic centre work I was keen to see if I could make a detection in M17. My attempts in evening twilight around this time of year did not succeed though it was evident that something was there. We concluded that the hydrogen was so extended that we could not chop off it, so we came up with the idea of chopping in wavenumber space, which was implemented for the following year. Simply put, the solution was to mount a Fabry-Perot etalon in the beam feeding a photometric cryostat and measure the flux at selected spacings. Luckily there was a source of FPs that could be commanded by computer (Queensgate Instruments). We made an extensive map using an 18 arcsec beam.

Parallel Developments – Buildings

The minimum-sized dome turned out to be too minimal. An extension was added to house instrument preparation equipment and allow much of the computing and other noise and heat generating equipment to be moved from the control room (Fig. 2.9).

In the beginning we operated from a warehouse building near Hilo airport. This soon became cramped. At the time of negotiations for the mountain-top site and in the course of discussions on collaboration with CFHT, there was a gentleman's agreement at Head Office level that both would site their low level base in Waimea. We bought adjacent plots of land there in 1979. It became clear in the course of the first year or so that from many considerations, including the lives of most of the staff, remaining in Hilo was preferable. The University of Hawaii was keen to open a new section of the Hilo campus to accommodate astronomy facilities. In 1982 the decision was made to design and construct a permanent base for UKIRT there. Ground was broken in 1983. The facility was opened in 1985.

Fig. 2.9 Emission from rotation vibration transitions of shocked molecular hydrogen in M17

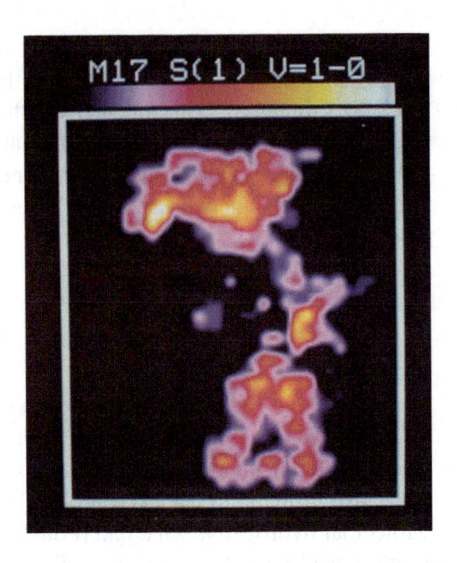

The huts at Hale Pohaku will remain long in the memories of older observers for days of interrupted sleep and limited privacy. We participated in the design details for the mid-level facilities which were completed later in the 1980s. With the mid-level facility came a power-line extending all the way to the summit to replace the diesel generators (Fig. 2.10).

Signs of Success

First early signs of success were that observing time was well oversubscribed, people wanted to come back. They published observations and made convincing cases. Later came independent signs external to the UK. The Japanese National Observatory had been impressed by UKIRT as an observatory that provided both telescope and instruments to the observing community. In Mitaka they sought in-depth advice on how to emulate us for a 6 m plus project. We also set up a collaboration between the two countries on UKIRT and the new Nobeyama mm telescope. After many years of selling single pixel detectors the Santa Barbara Research Corporation had come up with a plan to provide two-dimensional detectors for astronomy. As pilot project they chose UKIRT as one of their two IR array partners; the other was NOAO. It was indeed unusual for a US company to engage in this way with a foreign agency. Their assessment of our overall capability must have been high indeed. Ian McLean will tell all about this in his paper.

Fig. 2.10 Building projects

Tribute

Of the scores of people who contributed to the project and its early years, two stand out:

Jim Ring was a brilliant instrument scientist and a charming sincere person who brought and held the infrared community together. He convinced committees and attracted excellent researchers and students, many of whom still contribute to astronomy in key roles.

David Beattie brought us at UKIRT through some difficult times. As my deputy he was there when I was not and much more. With his great drive, enthusiasm, energy and sense of humour (at times macabre) he moved things along on the mountain and in Hilo. On any day or night he might take on the role of staff astronomer, instrument specialist, day work supervisor, contract negotiator, building design coordinator . . . He was my partner for 10 years in a great adventure.

Chapter 3
The IRCAM Revolution

Ian S. McLean

Abstract The advent of infrared "array" detectors in the mid-1980s completely changed the way infrared astronomy was done. The UK Infrared Telescope and the Royal Observatory Edinburgh played significant roles in this revolutionary development. From 1984 to 1986, I was involved in the development of IRCAM, the first common-user infrared camera for the 1–5 μm region to use a solid-state imaging device developed specifically for ground-based astronomy. Although that first 58 × 62 pixel array is small by today's standards, that remarkable little device, and its exploitation on the 3.8 m telescope, opened the flood gates and resulted in exponential growth for infrared astronomy. This article recounts some of the story behind the development and deployment of IRCAM.

Introduction

Following a series of significant discoveries in the 1960s, such as the detection of the Galactic Centre at near-infrared wavelengths by Eric Becklin and Gerry Neugebauer (1968), infrared astronomy began to grow during the 1970s. The Caltech Two Micron Sky Survey, using single-element PbS detectors, and the pioneering work at longer wavelengths with Ge bolometers by Frank Low and colleagues (e.g. Kleinmann and Low 1970), led the way. Developments in the 1970s culminated in the opening of two telescopes on the summit of Mauna Kea, Hawaii, dedicated to infrared observations in 1979. One of these was the 3.0 m NASA Infrared Telescope Facility and the other was, of course, the 3.8 m UK Infrared Telescope (UKIRT). Instrumentation for these telescopes was based on bolometers, or photometers using single-element photovoltaic or photoconductor

---

I.S. McLean (✉)
Department of Physics and Astronomy, University of California, Los Angeles, CA 90095, USA
e-mail: mclean@astro.ucla.edu

detectors. I joined the Royal Observatory, Edinburgh in November 1979 from my post-doc position at the University of Arizona in Tucson, but before returning to Scotland I went to visit UKIRT and made my first trip to the 14,000 ft summit of Mauna Kea.

The 1970s was also the era of the charge-coupled device (CCD) which was invented in 1969 by Boyle and Smith and first introduced into astronomy in 1974. The term "pixels" became part of the lexicon and CCDs changed everything. From 1977 to 1979, I had been working as a post-doc at the University of Arizona on solid-state array detectors, including the Reticon, the Charge Injection Device and finally, the Charge-Coupled Device or CCD. On arriving at ROE, my first task was to work on a CCD-based imaging spectropolarimeter with Ray Wolstencroft, but Director Malcom Longair and Head of Technology Malcolm Smith soon asked me what I knew about the development of infrared "array" detectors. Over the years I have written extensively in articles and books about the development of detectors for astronomy, especially regarding the remarkable impact of the CCD and infrared array. The latest edition of my book (McLean 2008) contains many historical references, and provides a chronicle of the key developments through this transitional period.

Thus it was that in 1982, I undertook a thorough survey of infrared detector array technology for ROE. I was surprised to find out that people had already tried to make CCDs using semiconductors with lower band-gaps than silicon; there was one example of a CCD made from indium antimonide (InSb). These efforts were not successful due to the difficulties of processing such materials to the level of purity already achieved for silicon. Nevertheless, infrared array devices using techniques other than charge-coupling were possible, and some were already in use for classified military applications. A few of these devices had "slipped out" into the US astronomical community. Of course, all of these early infrared arrays had small formats compared to existing CCDs and they were really intended for high-background, real-time applications. Devices that had made it to a telescope had been disappointing for faint-object astronomy. Thus, prospects seemed very bleak at first.

Development of IRCAM

Using contacts supplied by Donald Pettie, manager of the ROE Technology Unit, I made contact with Dr. Ian Baker at Mullard in Southampton. One of the technologies pioneered there was the use of mercury-cadmium-telluride (HgCdTe or MCT; actually known as CMT in the UK where this ternary semiconductor alloy was invented). The beauty of HgCdTe is that the band-gap is controlled by the ratio of mercury to cadmium, and thus the long-wave cut-off can be "tuned". For UKIRT, our goal was 1–5 μm imaging, but the Mullard work emphasized HgCdTe with 10–15 μm cut-offs. Consequently, Ian Baker directed me to the Rockwell Science Center in Thousand Oaks, California for shorter wavelength HgCdTe. Meanwhile, I became aware of the possibility that Santa Barbara Research Center (SBRC)

Fig. 3.1 A Sensor Chip Assembly or SCA composed of two slabs, one containing the infrared detector array and the other containing an array of silicon transistors, connected via columns of indium (bumps) and pressed together

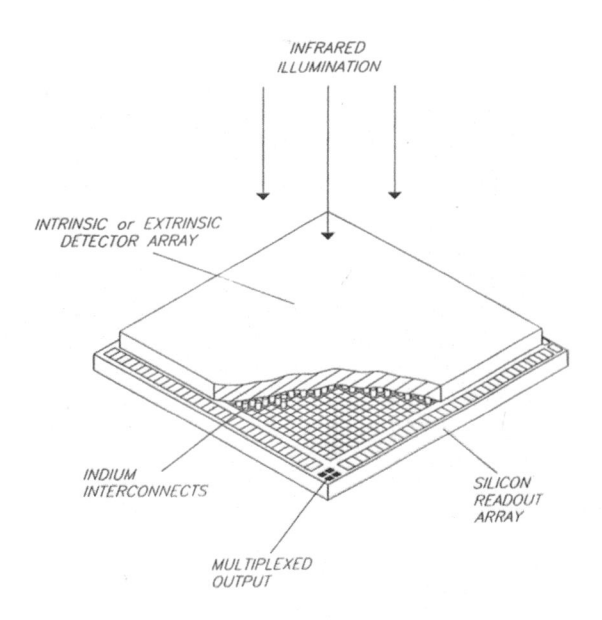

in Goleta, CA, from where UKIRT had obtained its individual InSb detectors, could make arrays of InSb or HgCdTe. In October 1982 I visited both of these vendors in the United States. Jon Rode of Rockwell estimated that it would take over one million dollars to develop HgCdTe arrays for low-background astronomy applications. Subsequently, when NASA funded the development of the NICMOS instrument for HST, astronomical devices using HgCdTe became possible and popular. At SBRC however, I encountered Dr. Alan Hoffman, an astronomer who had in fact already loaned a small format InSb array with a silicon CCD readout device to the University of Rochester. As a result of our interest, Alan convinced his upper management to take a closer look at the ground-based astronomy applications. In 1984 ROE selected SBRC to develop an InSb array specifically for near-infrared astronomy. As illustrated in Fig. 3.1, this device was not a CCD but a "hybrid" device composed of an array of 58×62 InSb photodiodes "bump-bonded" to a custom, matching, silicon readout circuit designed to sequentially access each pixel and feed the output to an output amplifier.

The development contract for the 58×62 InSb array was sealed in March 1984 during a spectacular visit to Mauna Kea by SBRC management while lava erupted from Mauna Loa. Terry Lee (Astronomer in Charge of UKIRT) drove me and the group to the summit. While Terry mended a flat tire in the twilight, I gave the group a quick tour of the telescope. In May 1984, I followed up with a visit to Santa Barbara, California accompanied by Tim Chuter who would soon become the engineering manager for the IRCAM project. On Monday, July 2, 1984 we held our first project meeting at ROE and UKIRT's first infrared camera got started officially.

Three years later we invited SBRC management back to see IRCAM in action on the telescope. The picture in Fig. 3.2 was taken during that visit.

Fig. 3.2 The author and Terry Lee flank the Vice President of SBRC (*centre*). David Beattie, Malcolm Smith and Malcolm Longair are *on the right*, and two other members of the SBRC group are *on the left*. SBRC is now a part of Raytheon

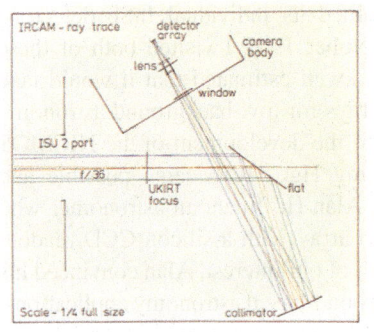

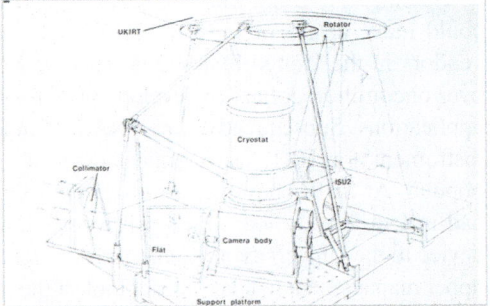

Fig. 3.3 *Left*: The optical layout of IRCAM, illustrating how the beam from the telescope was collimated outside the cryogenic body. *Right*: A beautiful hand-drawn sketch by ROE mechanical engineer Ron Beetles showing how IRCAM would be mounted on UKIRT

Construction of IRCAM was challenging. We knew that we wanted to provide spatial resolution that matched seeing, but it was not certain what the best image scale should be. I suggested that we provide three different camera lenses, each of which could be used with the same fixed collimator (Fig. 3.3).

We decided to collimate the beam externally using gold-coated mirrors and pass the collimated beam through the entrance window into the cryogenic filter wheel and camera section.

This decision not only kept the entrance window relatively small, but also enabled us to place a Fabry-Perot etalon just outside the window in the collimated beam. It was clear from the outset that a custom vacuum chamber would be needed to enclose the preferred optical design, rather than trying to fit the design inside

3 The IRCAM Revolution

Fig. 3.4 IRCAM is seen mounted on an optical test bench at the Royal Observatory, in June 1986

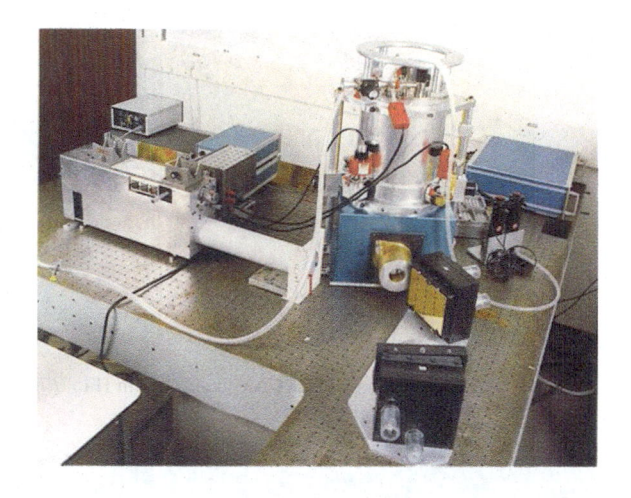

a standard IR Labs dewar. A dual filter wheel and a detector focus mechanism were also provided. These mechanisms were driven by stepper motors using shafts through the vacuum chamber. Unlike some previous infrared instruments that used hand-operated controls, IRCAM was fully automated and computer-controlled. One unique feature of the project was the up-front emphasis that was placed on making it easy to take and display images, as well as provide some quick look data analysis in real time.

Because the InSb detector needs to operate at 30 K, cooling would require liquid nitrogen for the optics and liquid helium for the detector. We separated the camera chamber with the optics and mechanisms from the cooling system, which was based on an Oxford Instruments LN2/LHe cryostat. Years later this approach paid dividends when it became possible to retrofit closed-cycle refrigerators and eliminate the liquid cryogens. Figure 3.4 shows the final instrument under test at ROE in June 1986. The module on the left with the white tube is an f/36 telescope simulator.

First Light

I had already moved my family to Hawaii in August and then I returned for 6 weeks to prepare IRCAM for shipping. The completed instrument was shipped out from Edinburgh in September 1986, 12 crates in total, for the long journey to Hawaii. Upon arrival, we set up the instrument in the lab at the Joint Astronomy Centre in Hilo for complete end-to-end testing before taking it to Mauna Kea.

After IRCAM was shipped, I was joined almost immediately in Hawaii by my colleague Colin Aspin, the IRCAM post-doc who was largely responsible for the camera software. Later, our two Edinburgh University Ph.D. students, Mark McCaughrean and John Rayner, also came out for planned observing runs. As shown in Fig. 3.5, by October 21, 1986 the entire system was mounted on the telescope and ready to go (Fig. 3.6).

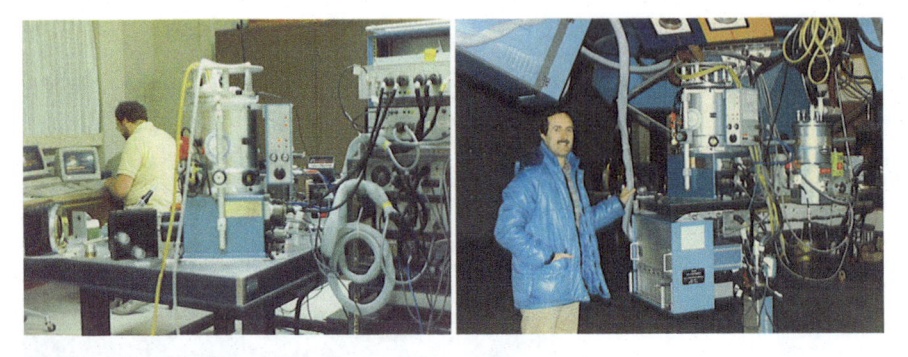

Fig. 3.5 *Left*: Colin Aspin with IRCAM under test in Hilo. *Right*: The author alongside IRCAM already mounted on UKIRT (October 21, 1986)

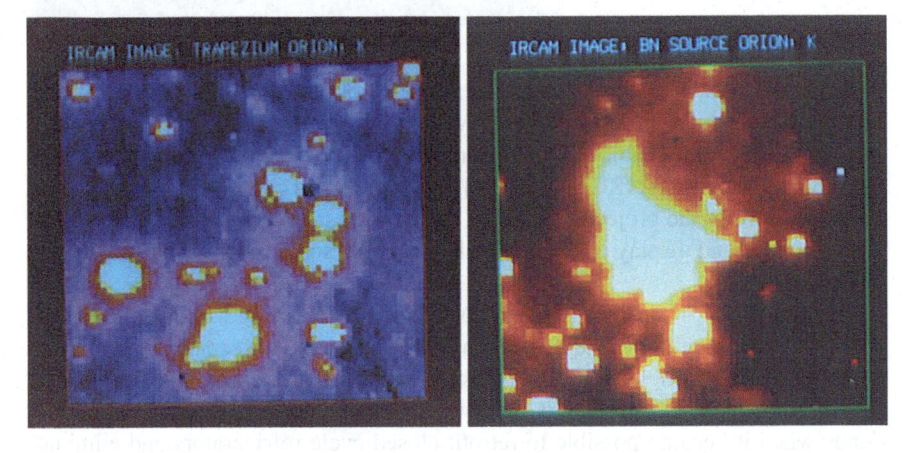

Fig. 3.6 Screen captures of the "first light" images with IRCAM on October 23, 1986. *On the left*, the Trapezium star cluster in Orion and *on the right* the Becklin-Neugebauer (BN) infrared object located a few minutes of arc to the north. The wavelength is the K-band

We were allocated "morning" observing time, beginning at 8 am on October 23. At the telescope was Colin Aspin, Gillian Wright, Dolores Walther and me. Our first target of the twilight was the Trapezium star cluster in Orion. As that first image came in, Colin and I exchanged smiles, and then from behind me I heard Gillian's voice say, "I don't believe you two!" She meant of course that a momentous occasion had just occurred and Colin and I had passed it off almost as if nothing had happened. Of course, Colin and I had taken hundreds of images before the camera even left Edinburgh and so we knew that it worked. Nevertheless, Gillian was right. This was a moment to be celebrated. We all laughed and shook hands, and then we took an image of a true infrared source, the Becklin-Neugebauer (BN) object, but it was only the beginning. In the weeks that followed, the flood gates opened.

3 The IRCAM Revolution

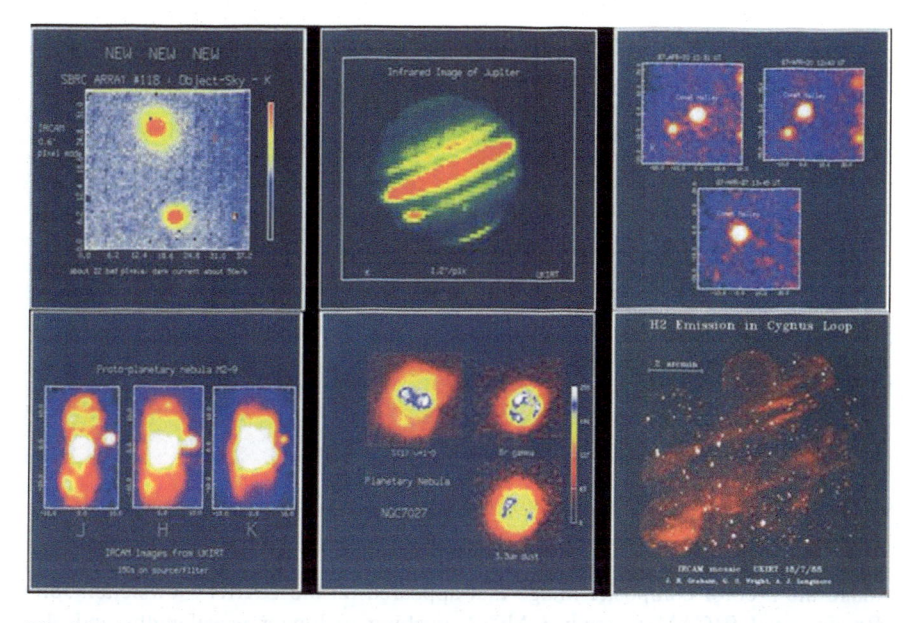

Fig. 3.7 A selection of early images obtained with IRCAM which illustrate the range and power of the new camera to resolve detail. Most of the images use the 0.6″ scale, but the Jupiter image has the 1.2″ scale and the Cygnus loop was done with 2.4″ per pixel

Early in 1987 we replaced the detector with a new and better device. This chip, FPA 118, had about 22 bad pixels and about 50 electrons/s/pixel dark current. Runs were planned that used all three of the IRCAM image scales, 0.6, 1.2 and 2.4 arcsec per pixel, and as Fig. 3.7 illustrates, targets ranged from solar system objects like Jupiter and Comet Halley, to planetary nebulae and supernova remnants.

We also perfected the method of constructing "mosaics" comprising many overlapping tiles so that large fields of view could be displayed as if the detector had many more pixels. Figure 3.8 shows the Antennae galaxies and M51. The Antennae is a two-frame mosaic in H-band with a 400 s exposure using the lowest spatial resolution scale of 2.4″/pixel. But perhaps our most exciting extragalactic image was the one of NGC3690 obtained by Gillian Wright and Bob Joseph while I was also conducting a tour of the UKIRT for the Vice-President of Santa Barbara Research Center. When the nuclei of both colliding galaxies was resolved and the screen image clearly showed that nucleus B was double, the astronomers erupted in jubilation and our visitors experienced the full impact their technology held for astronomy. The ability to obtain good images of extragalactic objects was very gratifying, especially as some people had been skeptical as to whether the arrays would have enough sensitivity.

One of the largest and most significant mosaics was created by Mark McCaughrean for his thesis. This was a mosaic of the Orion Nebula. Over 120 overlapping frames, in three colors (J, H and K), were obtained and combined to

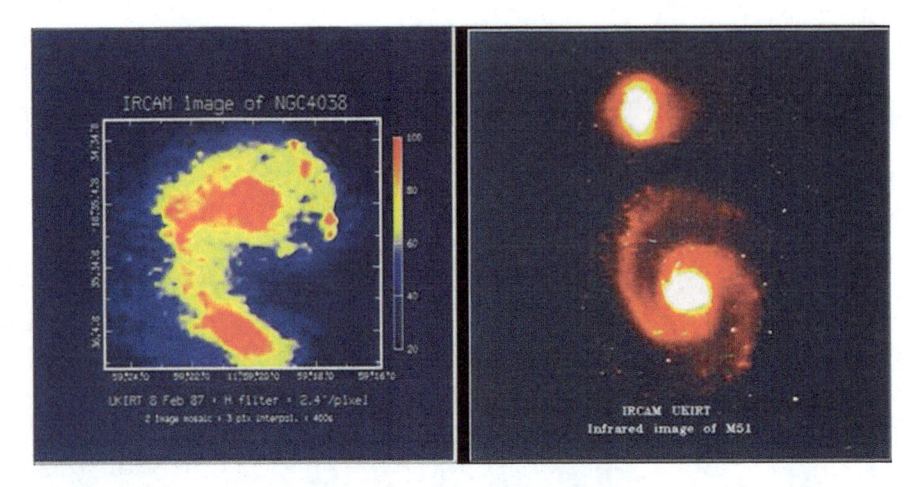

Fig. 3.8 *Left*: A mosaic of NGC4038/4039 (the Antennae galaxies). *Right*: A mosaic of M51

make the 3-color composite image that appeared on the UKIRT Newsletter. John Rayner used IRCAM to study OMC-2, pushing to longer wavelengths and also using the polarimetry mode.

The Turning Point

Many results from IRCAM were presented at the March 1987 conference in Hilo called "Infrared Astronomy with Arrays". The local organization of this meeting was handled by David Beattie and me, and proceedings were edited by Gareth Wynn-Williams and Eric Becklin (1987).

There is no doubt from the reviews at that meeting that the quality of the IRCAM images caught everyone by surprise. I am often asked whether or not I knew that the world of infrared astronomy had changed. I think we did have a good sense that something major had just occurred and that UKIRT was right at the centre of it. Based on comments from Wayne van Citters (NSF) and Don Hall (UH) at the close of the Hilo meeting, I think the community also realized that a turning point had been reached. Wayne van Citters said: "I don't think there's any doubt that we are on the verge of a promised land in optical and infrared astronomy." Referring specifically to IRCAM, Don Hall summarized the conference by saying, "Things that seemed in the future and beyond my grasp for so long are clearly here."

The buzz and excitement of the conference bubbled over and was captured in a moment that will live with me forever. In the middle of my talk I showed an IRCAM mage of OMC-2 obtained by John Rayner and me (Fig. 3.9). As I displayed the polarization map of the reflection nebula and then a set of images at progressively longer wavelengths to illustrate the changes in appearance with less reddening I said,

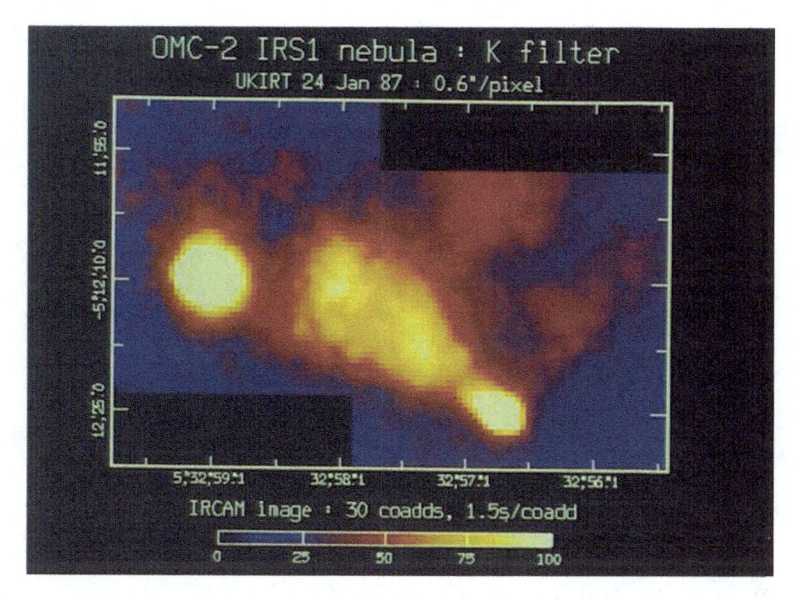

Fig. 3.9 This is a two-frame mosaic of the OMC-2 IRS1 nebula in K-band with 0.6″ per pixel

"If this is what infrared astronomy is going to be like from now on, then all I can say is – I like it!" To my astonishment I received spontaneous applause in the middle of the talk. Clearly, I struck a nerve, and I think I simply stated what everyone was feeling. It was a great moment.

The results kept coming. Among the many important new areas opened up by IRCAM was the ability to study the Galactic Centre in unique ways. For example, IRCAM was used by Andy Longmore and others to obtain high-speed occultation snapshots during the lunar occultation in April 1987. Later, using a Fabry-Perot etalon in front of IRCAM, Gillian Wright and others were able to create velocity maps in Brackett gamma at 2.17 μm.

Then and Now

In the years since those pioneering days of the late 1980s, much has happened. Infrared astronomy has benefited from the advent of adaptive optics, as well as the continued and successful growth of detector array technology. In 1984 I was working with an array with 58×62 pixels, each pixel 60 μm in size. Today, $2\,k \times 2\,k$ arrays with 18 μm pixels are standard, dark currents are less than 0.01 electrons/s/pixel and readout noise levels are typically around 5 electrons rms or better with multiple sampling.

Figure 3.10 shows an image that I obtained in 1987 of the Galactic Centre with IRCAM alongside a diffraction-limited image of the same region obtained by the

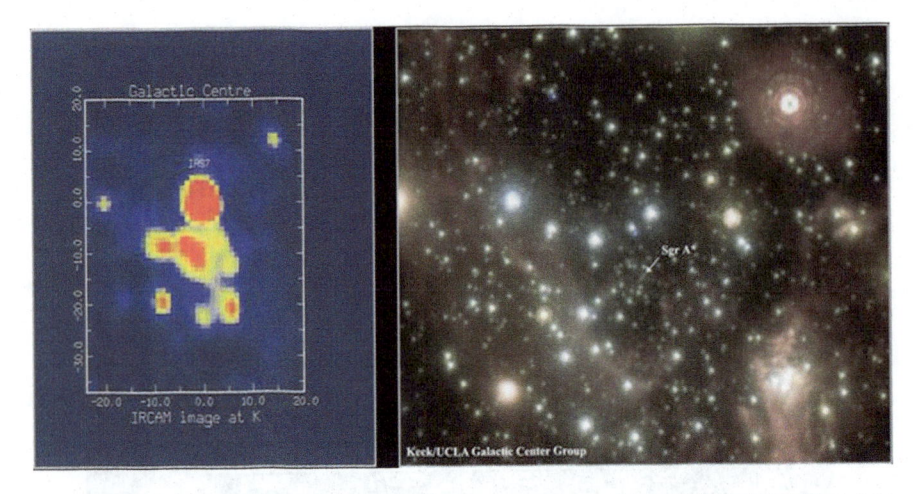

Fig. 3.10 *Left*: A K-band image of the Galactic Centre obtained with IRCAM in 1987. The bright source is IRS7. *Right*: A diffraction-limited infrared image obtained by the UCLA group in 2007 using the laser guide star adaptive optics system on the Keck 10-m telescope

UCLA group using one of our current IR cameras on the Keck 10-m telescope. Twenty years of hard work and dedication by a huge number of people is illustrated here, but I am certain that the decision to develop IRCAM for UKIRT was a key part of the stimulus that led to where we are today.

Acknowledgments It is a pleasure to acknowledge Tim Chuter and the entire IRCAM team at the Royal Observatory Edinburgh and at UKIRT, as well as Mark McCaughrean and John Rayner. The IRCAM revolution could never have happened without this wonderful team. It is also a pleasure to acknowledge the late Donald Pettie for his personal support and encouragement.

References

Becklin, E.E., Neugebauer, G.: Infrared observations of the Galactic Center. Astrophys. J. **151**, 145–161 (1968)
Kleinmann, D.E., Low, F.J.: Observations of infrared galaxies. Astrophys. J. **159**, L165–L172 (1970)
McLean, I.S.: Electronic imaging in astronomy, 2nd edn. Springer, Berlin/Heidelberg/New York (2008)
Wynn-Williams, C.G., Becklin, E.E. (eds.): Infrared astronomy with arrays. The Institute of Astronomy, University of Hawaii, Honolulu (1987)

Chapter 4
Continuum Submillimetre Astronomy from UKIRT

Ian Robson

Abstract We review the development of submillimetre continuum astronomy on UKIRT between the telescope's inauguration and the handing over of the baton to the JCMT in 1997.

Introduction

Submillimetre continuum astronomy means getting above most of the atmospheric water vapour and that translates to high altitudes for ground-based sites. The Queen Mary College group had spent a number of years sampling telescopes and sites, such as the Pic du Midi at around 10,000 ft and Mount Evans at 14,260 ft in the Colorado Rockies. However, the best combination of telescope size and altitude by far was the University of Hawaii's 88-inch telescope on Mauna Kea, at almost 14,000 ft above sea-level. Although our experience of using the telescope and weather had not been wonderful (Eve et al. 1977), nevertheless, it was clear that for truly submillimetre (as opposed to near-millimetre) astronomy, Mauna Kea was the best developed site at the time. Therefore, when I joined the UKIRT Oversight Committee in the mid-1970s, submillimetre (or far infrared astronomy as it was often called then) was in it infancy (Robson 1979) and we were very determined to maximise the effectiveness of the telescope for our work. Additionally, I could bring first-hand experience of observing on Mauna Kea to the debates.

My first contribution was to remove the 'folded-prime' configuration that had been selected as the location for the submillimetre photometer in order to retain a low f-number. While this would allow the detector to be more easily optically fed, it

---

I. Robson (✉)
UK ATC, Royal Observatory, Blackford Hill, Edinburgh EH9 3HJ, UK
e-mail: ianrobson@stfc.ac.uk

meant that the instrument would sit above the primary mirror and access (and helium fills) would be through the Cassegrain hole. This was less than a practical way forward and I could see that it would not help in getting telescope time, so we agreed that scrapping the folded prime and working at Cassegrain would be an acceptable option. This was a very popular decision and as a result I suspect someone in the project was the recipient of a 1 m diameter perfectly polished flat piece of glass – a brilliant coffee table no doubt. This then enabled the submillimetre to be accepted as part of the instrument suite: Cass-mounted, readily accessible and 'part of the general system'. My second contribution was to help persuade SERC to have the UKIRT headquarters in Hilo, rather than Waimea, in spite of the fact that land had already been purchased there. My third contribution was to agree on the colour of the telescope and mount with Gordon Carpenter, the Project Manager.

Submillimetre Photometry

One aspect that had been clear from our extensive experience on many telescopes was that in order to control background radiation (in the absence of chopping secondaries, which were not yet developed) we needed a purpose-built photometer, with careful consideration of tertiary chopping and background suppression. Our design (Robson et al. 1978) was tested on the Flux Collector at Izaña and this produced a large improvement, albeit the site was not conducive to submillimetre observing.

With the opening of UKIRT we were one of the first groups to gain access to the telescope and mounted our submillimetre photometer at the $f/9$ Cassegrain focus and began opening up the discovery space. The photometer was directly fed at the Cassegrain focus and came equipped with a large field-of-view offset guider (eyeball driven), which got around the problem of poor telescope pointing and the ability to guide in regions of the sky such as dark clouds, where there were few guide stars. However, the QMC photometer (as it was called) required someone (usually a research student) to guide the telescope from the Cass focus, often balanced precariously on the top of steps in the pitch dark (and freezing conditions) of the dome (health and safety was different in those days). The 'real' astronomers sat in the control room in the relative warmth, trying to manage the observing and data taking, get the data out of the PDP11/40 computer and the signals on the strip chart (no on-line data analysis) and try and make sense of whether it was worthwhile integrating further or go to another target (Fig. 4.1). Communicating with the 'guider' in the dome was via radio mikes and headsets. On the other hand, guiding through a large eyepiece on a 4 m telescope produced some spectacular views that we will all remember (Fig. 4.2).

In those early days the continuum detectors were still quite primitive, being liquid helium cooled doped germanium bolometers with rudimentary broad-band filtering. The end-product was that progress at the time seemed disappointingly slow and was limited by telescope, weather and instrument problems, with the result that only the brightest submillimetre objects were detected, and usually only at 800 μm.

Fig. 4.1 The observing room at UKIRT in the early days – the author and Peter Ade

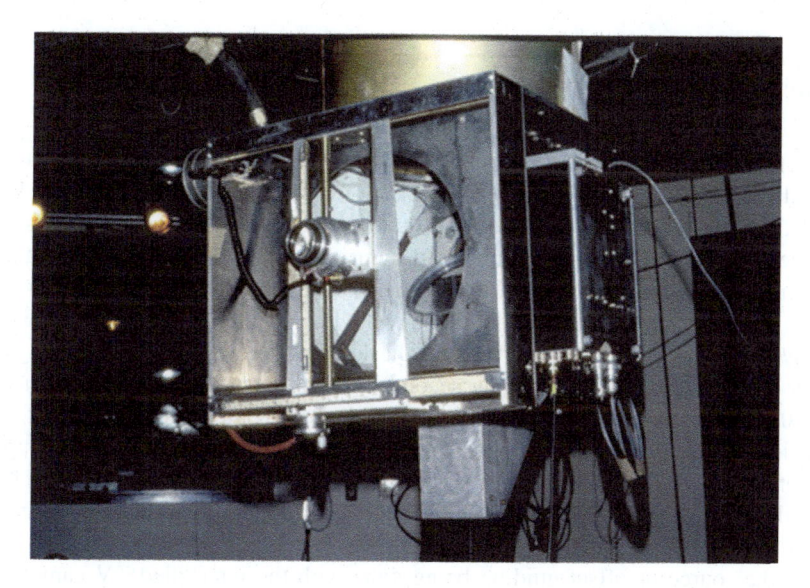

Fig. 4.2 The QMC Photometer on UKIRT showing the offset guiding side

Nevertheless, looking back, this was a time when valuable knowledge and experience of understanding the 'observing system' (filters, atmospheric attenuation, noise sources) were gained, all of which would lead to improvements and ultimate breakthroughs. Throughout this time, calibration was always difficult. Mars was the primary calibration source and there was very little else. From this early era one paper stands out: 'The submillimetre spectra of planets: narrow-band photometry' (Cunningham et al. 1981).

Nevertheless, in spite of this UKIRT was well up with the state-of-the-art in submillimetre astronomy at the time, with other groups (notably Caltech, Chicago, Arizona) making similar observations from various telescopes and the KAO. The challenges facing us on UKIRT and all the ground-based telescopes were overcoming the high and variable sky-background power with the resulting baseline instability and of course the sky-noise itself. Our very fast 110 Hz chopping removed the high-frequency component well enough but still left major baseline drifts. Also, as wavelength filtering was in its infancy, understanding the effects of an unknown spectral index of the source and potential high- and low-frequency filter leaks were other factors that had to be taken into account. Improvements were targeted at: better control of filtering; improved detector sensitivity; and crucially, having a chopping secondary on the telescope.

The Game Changers

The first two avenues were completed during 1980–1981 with new and narrower-band filters possessing excellent leak rejections. This was the speciality of Peter Ade, who went on to become the filter supplier to the Universe of astronomers. The detector sensitivity was vastly improved by using helium-three cooled germanium bolometers; being cooled to 350 mK rather than 1.2 K with pumped helium-four. This came about through collaboration with Ira Nolt and Jim Radostitz of the University of Oregon, who developed the helium-three systems. The really big step came with the $f/35$ chopping secondary being introduced on UKIRT in late 1980 and becoming the only top-end available a year later. From this point until the introduction of UKT14, improvements would be incremental. The chopping secondary came with a new Cassegrain mounting unit (the famous ISU 1, or 'gold dustbin') incorporating a 'straight-through' optical TV guider, which was sensitive enough that all future observing would be done from the control room with the source or offset guiding being done with the intensified TV camera. At a stroke the QMC photometer was made redundant – the submillimetre radiation was now fed directly into the detector cryostat from the tertiary dichroic and this arrangement became known as the 'QMC-Oregon' photometer (Ade et al. 1984). It was essentially a compact cryostat with all the filtering done internally via a filter wheel at 4.2 K (Fig. 4.3).

The chopping secondary produced huge improvements in baseline stability. The chopping frequency was usually around 12.5 Hz and the two beams, each of about

Fig. 4.3 (*Left*) Peter Ade and Jim Radostitz with the QMC-Oregon photometer and (*right*) the photometer itself

64 arcsec diameter at 1.1 mm, were separated by 120 arcsec on the sky. The new wavelength filtering also better isolated the atmospheric windows and for the first time separated out the 450 and 350 μm windows. Although this resulted in a small reduction of flux from the source, the benefit of removing a large and varying atmospheric emission power was dramatic and far outweighed the loss of flux in terms of detectivity. This was a crucial feature for UKIRT because most other groups continued to use very wide-band filters for many years.

The results of all these improvements led, by 1982, to an improvement in the noise-equivalent flux density from around 200 Jy Hz$^{-0.5}$ to around 5–10 Jy Hz$^{-0.5}$ at 1.1 mm. This was the breakthrough we had all been waiting for and the discoveries followed thick and fast. A further benefit was that on-line data analysis was now available for the first time to the submillimetre continuum community – but this was another story, which started off with a bang (see Radio Quiet Quasars below).

The trend to design a photometer system that basically fitted onto the telescope like any other facility instrument was also used by other submillimetre groups headed up by the University of Chicago with Roger Hildebrand and involving astronomers from various institutes in California. UKIRT observations of a range of evolved stars (such as IRC + 10216, CRL2688, NGC7027 and VYCMa – which would all go on to become well-observed objects from UKIRT and subsequently the JCMT) in 1981 (Sopka et al. 1985) using the Chicago submillimetre photometer (Whitcomb et al. 1980) demonstrated the success of this technique. Again, these groups began with He-4 and like ourselves, moved into He-3 for greater detectivity.

Atmospheric Attenuation

The move to narrower-band filters not only helped the photometry and noise reduction but also enabled atmospheric attenuation to be better understood. Extensive work was undertaken on this by Bill Duncan and myself, including experiments with two- and three-position chopping/nodding techniques but regrettably this was never fully written up. As well as the experiments in sky-background/noise cancellation it was also vitally important to determine the attenuation as accurately as possible because certainly in the submillimetre the attenuation coefficient was high and so any change of airmass between calibrator and source could mean a very significant difference in flux estimation. The problem of the changing atmospheric attenuation would plague submillimetre astronomy, limiting the accuracy of observation, for many years until SCUBA on the JCMT along with sky emission monitors essentially solved the problem.

In the UKIRT days the best that could be done was to make frequent measurements of calibration sources and produce secant plots of flux versus airmass. Of course this was a time-consuming task as it had to be done in each filter, which added overhead to the astronomical programme. For stable nights this gave good results, allowing an attenuation coefficient to be derived at each filter and so some degree of confidence in astronomical fluxes. For many nights, however, secant plots were anything but stable, leading to uncertainty in the attenuation and final data. It also seemed that there were never enough stable nights to be able to determine accurately the attenuation coefficient as a function of precipitable water vapour, or even know what this latter value was (apart from through models). Nevertheless, most of us accepted these observing overhead in the quest for best photometric accuracy of the data.

Steady-State Submillimetre Progress

Some of the first submillimetre measurements from UKIRT in late 1980 dealt with emission from the planets, the thrust being to use some of these as calibration sources, but also to understand the atmospheric emission models and for the first time the contribution of Saturn's rings to its total emission (Cunningham et al. 1981) and hence dust particle size distributions. Mars was the primary calibrator for all submillimetre observations due to it having essentially no atmosphere, a solid surface, and had been extensively studied by the Mariner 6 and 7 probes. This allowed models of the far-infrared emission to be developed (Wright 1976), which became the basis for all subsequent calibration. The work on the planets continued throughout the 1980s, with observations making more inroads on the physics of planetary atmospheres as the detectivity improved along with the narrow-band filters. Highly detailed measurements of Jupiter were made by Griffin et al. (1986) who found that from 350 to 450 μm observations the Jovian atmosphere is best described by the addition of ammonia ice particles of size between 30 and

100 μm. However, the authors also pointed out that if the absolute calibration scale (due to Mars) was increased by only 5 %, then there would be no need to incorporate additional absorbers in the Jovian atmosphere.

A novel collaboration took place between groups working on the IRTF and UKIRT. It was clear that the IRTF was better at the thermal infrared, especially 10 and 20 μm, and so it was agreed to undertake joint programmes of observations utilising the best features of both facilities. UKIRT provided the submillimetre and both telescopes the near-IR as suited. The programmes covered a range of objects but one of the first observations was of the asteroid, (10) Hygiea (Lebofsky et al. 1985). This was the first detection of an asteroid in the submillimetre and along with the thermal- and near-IR data provided the output spectrum over a crucial range of wavelengths and enabled a discussion of the temperature and surface properties to be undertaken. Subsequently (1) Ceres was later also detected at all submillimetre wavelengths.

The Arrival of UKT14

The popularity and productivity of continuum submillimetre astronomy had resulted in the procurement of a common-user photometer from the ROE. This was called UKT14 (number 14 in the series of ROE photometers) and although originally designed primarily for UKIRT, was capable of operating on the JCMT with a minor modification. Bill Duncan was the Project Scientist and I was the external Project Scientist. The photometer was designed to incorporate all the lessons and advances that had been learned from the years of work on UKIRT with the QMC systems (Duncan et al. 1990). The detector in UKT14 was a composite germanium bolometer, cooled by helium-3 to 350 mK. The radiation was channelled to the detector through a variable aperture stop, stray-light minimising optics, a 10-position filter wheel and a low-pass filter. This represented a step-change, not so much in raw detectivity, but in noise suppression and bandpass filter control, the latter being upgraded to the latest state-of-the-art filters in mid-1988.

UKT14 was the pinnacle of single-pixel devices and it would be a hugely successful instrument, doing excellent work on UKIRT and then the JCMT. The impact of UKT14 is often forgotten following the advent of SCUBA, but it laid the foundations for a community of astronomers and much of the work that SCUBA would revolutionise (Fig. 4.4).

The Link with IRAS

One of the key events in far infrared astronomy was the IRAS mission. This promised to provide a wealth of new discoveries and we were particularly keen to capitalise on this by undertaking follow-up observations as early as possible.

Fig. 4.4 The author with UKT14 (*red*) on UKIRT

Funding was awarded for a 6-month programme of submillimetre IRAS Follow-Up observations on UKIRT using the early morning twilight hours – also necessary due to the IRAS orbit and observing pattern. Unfortunately, this was not successful due to a number of reasons-technical, logistic and the information coming out of the IRAS data centre at RAL-and was eventually curtailed after about six weeks.

However, IRAS follow-up observations continued through regular observing sessions and a number of discoveries were made. I will pick just one example, the measurement of the submillimetre spectrum of Arp220 in May 1984 (Emerson et al. 1984). The 350 and 760 μm observations were augmented by 20 μm measurements, the latter showing that all the emission came from within the 4-arcsec beam. The data, along with the IRAS fluxes gave a dust temperature of 61 K and a luminosity of 10^{12} L$_\odot$. This is around 100 times that of the nearby starburst galaxy M82 and pioneered the class of super-starburst galaxy referred to at the time as 'IRAS high luminosity galaxies' (to become LIRGS and ULIRGS in due course).

Galactic Observations

IRAS also played a part in galactic observations from UKIRT. These continuum studies had lagged some way behind their extragalactic counterparts for one very good reason, they were mostly extended and mapping with a single pixel detector was a non-trivial process, being both time-consuming (requiring stable weather conditions over many hours) and extensive data reduction by hand as there were no software packages available at the time. Although a number of sources were observed, very few were ever written up apart from in Ph.D. theses. Galactic star forming regions were undoubtedly the realm of the heterodyne instruments, which made very good progress on UKIRT.

Nevertheless, one of the first measurements was the submillimetre emission from the planetary nebula NGC7027 in 1982 and 1983 by Gee et al. (1984). This was at 350 μm and showed that the emission was above an extrapolation of the free-free emission, being almost certainly due to thermal emission from dust with a temperature of order 20 K. This was the first detection of cold dust in the halo of a planetary nebula and the mass of dust was consistent with the gas mass deduced from CO observations. Another example of Galactic (almost point-source observations) was GL490, a deeply embedded bipolar far infrared outflow source. This was part of the IRAS follow-up work and the UKIRT 350 μm mapping observations (Gear et al. 1986b) showed that the dust was most likely in the form of a disc, but that it was incapable of channelling the outflow, which must be collimated on a much smaller size-scale.

A more extensive programme of submillimetre observations of a sample of compact HII regions was undertaken by Gear et al. (1988), who used the newly arrived common-user photometer, UKT14, early in 1986 to determine the emissivity index, β, of the dust emission using matched beamsizes at 350, 800 and 1,100 μm. The result was a mean value of $\beta = 1.75 \pm 0.20$. Indeed, the arrival of UKT14 in 1986 allowed some major submillimetre mapping programmes to be undertaken on selected regions. It also turned out that some of the last submillimetre continuum observations to be done on UKIRT were galactic, dark-cloud mapping programmes. Ward-Thompson and Robson (1990) undertook IRAS follow-up observations by mapping the W49 complex at 350, 800 and 1,100 μm in 1988 revealing extensive cool dust of $T_d \sim 50$ K and an emissivity index, β of 1.8. The far infrared luminosity of 2.7×10^7 $L_\odot$ made this complex one of the most luminous star-forming regions in the Galaxy and the data showed that the dust cloud must be fragmented on small-scales, possibly indicating continuous star formation. This was followed up by mapping of the ρ Ophiuchi dark cloud complex, again in 1988, by Ward-Thompson et al. (1989). The submillimetre mapping was combined with IRAS maps to reveal the extent of the cool emission and the highlight was the discovery of a new source, labelled SM1, which was very cold ($T_d \sim 15$ K) and of sufficient mass that it was a candidate for a new protostellar object. However, the very nearby (less than 1 arcmin) Class-0 discovery source, VLA1623 would have to await observations using UKT14 on the JCMT, a few years later.

The Radio Quiet Quasar Story

Turning to the extragalactic submillimetre programmes on UKIRT, a group from Europe had produced a paper claiming a high ratio of detections at 1 mm wavelength of a sample of radio-quiet (actually radio-weak) quasars using the prime focus of the 3.6 m ESO optical telescope at La Silla, Chile (Sherwood et al. 1982). This was exciting because if the observations were correct, then this meant a possible new source of emission as the 1 mm point lay way above the radio extrapolation and the radio-1 mm spectral slope was too steep for thermal emission from dust. We set about checking this claim in 1982 and with the first QSO we obtained an instant detection – a 5-sigma result in just a few tens of minutes. This repeated with the next two sources, but then caution set in, this seemed far too good to be true. Sure enough, looking at the raw data quickly revealed that there were as many negative signals as positives, which prompted immediate suspicion and a quick check on a HP45 calculator showed zero detection.

So, what had gone wrong? The puzzle was quickly solved; a software bug in the brand-new, on-line facility data analysis package that was being used for the first time had allowed the modulus of the signals to be used rather than their absolute values. Hence we were doomed to detect anything we looked at, even blank sky. Of course for bright sources such as the planets, which we had been observing for calibration and set-up (pointing/focus) purposes, this had no effect.

Euphoria rapidly turned to disappointment when over the next two nights every one of the quasars in our sample showed zero detection, well below the claimed detected values. We extended the sample and wavelength range and eventually published the result (Robson et al. 1985) removing the possible new physics and the earlier claim was quietly forgotten, showing the perils of the early days of the field. One of the lessons we had learned was that we had to be extremely careful to be sure that we had taken into account all possible sources of noise and systematics in determining the putative detection of these hard-to-observe sources that were on the threshold of detectivity, or below. In fact, detecting radio-quiet quasars in the submillimetre was to remain a challenge for almost the next decade.

Extragalactic Observations

As noted above, point-like objects were always easier targets than very extended objects and so the extragalactic field opened up rapidly, albeit always fighting sensitivity challenges. One of the first successes was, looking back on it, an amazing detection of the nearby active galaxy NGC5128 (Centaurus-A). This was clearly done on a very dry night as the elevation of Cen-A from Hawaii is only 27° at transit! Also, this was only 1981 and so very early in the lifetime of submillimetre astronomy. The 370 μm flux was 59.5 Jy (a 4σ detection, which gives an idea of the sensitivity at the time). Detections were also made at 770 and 1,070 μm which allowed the thermal emission from dust to be well characterised (Cunningham et al. 1984).

This observation was followed by a multi-wavelength study from UKIRT ranging from the near IR through to the millimetre. In this case the target was another AGN, NGC1275 (Perseus A). Unlike Cen-A, the observations did not unambiguously show thermal emission from dust, but rather a smooth spectrum extending throughout the wavelength regime. The observational programme was undertaken on three occasions from 1980 to 1982 and so variability was always a concern. Nevertheless, this was one of the first sets of observations of its type from UKIRT (Longmore et al. 1984). Later observations of NGC1275 (Gear et al. 1985b) would go on to show that while the millimetre through near-IR emission was dominated by synchrotron radiation from a very compact component, thermal emission from heated dust, very similar to NGC1068, was also present. Furthermore, the material responsible for the star-formation causing the far-infrared emission was undoubtedly derived from the in-falling, cooling-flow from the Perseus cluster itself.

Radio-loud, flat-spectrum AGNs had been a target of the group for some time, using Kitt Peak prior to UKIRT (Rowan-Robinson et al. 1978). These programmes continued and an early success was with the radio-loud quasar 3C273, which had been part of the multi-wavelength (millimetre through infrared – and IRTF thermal IR observations as noted above). We had managed to obtain sufficient observations that the 'baseline', or 'quiescent' emission spectrum of 3C273 was known (Gear et al. 1984) and so when the object had a major flare in early 1983 we were able to monitor the emission and subtract this baseline to obtain the spectrum of the emission component (Robson et al. 1983). The data showed that the flare propagated to longer wavelengths while decaying at shorter wavelengths and the timescale was suggestive of an event in the central 0.1 pc of the source. The flare was modelled by Marcher and Gear (1985) as a shock-in-jet, and this was to form the basis for the physics of the high frequency variability in Blazar-jet models for the future.

Continued monitoring of 3C273 eventually revealed a new infrared component in the emission, which was believed to be from free-free emission from the broad-line clouds (Robson et al. 1986). The work on 3C273 opened up what would become a long-standing collaboration with Thierry Courvoisier and resulted in the first of a truly multi-wavelength (radio-through X-ray) study as reported in Courvoisier et al. (1988). This work focused on the origin of the various components in the central engine of these AGNs, especially the X-ray emission (synchrotron-self-Compton etc.), and hinged on the temporal variations of the longer or short-wavelength components. The later data in these papers were all from the JCMT.

Blazar monitoring from the near-mm through IR became one of the well supported programmes on UKIRT for many years, producing a new generation of submillimetre astronomers in the process along with a wealth of papers. The 'Multifrequency observations of Blazars' series, from paper I through IV (Gear et al. 1985a, 1986a; Brown et al. 1989a, b) made breakthroughs in terms of the jet physics from blazars and further demonstrated the viability of the Marscher and Gear model. This series of observations essentially spanned the most productive epoch of submillimetre observing on UKIRT, mostly with UKT14.

Although the radio-loud sources were the 'bread and butter' targets, strenuous attempts were made to detect submillimetre emission from the 'starburst' galaxies and in 1985 the southern object, NGC253, was 'mapped' with the QMC photometer (Gear et al. 1986c) showing that the thermal emission extended along the major axis of the galaxy. This provided an estimate of the interstellar mass in the central kpc of around 3×10^8 $M_\odot$, which, along with CO data suggested that the material is in the form of massive clouds, similar to what is seen in our galaxy but unlike what was thought to be the case for M82 (large ensemble of small clumps).

The Remote Eavesdropping/Observing Experience

Before closing it is worth just mentioning for the record the work on remote eavesdropping and observing that we participated in extensively as part of the monitoring programme. The blazer monitoring was an ideal test-programme in that it was routine, had experienced observers and used standard techniques. It was very clear that although the technology was rudimentary, not only was the dedicated telemetry line a requirement but voice communication between Edinburgh and the summit of Mauna Kea was also a requirement. The experiments were way ahead of their time and showed that careful planning was crucial but that in principle, with well understood instruments at the telescope, the algorithms for observing could be written in advance. However, the key was that solid data reduction pipelines would need to be operating both at the summit (in case of line dropout) and for the remote observers in order that the programmes could be carried out efficiently in real-time. This also meant having a good handle on the atmospheric extinction, which for the submillimetre, posed the greatest drawback to remote observing and confidence building. Nevertheless, these experiments were critically valuable for the data analysis work that would eventually lead to the pipelines and queue-scheduled, priority-based, weather-banded scheduling and observing, that became the feature for SCUBA on the JCMT, UKIRT itself with UKIDSS and now in common use on the major telescopes of today.

So in conclusion, not only did UKIRT provide fantastic discovery science and set ground-based submm astronomy on a firm footing. It also produced a whole generation of submm astronomers who would go on to have highly distinguished careers and leaders in their fields. An amazing achievement for a cheap, flux-collector and a great legacy.

References

Ade, P.A.R., et al.: The Queen Mary College/University of Oregon photometer for submillimetre continuum observations. Infrared Phys. **24**, 403 (1984)

Brown, L.J.M., et al.: Multifrequency observations of Blazars III: the spectral shape of the radio to Xray continuum. Astrophys. J. **340**, 129 (1989a)

Brown, L.J.M., Robson, E.I., Gear, W.K., Smith, M.J.: Multifrequency observations of Blazars IV: the variability of the radio to ultraviolet continuum. Astrophys. J. **340**, 150 (1989b)

Courvoisier, T.J.L., et al.: Rapid infrared and optical variability in the bright quasar 3C273. Nature **335**, 330 (1988)

Cunningham, C.T., et al.: The submillimetre spectra of planets: narrow-band photometry. Icarus **48**, 127 (1981)
Cunningham, C.T., Ade, P.A.R., Robson, E.I., Radostitz, J.V.: The submillimetre and millimetre spectrum of NGC5128. Mon. Not. R. Astron. Soc. **211**, 543 (1984)
Duncan, W.D., et al.: A common-user millimetre-submillimetre continuum photometer for the James Clerk Maxwell Telescope. Mon. Not. R. Astron. Soc. **243**, 126 (1990)
Emerson, J.P., et al.: IR observations of the peculiar galaxy Arp220. Nature **311**, 237 (1984)
Eve, W.D., Sollner, T.C.L.G., Robson, E.I.: Submillimetre lunar emission. Astron. Astrophys. **59**, 209 (1977)
Gear, W.K., et al.: Millimetre wave observations of flat spectrum radio sources. Astrophys. J. **280**, 102 (1984)
Gear, W.K., et al.: Multifrequency observations of Blazars I: the shape of the 1 micron to 2 millimeter continuum. Astrophys. J. **291**, 511 (1985a)
Gear, W.K., Gee, G., Robson, E.I., Nolt, I.G.: Thermal and non-thermal emission from NGC1275(3C84). Mon. Not. R. Astron. Soc. **217**, 281 (1985b)
Gear, W.K., et al.: Multifrequency observations of Blazars II:the variability of the 1 micron to 2 mm continuum. Astrophys. J. **304**, 295 (1986a)
Gear, W.K., et al.: Submillimetre observations of the disc around the embedded source GL490. Mon. Not. R. Astron. Soc. **219**, 835 (1986b)
Gear, W.K., et al.: Submillimetre continuum observations of NGC253. Mon. Not. R. Astron. Soc. **219**, 19p (1986c)
Gear, W.K., Robson, E.I., Griffin, M.J.: Millimetre and submillimetre observations of emission from dust in compact HII regions. Mon. Not. R. Astron. Soc. **231**, 55p (1988)
Gee, G., et al.: Submillimetre observations of the cold dust halo of NGC7027. Mon. Not. R. Astron. Soc. **208**, 517 (1984)
Griffin, M.J., et al.: Submillimeter and millimeter observations of Jupiter. Icarus **65**, 244 (1986)
Lebofsky, L.A., et al.: Submillimeter observations of the asteroid 10 Hygeia. Icarus **63**, 192 (1985)
Longmore, A.J., et al.: The continuum emission from the nucleus of NGC1275. Mon. Not. R. Astron. Soc. **209**, 373 (1984)
Marscher, A.P., Gear, W.K.: Models for high-frequency radio outbursts in extragalactic sources with application to the early 1983 millimeter to infrared flare of 3C 273. Astrophys. J. **298**, 114 (1985)
Robson, E.I.: Far infrared astronomy. Sci. Prog. Oxf **66**, 119 (1979)
Robson, E.I., et al.: Submillimetre multicolour photometry of galactic sources. Infrared Phys. **18**, 78 (1978)
Robson, E.I., et al.: The variability of 3C273: a flare in its millimetre to infrared spectrum. Nature **305**, 194 (1983)
Robson, E.I., et al.: Millimetre observations of optically selected quasars. Mon. Not. R. Astron. Soc. **213**, 355 (1985)
Robson, E.I., et al.: A new infrared spectral component of the quasar 3C273. Nature **323**, 134 (1986)
Rowan-Robinson, M., Ade, P.A.R., Robson, E.I., Clegg, P.E.: Millimetre observations of planets, galactic and extra-galactic sources. Astron. Astrophys. **62**, 249 (1978)
Sherwood, W.A., Schultz, G.V., Kreysa, E., Gemund, H.: In: Heeschen, D.S., Wade, C.M. (eds.) Proceedings of IAU symposium 'Extragalactic Radio Sources', p. 305. Reidel, Dordrecht (1982)
Sopka, R.J., et al.: Submillimeter observations of evolved stars. Astrophys. J. **294**, 242 (1985)
Ward-Thompson, D., Robson, E.I.: Dust around HII regions – II: W49A. Mon. Not. R. Astron. Soc. **244**, 458 (1990)
Ward-Thompson, D., et al.: Infrared and submillimetre observations of the Rho Ophiuchi Dark Cloud. Mon. Not. R. Astron. Soc. **241**, 119 (1989)
Whitcomb, S.E., Hildebrand, R.H., Keene, J.: Publ. Astron. Soc. Pac. **92**, 863 (1980)
Wright, E.L.: Recalibration of the brightness temperature of the planets. Astrophys. J. **210**, 250 (1976)

Chapter 5
CGS4: A Breakthrough Instrument

Phil Puxley

Abstract CGS4 was the fourth in a series of Cooled Grating Spectrometers built for the United Kingdom Infrared Telescope (UKIRT). This paper reviews the impact of CGS4 on the scientific productivity of UKIRT, examines the origins of this success and concludes with a few personal science highlights.

The Impact of CGS4

CGS4 was designed and built at the Royal Observatory, Edinburgh (ROE) in the late 1980s and delivered to UKIRT in 1991. It represented a major advance in capabilities for infrared spectroscopy; the precursor instruments in use at the time included CGS2 with its 7-element array and the single-channel spectrophotometers UKT6 and UKT9. In contrast, CGS4 offered a low-resolution grating that covered in one setting the whole of an atmospheric window for line ratio and detection experiments, a high-resolution (15 km/s) echelle grating for velocity studies, a wide slit that was well-matched to the telescope image quality and provided high throughput, and a long slit enabling background subtraction and optimal extraction techniques derived from optical spectrographs. These capabilities allowed CGS4 to expand the reach of near-infrared spectroscopy into the wider astronomical community e.g. observational cosmology.

CGS4 frequently dominated the telescope schedules, garnering as much as two-thirds of the time in some semesters, a figure limited perhaps by the available support resources. The use of the instrument is reflected in its science output, for example as measured by the publication statistics illustrated in Fig. 5.1. The data show that CGS4 contributed about half of all the science derived from UKIRT for

P. Puxley (✉)
Directorate for Mathematical and Physical Sciences, Division of Astronomical Sciences, National Science Foundation, Arlington, VA 22230, USA
e-mail: ppuxley@nsf.gov

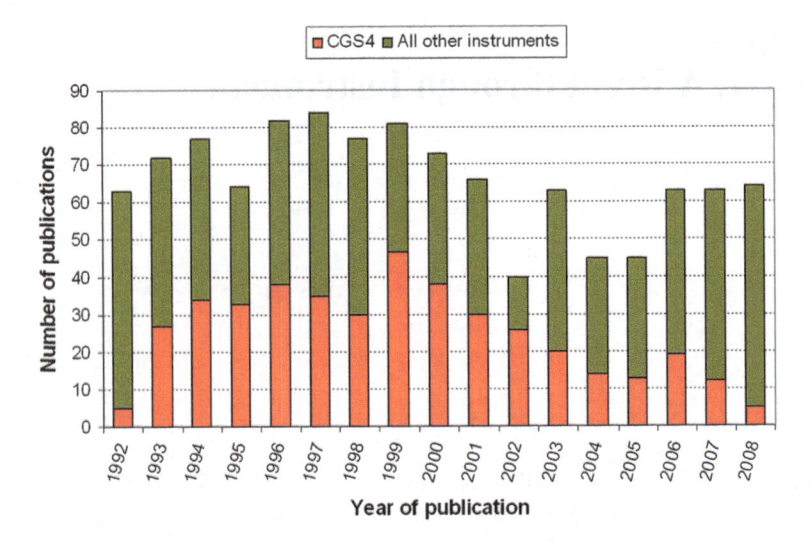

Fig. 5.1 Publications for CGS4 and all other UKIRT instruments combined

the decade or more after its commissioning, through an upgrade to a 256 × 256 pixel detector in 1996, until the publication statistics became dominated by UFTI and UIST in about 2003 and then by WFCAM in 2007. In total CGS4 contributed more than 400 papers.

Beyond the science impact, CGS4 was a forerunner of the ROE/Astronomical Technology Centre (ATC) approach to engineering that was later widely employed for large, complex instruments on 8 m-class telescopes.

CGS4 Engineering Innovations

CGS4 was a breakthrough because of a number of engineering and technical innovations, its science and engineering management and team, and its development as an integrated system that included the telescope and data processing. Each of these aspects came with its own challenges.

A schematic of the CGS4 optical path that illustrates many of the engineering innovations is shown in Fig. 5.2. It is worth noting that the instrument thus designed and built differs markedly from the original specifications proposed by Richard Wade: sized to fit in an HD3-8 (CGS2-like) dewar, a weight of 100 kg, mounting on the UKIRT Instrument Support Unit, a cost of under 500 thousand pounds and delivery within 4 years. The requirements were changed due to the principal science drivers whose flowdown required a 3 arcsec slit for high throughput and the range of spectral resolution described earlier. In fact only the last of the original requirements was close to being achieved. Delivered in ∼5 years, CGS4 had a size of ∼1 m,

5 CGS4: A Breakthrough Instrument

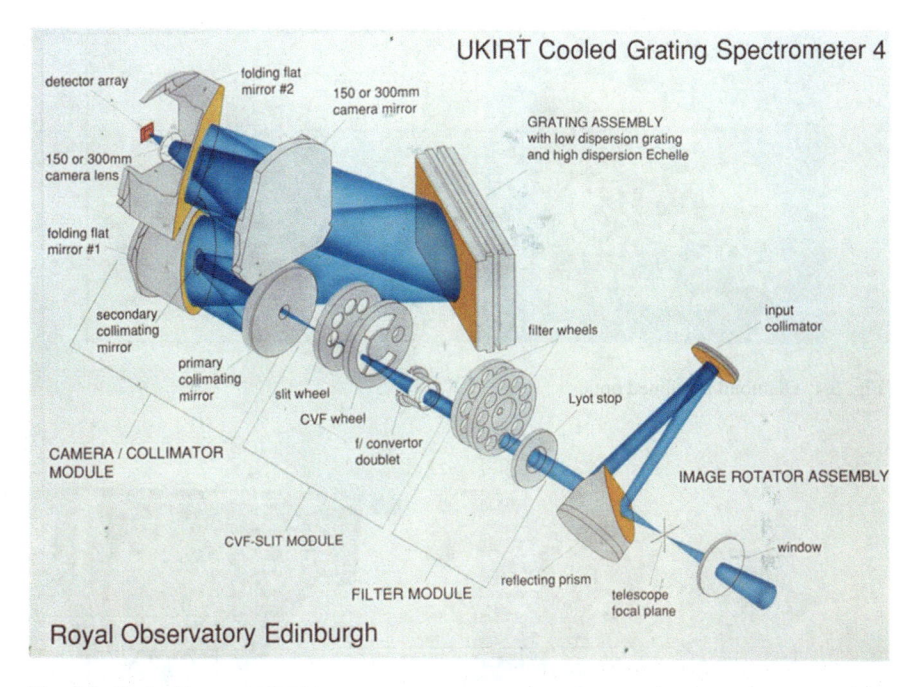

Fig. 5.2 Optical layout of CGS4

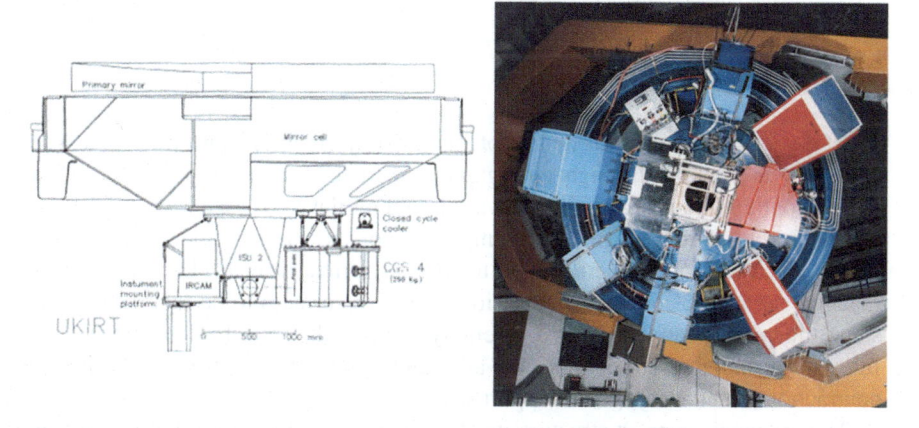

Fig. 5.3 Schematic (*left*) of CGS4 and the UKIRT mirror cell, instrument support unit and IRCAM. The scale bar shows 1 m. CGS4 and its two electronics cabinets (*red*) on the telescope (*right*)

weighed 250 kg, was mounted directly on the UKIRT mirror cell and cost in the region of three million pounds. As noted by Bob Joseph when researching this article, sometimes "it does cost money to do things right". A schematic and photo of CGS4 mounted on UKIRT are shown in Fig. 5.3.

Fig. 5.4 Diamond machined optics

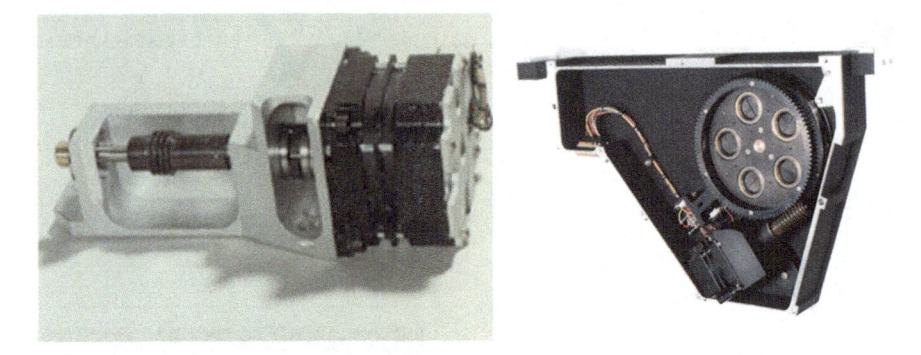

Fig. 5.5 Cryogenic mechanisms and motors

CGS4 was pioneering in its use of diamond-machined metal optics that were necessary to achieve the high throughput and image quality in a compact (fast) folded configuration. Figure 5.4 shows the folding flat mirror #1 with its included secondary collimating mirror and components of the camera module. The optics incorporated a planned upgrade path to a larger format detector. Use of diamond-machine surfaces also enabled precision reference surfaces to be fabricated at the same time and greatly facilitated assembly to the tight optical tolerances with no further adjustments. Furthermore the all-metal optics ensured predictable thermal contraction with the cryogenic support structure.

The cryogenic support structure itself was an innovative design fabricated from a single aluminium casting.

CGS4 was also pioneering in its use of stepper motors at cryogenic temperatures that eliminated the need for vacuum feedthroughs and thereby made the instrument more compact and reliable. The large cryogenic mechanisms saw several technical developments including vespel worm gears and also the first ground-based use of space-qualified bearings. Figure 5.5 shows two of the cryogenic mechanisms.

In the area of engineering design, while AutoCAD was first used at ROE for IRCAM, CGS4 would probably have been impossible without it. Likewise, Finite

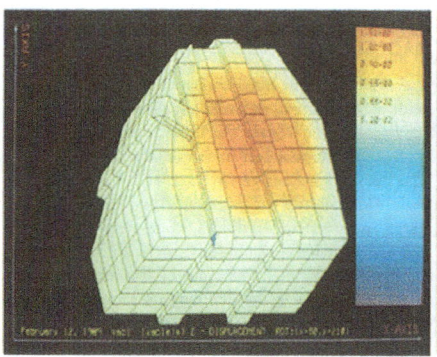

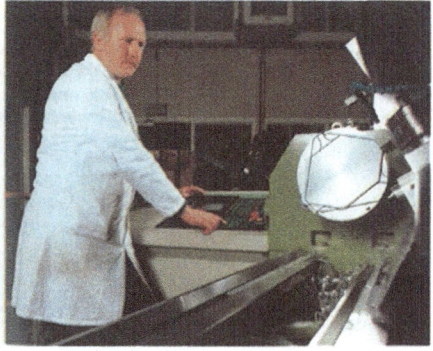

Fig. 5.6 An early example of Finite Element Analysis of CGS4 and computer numerically controlled machining of a mirror blank

Element Analysis was found to be crucial and the combination of numerical modeling with Code V optical design and Monte-Carlo tolerance analysis presaged what became the widespread use of systems engineering techniques in later instruments. These new design techniques were echoed on the manufacturing side by use of CNC tools and multi-axis measuring machines. Figure 5.6 shows an example FEA of the cryostat lid deformation under vacuum and CNC manufacture of one of the mirror blanks prior to diamond machining.

At the heart of all astronomical instruments is the detector and CGS4 was designed to use the latest generation of 58×62 pixel array detectors (see Fig. 5.7) and also to have an upgrade path to the 256×256 detectors that were then in development. The relatively noisy infrared detectors were adapted for lower spectroscopic backgrounds using a new multiple non-destructive readout technique that sampled the signal up the integrating ramp and provided superior performance to other sampling schemes. To maximize wavelength coverage with the relatively small format detector, a novel detector translation mechanism was implemented to enable the required Nyquist and finer spectral sampling.

CGS4 Management and Team

CGS4 was also innovative in the application of engineering and management processes by the team. Earlier instruments had largely followed the one scientist-one engineer model e.g. the CGS1 & 2 team of Richard Wade and David Robertson and the IRCAM 1 & 2 team of Ian McLean and Tim Chuter. In contrast, the much more complex CGS4 had Matt Mountain (Project Scientist) and David Robertson (Project Manager) in protagonist roles, to provide the "creative tension" sought by ROE's Chief Engineer Donald Pettie, a team of specialist scientists and engineers and what would now be called a "systems engineering" interaction between the various disciplines. The instrumental complexity and engineering challenges described

Fig. 5.7 The CGS4 58 × 62 pixel array

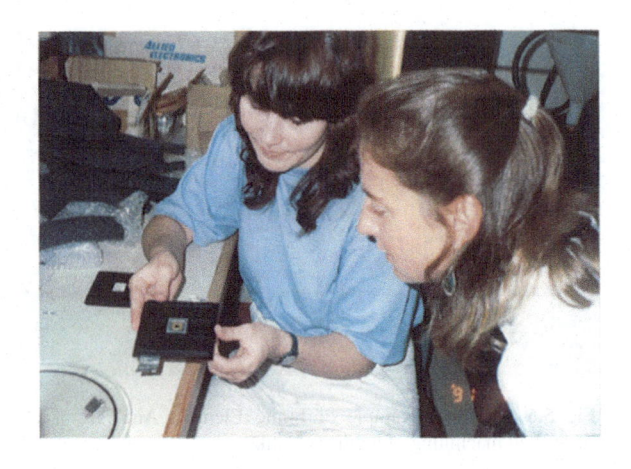

above demanded "real" project management ("...after a time...") notes Matt Mountain) which included the then novel concepts of a written specification, budget and formal oversight. Formal project scheduling techniques and error budget and configuration control were also applied. The project plan made allowance for comprehensive testing, including provision of a built-in calibration unit, and formal pre-shipping commissioning and acceptance test plans. Design and execution of the latter also took good advantage of the experience of the front-line UKIRT staff.

Repeatedly in researching this article it was remarked that central to CGS4's success was the strong engineering and science team shown in Fig. 5.8 with the completed instrument (and, inset, a faithful representation of the Project Scientist assembled by junior team members during a break in commissioning).

CGS4 as an Integrated System

CGS4 was perhaps the first large astronomical instrument to be developed as an integrated system incorporating telescope motion and data processing. The CGS4 control system coordinated telescope sequencing (offset motion and chopping), detector translation, calibration sources and exposures available through a simple but powerful user interface (the CONFIG(uration) and EXEC(utable) script system) for astronomers. The equally innovative CGS4DR data processing system (building on underlying Figaro routines) 'understood' the execution sequence and target identifiers (calibration, dark, arc, flat, object, sky) in a configurable processing sequence. CGS4DR presented the astronomer with real-time 1D- and 2D-spectra optionally with background subtraction, wavelength calibration, atmospheric correction and so forth. Figure 5.9 shows the CGS4DR system in use at the telescope and an early example of the 2D spectra and image display.

In combination, the control system and CGS4DR resulted in very efficient data collection and real-time update of accumulated data that enabled signal-to-noise ratio and detection assessment by the user.

5 CGS4: A Breakthrough Instrument

Fig. 5.8 The team and completed instrument

Fig. 5.9 The CGS4 data reduction during observations at the telescope and an example display from instrument commissioning

Some Science Highlights

The science derived from CGS4 is described more fully elsewhere in this meeting but a few, personal favourite examples serve to illustrated the breadth of research that was enabled by this instrument.

The astronomical discovery by Oka and Geballe and subsequent analysis of H_3^+ was an early and continuing highlight of CGS4 results, including understanding its

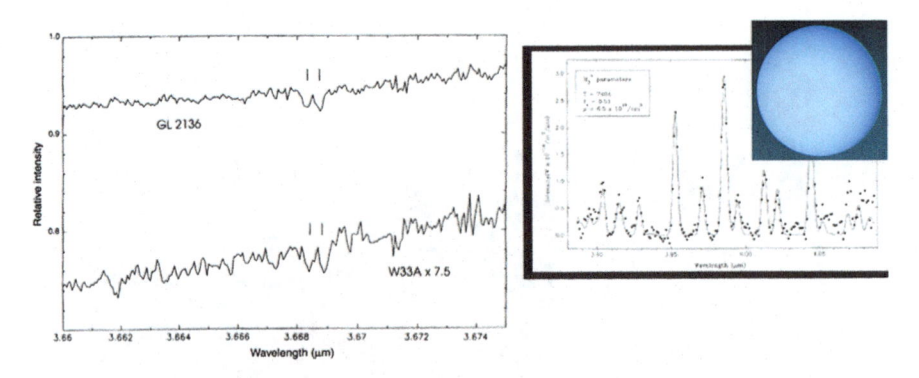

Fig. 5.10 H_3^+ in absorption and emission

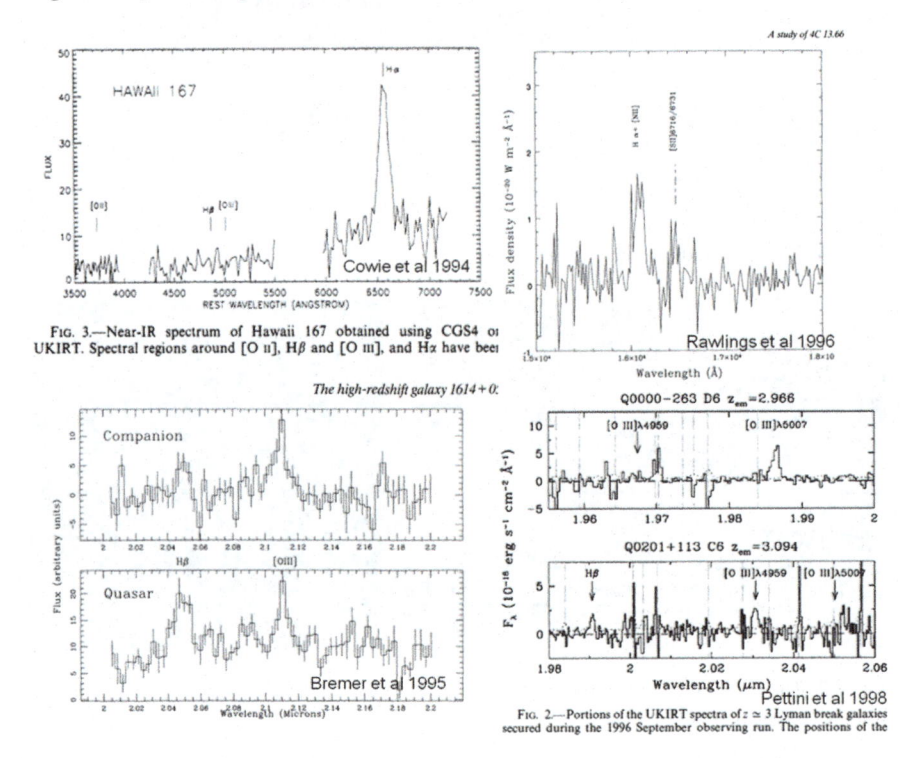

Fig. 5.11 Example spectra of high-redshift galaxies

pivotal role in gas phase chemistry and the formation of CO, H_2O, NH_3, CH_4 and other simple molecules, and measurement in such diverse sources as the diffuse interstellar medium and gas giant planets (see Fig. 5.10).

One of the first cosmological experiments with CGS4 was observation of the radio galaxy B2 0902 + 34, at z = 3.4 one of the highest red-shift galaxies known in the late 1980s, to determine whether the 'red bump' attributed at the time to PMS

5 CGS4: A Breakthrough Instrument

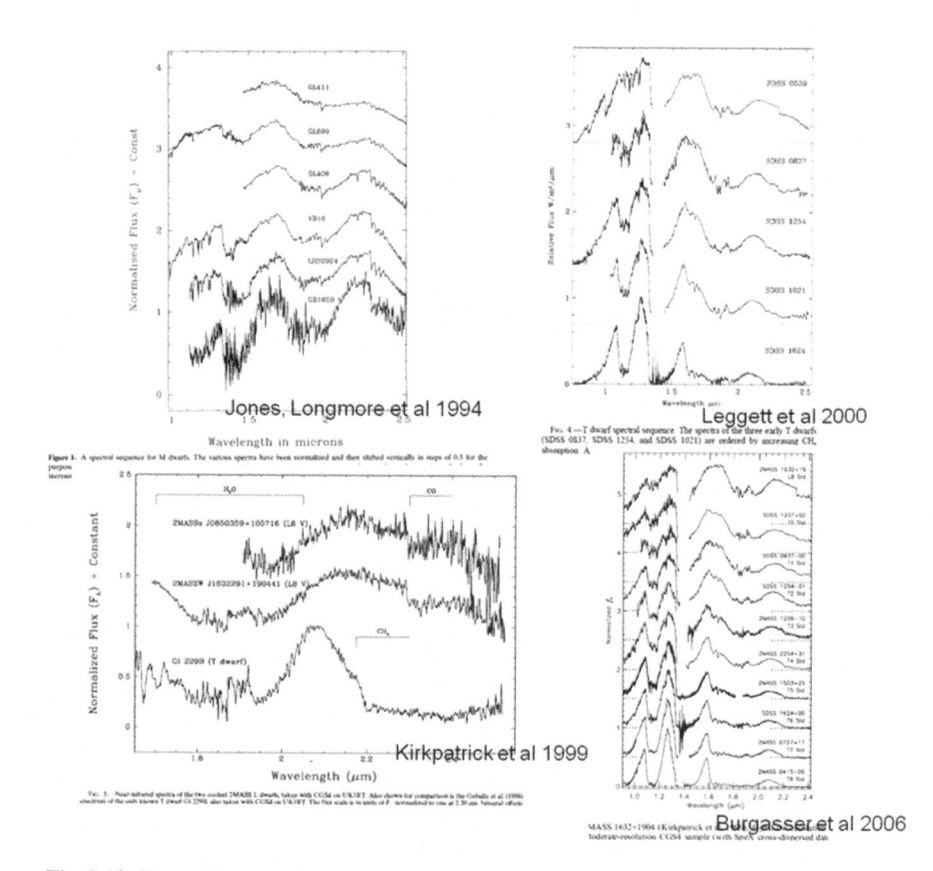

Fig. 5.12 Example spectra of very low mass stars

stars might be confused by redshifted [OIII] lines and therefore affecting assessment of its age and stellar population. Detection of the line emission and this clear demonstration of CGS4's sensitivity spurred many further studies of cosmologically interesting objects e.g. Fig. 5.11.

CGS4's broad wavelength coverage was ideal for studying the spectra of stars at the bottom of the IMF and these spectral sequences became one of the most highly-cited areas of UKIRT's scientific output e.g. Fig. 5.12.

Conclusions

CGS4 is considered to have been a breakthrough instrument because of the engineering and technical innovations driven by its science goals and the presence of a strong technical and science team encouraged to exploit new processes as necessary. This resulted in more than a decade of top-ranked science at UKIRT and defines CGS4 as a forerunner of the *current* generation of large, complex instrumentation.

Acknowledgements I am indebted to many members of the original project team and contemporaries from the time of CGS4's introduction for extensive comments, anecdotes and images: Alan Tokunaga, Bob Joseph, Bob Carswell, David Robertson, Tom Geballe, Gillian Wright, Andy Adamson, Ian Bryson, Jason Cowan, Peter Hastings, and David Montgomery, and in particular to Matt Mountain and Suzanne Ramsay for conversations and advice (and the "Snow Matt" picture!). I am also very grateful to the organizers of the meeting for the invitation and to the other participants for an excellent opportunity to reminisce, interact and plan.

Chapter 6
The UKIRT Upgrades Programme

Tim Hawarden[†]

Abstract Tim Hawarden presented this paper to the 30th anniversary workshop, just a month before his untimely death. The editors have done their best to convert his talk into this paper, and gratefully acknowledge the assistance of Nick Rees (a member of the Upgrades team, now at Diamond Light Source). Tim's discussion concerned the UKIRT Upgrades Project, which ran through the 1990s and transformed the telescope and made it truly competitive on the world stage for operation into the twenty-first century. The reference list at the end of the paper is comprehensive; some of these are referred to in the paper itself and some are included for completeness only.

Origins

As highlighted elsewhere in these proceedings, UKIRT was designed to be a very inexpensive telescope. Firstly, UKIRT had a thin (flexible!) primary, permitting very lightweight construction elsewhere. This mirror was figured far beyond the initial specification as Grubb Parsons had mastered the polishing process and persuaded the funding agencies to continue to achieve a high accuracy, almost optical quality mirror. Secondly, UKIRT was placed in the smallest possible dome, which although initially not ventilated by any means other than the normal slit, allowed additional ventilation to be easily installed. Finally, it had the lightest possible structure and so had lower thermal inertia than any similar sized telescope at the time or since.

[†]deceased

Overview

In UKIRT, Dunford Hadfields had designed the first truly modern high performance telescope, but despite its potential, UKIRT initially had many problems. The pointing was not very good, and once on target, tracking was unreliable at the level of arcsec. Further, the telescope was severely susceptible to windshake even in modest winds. Optical alignment was poor and hard to correct, and the design of the top end led to poor performance in the thermal infrared, a regime where telescopes on Mauna Kea should excel. Furthermore, the local seeing was poor, with thermal plumes being clearly visible in knife edge images.

The result was a telescope that was hard to use for programmes requiring high angular resolution. Indeed, early instruments were especially designed to be suited to low resolution applications (which actually led to some unique observations, e.g. global line strengths of galaxies). These intrinsic problems were not easy to fix. For example, Andy Longmore and Bill Parker worked on the telescope drive servo for years, but could not get good performance over the sky: if the servo was stiff to windshake in one part of the sky it tended to oscillate in others.

In September–October 1988 there was a 6 week shutdown for dome repairs and enhancements. The Devil makes work for idle hands, and the author spent much of that time thinking about the possibility of solving as many of UKIRT's problems as possible. A programme of work emerged from this, which was first presented to the UKIRT Users Committee in September 1991. The proposal was endorsed by the committee and over the course of 1992 the project coalesced and evolved into a multi-year collaboration involving four groups: ROE, RGO, MPIA (Heidelberg) and the JAC. At this point the stated goals of the project had been refined to be:

1. UKIRT must not degrade the delivered image quality by more than 0.25 arcsec (measured as FWHM) (this was the predicted image size for a tip-tilt corrected 4 m telescope at K in median MK conditions[1])
2. UKIRT should not degrade the delivered image quality by more than 0.12 arcsec (measured as FWHM) (this is the diffraction limit of UKIRT in the K-band)
3. Local seeing effects should not appreciably degrade image quality; no single element of the local seeing should degrade quality by more than the 0.12 arcsec criterion as in item 2, above.

For most of the programme the core group at the JAC were Nick Rees (optics, computing, control systems); Tim Chuter (electronics, optics, operational issues); Chas Cavedoni (JAC Project Manager, Mechanical Engineering); and myself (JAC project scientist, optics, AO, Science Operations, scientific direction). These individuals were helped at one time or another by most of the JAC technical staff and about half of ROE! From 1993 the overall project manager was Donald Pettie at the

---

[1] Assuming classical Kolmogorov turbulence, with a large fraction of seeing power in image motion (tip-tilt). With hindsight it is hard to find an observatory site on which this assumption is consistently appropriate.

ROE. More than 30 people were significantly involved at the four Institutes, and it is very hard to apportion fair credit. To recognise a major unofficial contribution, we freely plagiarised from the IRTF upgrades programme (starting by acquiring their engineer, Chas Cavedoni). For several years we let them make the mistakes first, but then (mistakenly?) overtook them and had to start making our own blunders.

Major Elements

To accomplish the goals of the programme, a set of technical changes were envisaged, as listed below and shown schematically in Fig. 6.1.

1. Fast two axis tip-tilt Secondary Mirror with accurate positioning, on a new, stiff, top-end, to correct image motion (esp. windshake!)
2. Sensitive CCD based fast guider (>100 Hz)
3. Active control of Primary Mirror aberrations
4. Control of local seeing by ventilating the dome and primary mirror
5. Actively cooling the primary

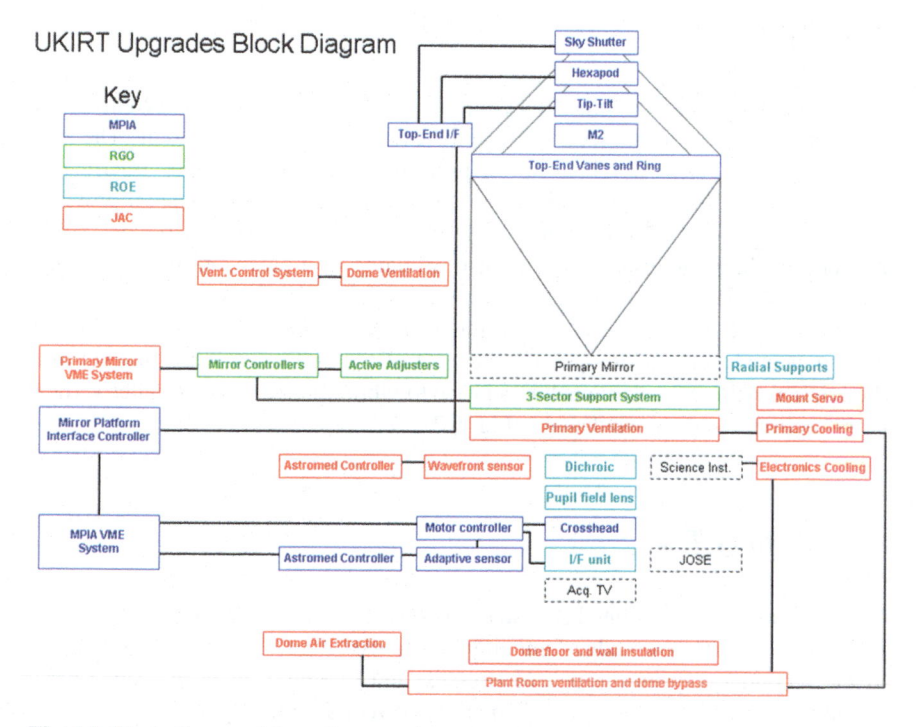

Fig. 6.1 Block diagram of the entire upgrades programme

Fig. 6.2 Secondary system attached to *top-end* structure and vanes (in lab at MPIA, Heidelberg). Note constrained-layer viscous damping panels on the vane sides!

New Top End

As can be seen from the block diagram, a major part of the MPIA contribution was a new top-end ring, secondary, and hexapod positioning system. The new secondary was a light weighted mirror produced by Prazisions Optik (Germany), optically tested through the back surface. It was controlled by a system provided by Physik Instrumente (Germany). Control was in two stages: a 6-axis "slow" hexapod positioning system capable of positioning to a precision of 1 μm, and a 3-axis "fast" piezoelectric tip-tilt system with a driven counterweight to minimise vibration transfer to the telescope structure (Pitz et al. 1994a, b). The new hardware was connected to the telescope via a new set of top-end vanes, much thicker than the original knife-edge and much less prone to vibration and with a smaller vertical cross-section to reduce wind shake (Fig. 6.2).

A Cautionary Tale

While the new light-weighted secondary performed well, it had a number of issues including a turned-down edge (amounting to a loss of 10 % of the functional area of the secondary), print-through of stresses, and thermally induced second-order trefoil at a level too severe for the primary to correct. Essentially this mirror was no good for AO use, which was still under consideration at that time. In addition to the

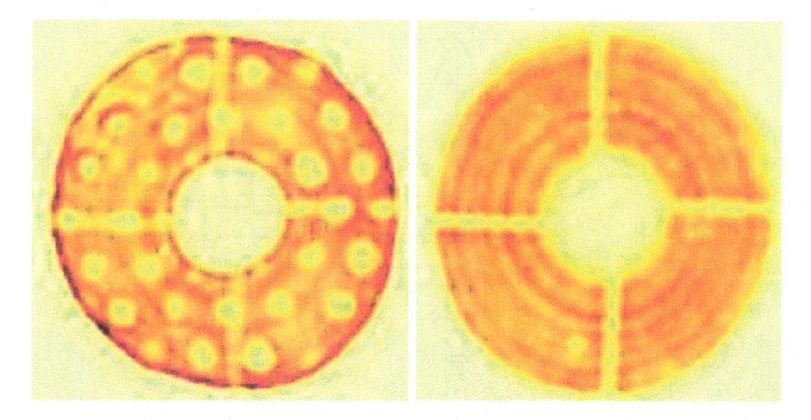

Fig. 6.3 *Left*: Original secondary, showing lightweighting print-through. *Right*: Revised secondary

UKAO programme, the IfA Adaptive Optics group had applied for an NSF grant for a UKIRT AO system on the basis of our initial success. The solution to these problems required manufacture of a new mirror by Prazisions Optik and funded by MPIA (Rohloff et al. 1999). It was slightly oversized and with the edge ground off to avoid turn-down; the back of the mirror was etched in hydrogen fluoride for stress relief after light-weighting, removing the print-through; and finally the mirror was given athermal mountings (designed by Richard Bennett from ROE, who had also diagnosed the cause of the problem), removing the trefoil (Fig. 6.3).

Bottom-End Modifications

At the Cassegrain focus, a number of modifications were made, including manufacture of a new, very stiff and highly accurate crosshead (by SKF), and a CCD fast guider, all provided by MPIA and installed in August 1996 (Fig. 6.4). After delivery this system was upgraded with a thinned, back-illuminated CCD in June 1998, employed a sophisticated Kalman filter (Yang et al. 2000), and was capable of guiding on an image with only ∼10 counts/cycle. The guider provided a wider field (30 arcsec) acquisition mode, and a Shack-Hartmann option for autofocus and measurement of seeing (Glindemann and Rees 1994a, b). Finally, a wavefront curvature sensor was provided for diagnostics and calibration of lookup tables for the primary mirror. This was initially described as an "interim" wavefront sensor since it was a very simple system built quickly in mid 1993. It also occupied one of the ISU2 instrument ports and the intention was to replace it with a permanently mounted wavefront sensor below the crosshead. However, it proved so successful that it was never replaced and proved critical for testing the performance of the telescope as the programme proceeded.

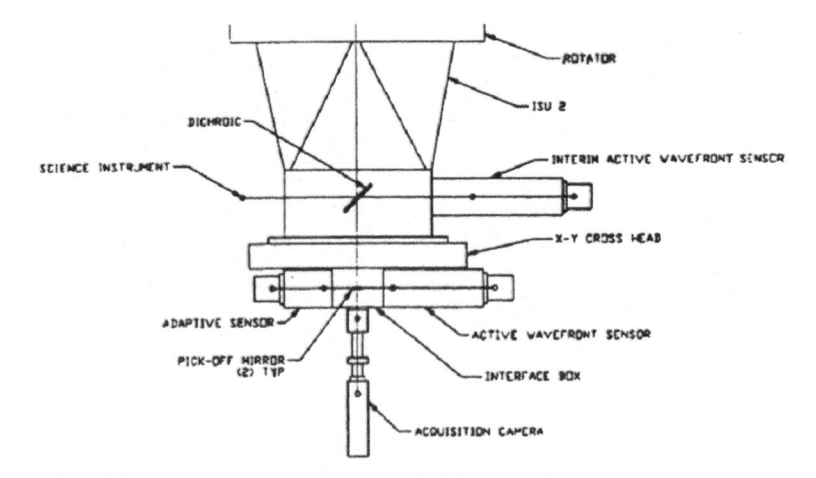

Fig. 6.4 The UKIRT *bottom-end* after upgrades

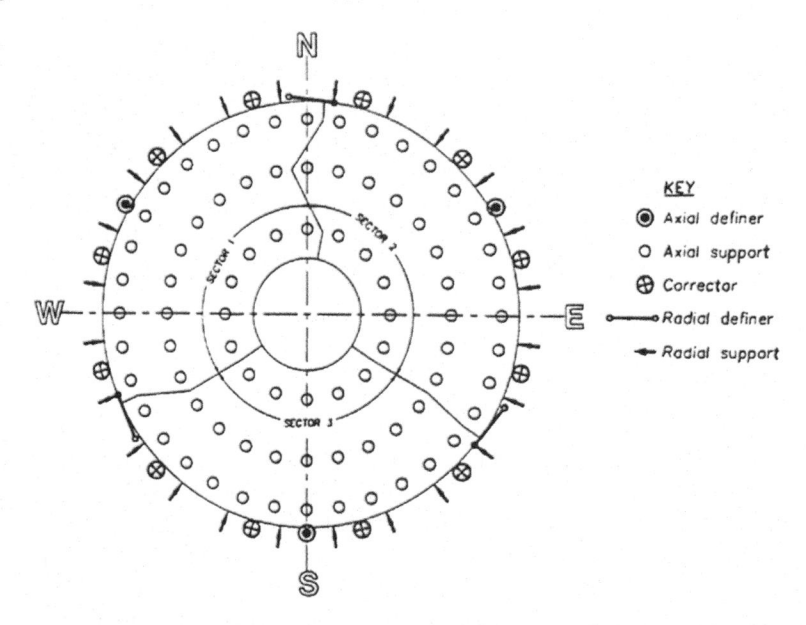

Fig. 6.5 Schematic of pneumatic axial support pads, edge supports, controllers and positioners

Mirror Support System

To solve the alignment and wavefront issues, the primary mirror support system was completely upgraded to a system capable of maintaining figure to <100 nm in all orientations of the telescope (Chrysostomou et al. 1998). The new passive support system comprised a three sector axial support utilizing the existing 80 pneumatic supports (Fig. 6.5) servoing to maintain zero load on three axial definers. The

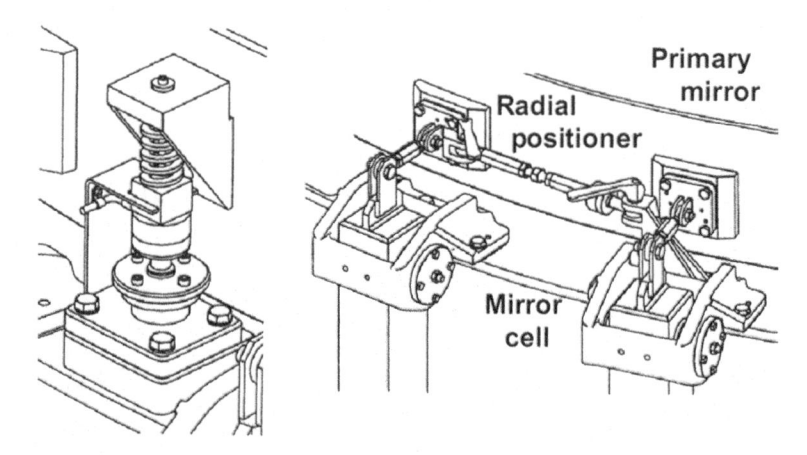

Fig. 6.6 Schematic of the various primary mirror adjusters

existing radial support consisting of 24 weights and levers (Fig. 6.6) was retained, but it was serviced with new bearings and new tangential-arm radial definers. The system was slightly unusual in that the all-new active support system was completely separate from the passive support and comprised a 12-point pneumatic support system applying axial loads on the mirror edge (Fig. 6.7). All of these systems were controlled over CANbus using the EPICS CAN driver. Most of the system was supplied by RGO (a good example of inter-observatory cooperation in what were trying times!), but the new radial bearings and linkages were from ROE, and much of the analysis was also done by Richard Bennett of ROE.

Local Seeing Control

Local seeing was to be controlled on a number of fronts. The dome was better insulated by insulating the crew room ceiling and the vestibule roof around the cargo entry area. Ventilation was added: passively by the provision of vents around the sides of the dome to allow airflow through the dome on opening (Neff et al. 1997), and actively flushing the dome using the Coudé room as a plenum chamber, with extractor fans in the plant room and taking advantage of pre-existing holes in the dome floor around the north and south columns and elsewhere (Cavedoni et al. 1997) (Fig. 6.8). Local sources of heat and emission were suppressed, for example by dispersing instrument heat below the primary mirror and the use of low emissivity paint on the top-end central boss. A servo'd mirror cooling system was designed, which would launch a cooling air flow over the primary from the central mirror lifting plug.[2]

---

[2]This system was the last of the upgrades to be implemented, we were able to report on it at the workshop itself. As an interim measure, cooling fans were mounted on the south column.

Fig. 6.7 The reality, showing all three types of active supports

A Second Cautionary Tale

We were about to place four inches of "inert" foam backed board over the whole dome floor, when a glycol fire at CTIO caused us to pause. Primary cooling was to use glycol. The flammability of various insulating materials was tested at altitude by the Keck observatory, and found to be higher than at sea level for almost all materials. As an interim measure, cooling fans were mounted on the south column.

Results

As well as improving thermal background stability (as intended), the installation of the new top-end systems removed non-linear effects ("flop"). In fact, just before we installed the new top end we were having emergency meetings to deal with rapidly deteriorating pointing. It turned out that the bolts that attached the truss to the top ring were still covered in hydraulic fluid from an ill-fated experiment with a hydraulic chopping secondary and they had worked completely loose. We discovered this when we installed the new top-end and a greatly improved pointing (to $\sim 1.3''$ RMS all sky) was the result. UKIRT became the best pointing equatorially mounted telescope in the world! Furthermore, the bottom-end fast guider provided phenomenal guiding performance. UKIRT was able to guide on a V $\sim$19.2 star at 40 Hz sampling rate (Balius and Rees 1997).

6 The UKIRT Upgrades Programme

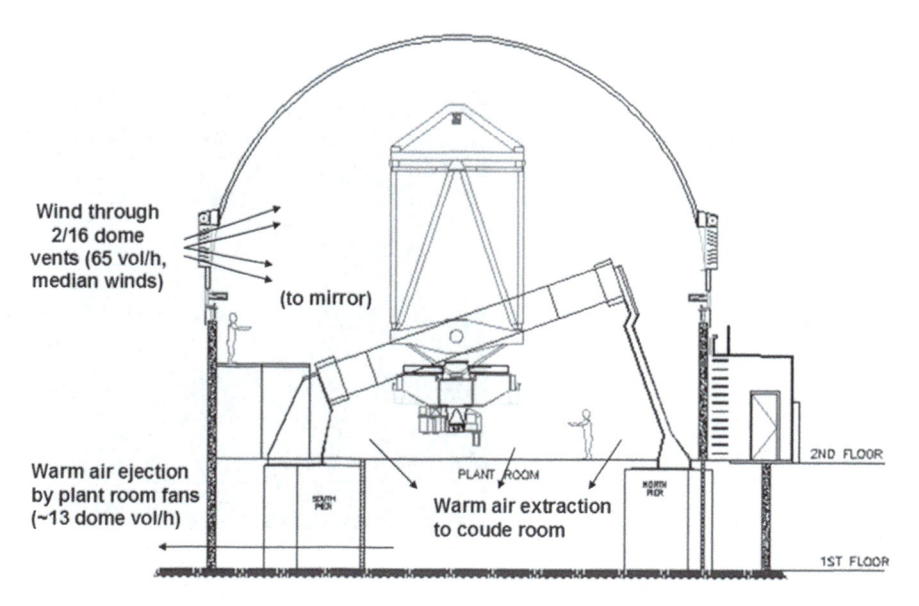

Fig. 6.8 Ventilating the UKIRT dome

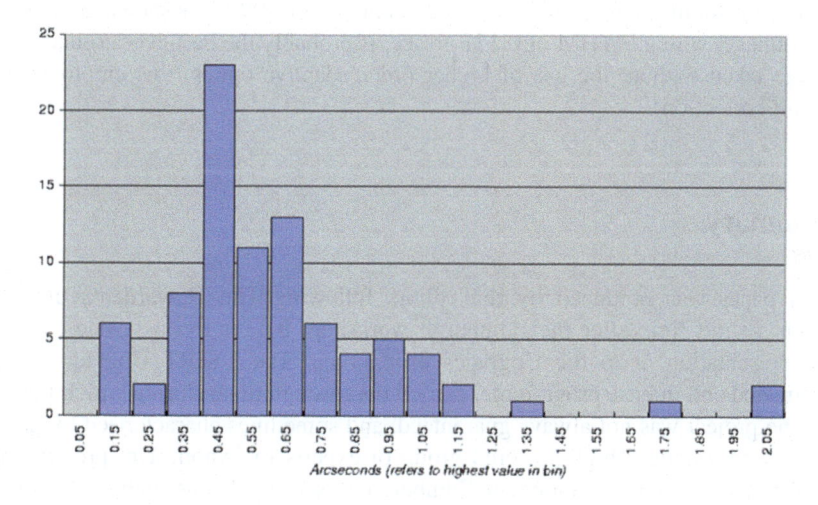

Fig. 6.9 Histogram of delivered seeing from UKIRT, post upgrades

Finally, the combination of dome ventilation, active dome flushing and tip-tilt system provided fantastic seeing performance. Figure 6.9 shows a histogram of delivered seeing after the Upgrades (all but primary mirror cooling) were in place.

With a peak below 0.5 arcsec, this brought UKIRT to a pinnacle of performance that had been unimaginable to the original designers of the telescope. The best image ever taken with UKIRT – using IRCAM3 and 5× optical magnifier during the

Fig. 6.10 Best ever image obtained with IRCAM on the upgraded UKIRT

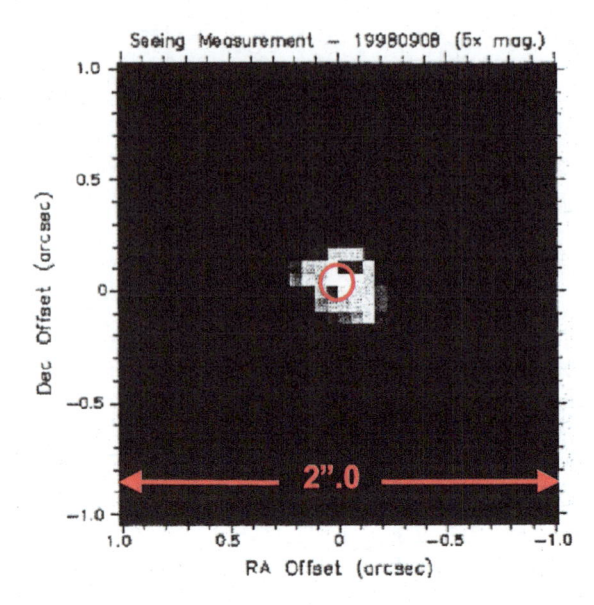

systematic monitoring campaign carried out in the late 1990s – is shown in Fig. 6.10. This image, with a FWHM of 0.171 arcsec, is probably the best ever ground-based image taken without the use of higher order adaptive optics (this employed just tip-tilt correction).

Summary

This paper was produced by the editors following Tim Hawarden's untimely death a short time after the Edinburgh workshop. It fails to do justice to either the presentation or to the Upgrades programme. The UKIRT Upgrades was a protracted and intense programme, carried out through times during which funding for the project was not always guaranteed, and sometimes characterised by heated discussion among a high-powered group of experts for whom Tim provided the perfect lead. His presentation in Edinburgh was every bit as enthusiastic, wide ranging and technically knowledgeable as expected and while we may have been able to reproduce the technical substance of it here, the passion was Tim's alone and not possible to transcribe. Whilst not all Upgrades papers have been cited in this paper, the editors have collated all known upgrades papers in the references (Hawarden et al. 1994, 1997, 1998a,b, 1999, 2000; Hawarden 1995; Rees and Hippler 1995; Seigar et al. 2002). As reflected in the rest of this volume, the upgraded telescope went on to provide some of the most exciting science to emerge from the UK's astronomical facilities over the first decade of the twenty-first Century, and Tim's instigation, championing and leadership of the Upgrades was critical to that success.

References

Balius, A., Rees. N.P.: UKIRT fast guide system improvements. In: Lewis, H. (ed.) Society of Photo-Optical Instrumentation Engineers (spie) Conference Series. Vol. 3112 of Society of Photo-Optical Instrumentation Engineers (spie) Conference Series, pp. 296–307 (1997)

Cavedoni, C.P., Hawarden, T.G., Chuter, T.C., Look, I.A.: UKIRT Upgrades Program: control of the telescope thermal environment. In: Ardeberg A.L. (ed.) Society of Photo-Optical Instrumentation Engineers (spie) Conference Series. Vol. 2871 of Society of Photo-Optical Instrumentation Engineers (spie) Conference Series, pp. 685–694 (1997)

Chrysostomou, A.C., Rees, N.P., Hawarden, T.G., Chuter, T.C., Cavedoni, C.P., Pettie, D.G., Bennett, R.J., Ettedgui-Atad, E., Humphries, C.M., Mack, B.: Active optics at UKIRT. In: Stepp, L.M. (ed.) Society of Photo-Optical Instrumentation Engineers (spie) Conference Series. Vol. 3352 of Society of Photo-Optical Instrumentation Engineers (spie) Conference Series, pp. 446–451 (1998)

Glindemann, A., Rees, N.: UKIRT 5-axis tip-tilt secondary – wavefront sensor simulations. In: Merkle, F. (ed.) European Southern Observatory Conference and Workshop Proceedings. Vol. 48 of European Southern Observatory Conference and Workshop Proceedings, p. 273 (1994a)

Glindemann, A., Rees, N.P.: Photon counting versus CCD sensors for wavefront sensing: performance comparison in the presence of noise. In: Stepp, L.M. (ed.) Society of Photo-Optical Instrumentation Engineers (spie) Conference Series. Vol. 2199 of Society of Photo-Optical Instrumentation Engineers (spie) Conference Series, pp. 824–834 (1994b)

Hawarden, T.: The upgraded UKIRT and other planetary systems. Astrophys. Space Sci. **223**, 158 (1995)

Hawarden, T.G., Pettie, D.G., Cavedoni, C.P.: The UK 3.8m Infrared Telescope (UKIRT): from light bucket to the diffraction limit. In: American Astronomical Society Meeting Abstracts. Vol. 30 of Bulletin of the American Astronomical Society, p. 1265 (1998a)

Hawarden, T.G., Cavedoni, C. P., Rees, N.P., Chuter, T.C., Pettie, D.G., Humphries, C.M., Bennett, R.J., Etad, E., Harris, J.W., Mack, B., Pitz, E., Glindemann, A., Rohlo, R.R.: UKIRT upgrades program: preparing for the 21st century. In: Stepp, L.M. (ed.) Society of Photo-Optical Instrumentation Engineers (spie) Conference Series. Vol. 2199 of Society of Photo-Optical Instrumentation Engineers (spie) Conference Series, pp. 494–503 (1994)

Hawarden, T.G., Cavedoni, C.P., Chuter, T.C., Look, I.A., Rees, N.P., Pettie, D.G., Bennett, E., Atad, R.J., Harris, J.W., Humphries, C.M., Mack, B., Pitz, E., Glindemann, A., Hippler, S., Rohloff, R.R., Wagner, K.: Progress of the UKIRT upgrades program. In: Ardeberg, A.L. (ed.) Society of Photo-Optical Instrumentation Engineers (spie) Conference Series. Vol. 2871 of Society of Photo-Optical Instrumentation Engineers (spie) Conference Series, pp. 256–266 (1997)

Hawarden, T.G., Rees, N.P., Cavedoni, C.P., Chuter, T.C., Chrysostomou, A.C., Pettie, D.G., Bennett, R.J., Ettedgui-Atad, E., Harris, J.W., Mack, B., Pitz, E., Glindemann, A., Hippler, S., Rohloff, R.R., Wagner, K.: Upgraded UKIRT. In: Stepp, L.M. (ed.) Society of Photo-Optical Instrumentation Engineers (spie) Conference Series. Vol. 3352 of Society of Photo-Optical Instrumentation Engineers (spie) Conference Series, pp. 52–61 (1998b)

Hawarden, T.G., Rees, N.P., Chuter, T.C., Chrysostomou, A.C., Cavedoni, C.P., Rohloff, R.R., Pitz, E., Pettie, D.G., Bennett, R.J., Atad-Ettedgui, E.: Postupgrade performance of the 3.8-m United Kingdom Infrared Telescope (UKIRT). In: Roybal, W. (ed.) Society of Photo-Optical Instrumentation Engineers (spie) Conference Series. Vol. 3785 of Society of Photo-Optical Instrumentation Engineers (spie) Conference Series, pp. 82–93 (1999)

Hawarden, T.G., Adamson, A.J., Chuter, T.C., Rees, N.P., Massey, R.J. Cavedoni, C.P., Atad-Ettedgui. E.: Thermal performance and facility seeing at the upgraded 3.8-m UK Infrared Telescope (UKIRT). In: Sebring T.A., Andersen, T. (eds.) Society of Photo-Optical Instrumentation Engineers (spie) Conference Series. Vol. 4004 of Society of Photo-Optical Instrumentation Engineers (spie) Conference Series, pp. 104–114 (2000)

Neff, D.H., Hileman, E.A., Kain, S.J., Cavedoni, C.P., Chuter, T.C.: UKIRT upgrades program: design and installation of the Dome Ventilation System (DVS). In: Ardeberg, A.L. (ed.) Society of Photo-Optical Instrumentation Engineers (spie) Conference Series. Vol. 2871 of Society of Photo-Optical Instrumentation Engineers (spie) Conference Series, pp. 710–721 (1997)

Pitz, E., Rohloff, R.R., Marth, H.: UKIRT 5-axis tip-tilt secondary – electrome-chanical and optical design. In: Merkle, F. (ed.) European Southern Observatory Conference and Workshop Proceedings. Vol. 48 of European Southern Observatory Conference and Workshop Proceedings, p. 267 (1994a)

Pitz, E., Rohloff, R.R., Hippler, S., Wagner, K., Marth, H.: Five-axis secondary system for UKIRT. In: Stepp, L.M. (ed.) Society of Photo-Optical Instrumentation Engineers (spie) Conference Series. Vol. 2199 of Society of Photo-Optical Instrumentation Engineers (spie) Conference Series, pp. 516–522 (1994b)

Rees, N.P., Hippler, S.: Controlling the UKIRT upgrades program. In: Wallace, P.T. (ed.) Society of Photo-Optical Instrumentation Engineers (spie) Conference Series. Vol. 2479 of Society of Photo-Optical Instrumentation Engineers (spie) Conference Series, pp. 2–10 (1995)

Rohloff, R.R., Pitz, E., Hawarden, T.G., Rees, N.P., Atad-Ettedgui, E., Kaufmann, H.W., Schmadel, L.: Lightweighted secondary mirror for the United Kingdom Infrared Telescope. In: Roybal, W. (ed.) Society of Photo-Optical Instrumentation Engineers (spie) Conference Series. Vol. 3785 of Society of Photo-Optical Instrumentation Engineers (spie) Conference Series, pp. 152–159 (1999)

Seigar, M.S., Adamson, A.J., Rees, N.P., Hawarden, T.G., Currie, M.J., Chuter. T.C.: Seeing statistics at the upgraded 3.8m UK Infrared Telescope (UKIRT). In: Quinn, P.J. (ed.) Society of Photo-Optical Instrumentation Engineers (spie) Conference Series. Vol. 4844 of Society of Photo-Optical Instrumentation Engineers (spie) Conference Series, pp. 366–375. doi:10.1117/12.460604 (2002)

Yang, Y., Rees, N.P., Chuter, T.: Application of a Kalman filter at UKIRT. In: Manset, N., Veillet, C., Crabtree, D. (eds.) Astronomical Data Analysis Software and Systems ix. Vol. 216 of Astronomical Society of the Pacific Conference Series, p. 279 (2000)

Chapter 7
Operational Innovations

Andrew J. Adamson

Abstract UKIRT evolved to its position as a world-leading wide-field astronomical observatory via a process punctuated by operational innovations which changed the way the telescope was scheduled and used. We review the some of these innovations and show how they came together to provide the UK with its first major infrared survey capability.

Introduction

It is often the case that operational innovation follows from technical innovation. As UKIRT the telescope benefitted from new technologies, e.g. in infrared arrays and tip/tilt correction, UKIRT Operations benefitted from new software tools and ways of working which would have been inconceivable at the start of operations in 1979. This paper describes changes in the science use of the telescope (particularly in recent years), and various innovations in scheduling and responsiveness implemented over the decades before. Most of the discussion here will refer to software and its impact. But it is worth noting that UKIRT was designed from the start with multiple instruments in mind, and a key decision was taken at the very beginning: to employ a tertiary pickoff mirror and side-looking cryostats. In early operations, this was useful but not crucial, as the telescope was classically scheduled. However with the implementation of flexible scheduling in the early 2000s, it became critical to the scientific productivity of the observatory.

A.J. Adamson (✉)
Gemini Observatory, Northern Operations Center, 670 N. A'Ohoku Place, Hilo, Hi 96720, USA
e-mail: aadamson@gemini.edu

Preparation and Execution

Ukirtprep

For many years, UKIRT was used in a highly classical way, with observations being both prepared and carried out at the summit. **ukirtprep** was the first move to require advance preparation of observations. "Advance" meant "in Hilo", but this was nonetheless a significant step forward and provided significant benefits including additional oxygen to the brain of the astronomer doing the preparation. **ukirtprep** was based on SMS (see Fig. 7.1), a menu system which ran on VAX computers and the then-ubiquitous VT100 terminals. A generation of UKIRT users became very familiar with the PF1 key. **ukirtprep** successfully moved the preparation stage off the summit and paved the way for later systems in which preparation would be done worldwide. It was standard across the facility instruments (e.g. CGS4, IRCAM), and the output was human (and machine) readable text files called "execs" and "configs"; the use of these files survives to this day, a testament to sound design.

ORAC

Developed jointly between the JAC and (at the time) ROE, and introduced initially with UFTI in 1998, ORAC was an ambitious control system which aimed to provide the astronomer with state-of-the-art tools for remote preparation of observations,

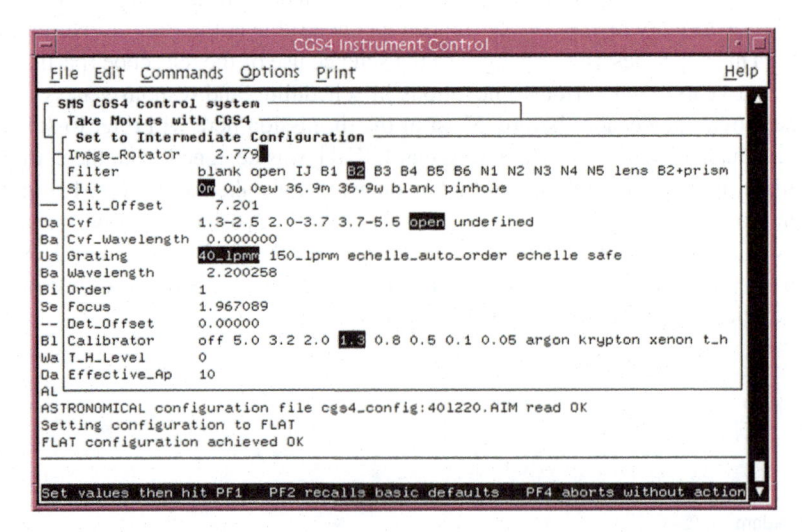

Fig. 7.1 CGS4's SMS menus, showing a part of the structure which not many observers dealt with directly (the intermediate configuration screen)

7 Operational Innovations

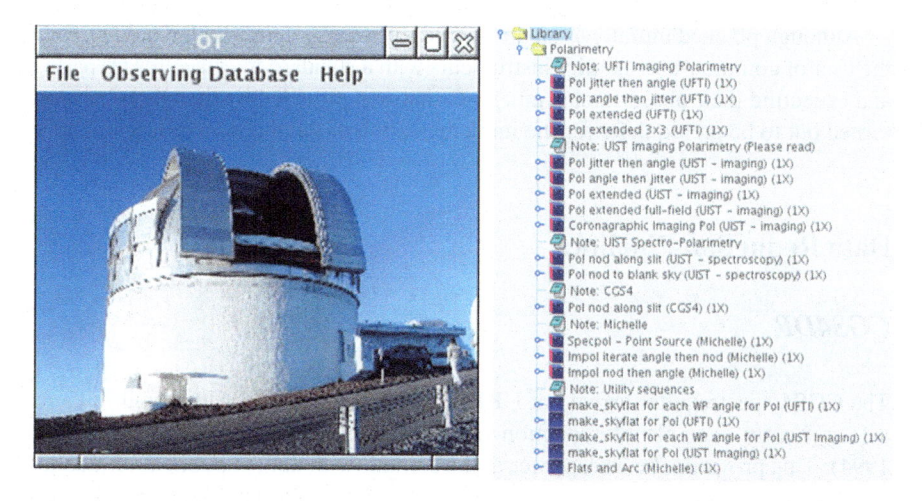

Fig. 7.2 The familiar UKIRT OT launch screen (*left*) and one of the numerous template libraries, this one catering for polarimetry (*right*)

and for seamless execution of those observations at the summit (e.g. Bridger et al. 2000). After the arrival of UFTI, and with the obvious success of the system, it was back-ported to CGS4 and IRCAM. ORAC remains the UKIRT standard interface (between astronomers, instruments and the telescope) to this day. Michelle, UIST and WFCAM, the mainstays of the UKIRT Development Programme, were all delivered ORAC-ready and slotted into the system with minimal fuss.

ORAC comprised the following elements: the Observing Tool (a variant on the Gemini OT which was under development at the time of the ORAC project and which provided a highly suitable infrastructure for UKIRT operations also), the Translator (a critical piece of software which translated the SGML or XML OT files into the execs and configs required by later systems in the sequence), and the Sequence Console via which observers interacted with execution at the highest level of detail (down to individual exposures). An observing database at the summit allowed PIs to submit their programmes from anywhere in the world. Finally there was the ORACDR data reduction pipeline which we will refer to in a later section.

The observing tool was given a particularly "UKIRT" feel by the implementation of "template libraries" – available via a menu in the OT main screen and providing template observations for all standard UKIRT observing modes (Fig. 7.2). Thus, libraries were provided for all the modes of the Cassegrain facility instruments, later UIST, Michelle and WFCAM, and for ancillaries like the Fabry-Perot and IRPOL. A fundamental link was made between these modes and library sequences and the data reduction recipes, which were referred to within the templates; the standard modes produced data which were guaranteed reducible with the specified DR recipe. Commissioning an instrument came to include the finalization of its template library and DR recipes, initial versions of which were generally provided as part of each instrument project.

Although planned initially as an up-to-date software system which would provide the user of common-user facility instruments with a standard interface for preparing and executing their observations, much of what was provided by the ORAC project turned out to be prerequisite for the implementation of flexible scheduling in 2003.

Data Reduction Pipelines

CGS4DR

The CGS4 spectrometer arrived at UKIRT in the early 1990s. Along with it came a (very) early example of reduction pipeline technology, CGS4DR (Daly et al. 1994). This provided the astronomer at the summit with almost publication-quality reduced data products, and the reduction process (selection of calibrations etc.) was customizable on the fly. It provided for queued, automatic reduction of the incoming data stream from the instrument, as well as for re-reduction of specific groups of data. Furthermore, CGS4DR was distributed to the user community with STARLINK so that it could be used at the user's home institution after their run was complete. CGS4DR broke much new ground in the provision of tools for optimizing the observing process: the astronomer could very easily assess their achieved signal-to-noise, allowing them to halt observations when acceptable and thus freeing up time for other targets which might otherwise have been missed.

ORAC-DR

The final element of the ORAC project was a data reduction pipeline which has since been used with every UKIRT facility instrument (imagers and spectrometers) and which provides the astronomer at the summit with real-time reduced data and quality-control information (e.g. Cavanagh et al. 2008). Results from the pipeline are almost publication-quality in almost all modes of observing, and the system is (by design) entirely hands-off – the observer may interact with the display of the results but has no interaction with the pipeline itself, short of starting it at the beginning of the night and stopping it at the end. Here the tight linkage between template observing sequences and data-reduction recipes is crucial, ensuring that the telescope and instrument carry out sequences of observations which will be reducible once the data reach the pipeline. Standard recipes were provided for all modes, and new ones developed over time as requests for alterations and new template sequences came in from observers. The pipeline is data-driven; the data files themselves "know" how they should be reduced. This information is encoded in the FITS headers, from which the pipeline takes its cue. The software system is hierarchical, broken down into two layers: the top-level "recipe", which is

Fig. 7.3 ORACDR output image from a UIST polarization measurement

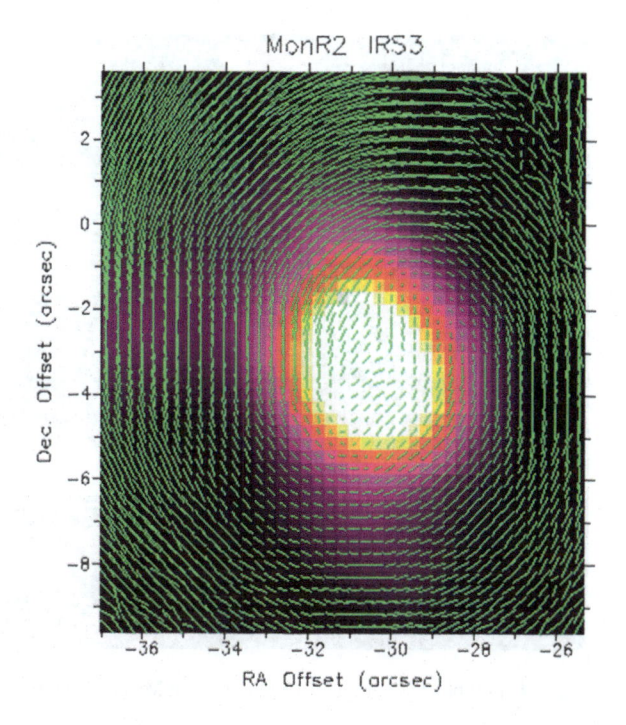

largely human-readable and amenable to modification by a user who knows what they are doing, and lower-level "primitives" which are called by the recipes and which are comprehensible to humans who can understand perl. Beneath all of this infrastructure lie the STARLINK monoliths, which provide the algorithm engine for all reduction steps.

Once again, ORACDR was designed in the context of classical observing but when the move to flexible scheduling was being considered it was quickly identified as a prerequisite. Without real-time feedback on the quality of data, it would be unreasonable, as UKIRT has now done for more than half a decade, to ask visiting observers to execute other people's observations, particularly in modes which the visitor knew little about. Figure 7.3 shows a good example: the ORACDR output image from a polarisation observation, with automatically superposed polarisation vectors.

The OMP

With the desire to implement a conditions-driven dynamic scheduling mode, came the requirement to provide a range of tools for project filtering and execution, plus communications with remote PIs who would not in all cases now visit the telescope to carry out their observing. The Observation Management Project (OMP; see,

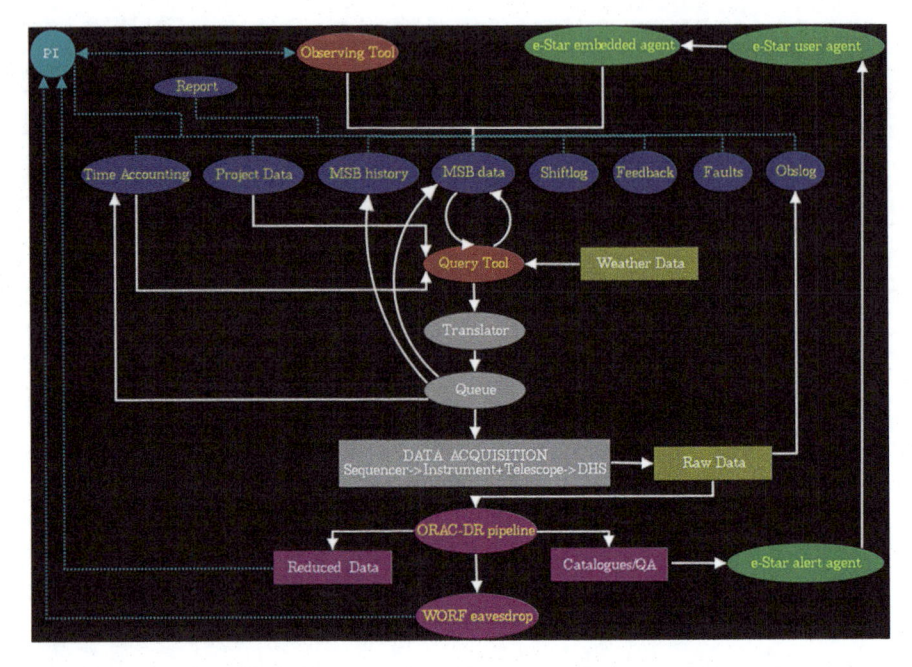

Fig. 7.4 Full block diagram of the Observation Management Project systems

e.g., Economou et al. 2002) provided the central database for observation storage, history, users, communications etc. which has been in use ever since, through into the UKIDSS sky survey and large-scale campaign projects which are now the norm at UKIRT. Figure 7.4 shows the OMP design in its entirety. Besides the preparation, submission and execution of programmes, the OMP also handles all communications between the observatory and the PI, some of which is automated (e.g. automatic emails following nights on which data are taken) and some of which is manual (e.g. communications over preparation issues). The OMP introduced a refreshingly small number of acronyms into the world, but one has become a central plank of UKIRT (and indeed JCMT) observing: the MSB (Minimum Schedulable Block). Everything you need to know about an MSB is there in its title: it represents a single block of observations on a given project, and it is Minimum in the sense that one never half completes an MSB.[1] Figure 7.5 shows the first tool employed by the summit observer – the "Query Tool" which allows the specification of observing conditions (seeing, cloud, atmospheric water vapour column) and selects only those projects which can be carried out in those conditions. It is from this tool that the observer selects the next MSB (or MSBs) to be executed.

---

[1] With only one exception – a SWIFT-triggered GRB event can cause the observer to stop and reject the currently-executing MSB.

7 Operational Innovations
81

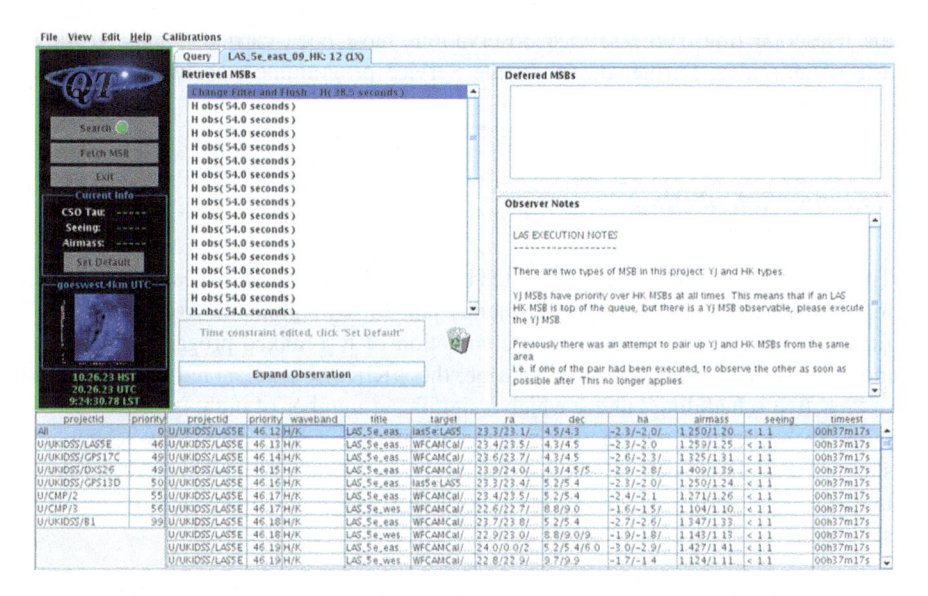

Fig. 7.5 The observer's interface to the summit database of observations: the Query Tool

Scheduling

Schedule management and diversity is a key to effective use of the available time at an observatory. UKIRT has led the way with many innovations in this area, culminating in the implementation of full flexible scheduling in the early 2000s.

Service Observing Programme

The ability to carry out short, sharply-focused projects on a short timescale compared to the usual semester rounds has proved to be highly important to the science productivity of UKIRT. Commenced in 1980s, the UKIRT observing programme was one of the first such programmes in the world and quickly established itself as indispensable to the user community. Programmes requiring less than 3 h were accepted into the programme and subjected to refereeing by a panel of typically five members; for many of these referees (including the author) the UKIRT service programme represented their first experience of peer-review panel membership. Deadlines for service programmes have come approximately monthly for two decades, and to date more than 1,700 projects have been carried out under the programme. In some years, service observing has produced up to 25 % of UKIRT publications. Observations are typically carried out by staff astronomers on each of a few nights per semester; but in recent years with WFCAM taking more of the telescope time some service observations have been carried out by visiting astronomers also. Further details of the Service programme were well described in a UKIRT Newsletter.[2]

Reactive Scheduling

The completion rate of projects accepted by time-allocation committees is a figure of merit by which observatories are frequently judged. With guidance from the UKIRT Board, which clearly identified the completion of top-ranked programmes as the most important, John Davies (at the time, the UKIRT scheduler) led a programme of "reactive scheduling" under which a significant number of nights were set aside in the schedule to allow reactive recovery of projects whose classically-scheduled summit time had been weathered out (Davies 1996, 1998). This programme, undertaken in 1999 and 2000, was successful and pointed the way to flexible scheduling.

Pair Flexing

Between 2000 and 2002, with the success of reactive scheduling clear and the Board's guidance unchanged, the emphasis moved toward full flexible scheduling as a way of maximizing time on a priority basis. This was considered to be a major psychological minefield; since no additional funding was to be provided (for example to allow for additional staff astronomers to carry out the observing) the visiting observer was expected to be asked to carry out more than just their own science projects (and potentially not to carry out their own project at all, if conditions required were never met). Would this work? A set of experiments, again run by John Davies, was carried out which mixed pairs of projects at the summit and asked the observers to switch between the two on the basis of conditions (e.g. Fig. 7.6). Results were generally positive; observers accepted the rationale of the system, and cooperated well with what were at the time quite unusual requirements on them. By the end of this programme, we were confident that full flexible scheduling could be implemented and made to work, and took a proposal to the UKIRT Board to do just that.

Full Flexible Scheduling

Under flexible scheduling, no observation is ever undertaken in conditions inappropriate for it. The success of the pair flexing experiment showed that observers

---

[2]http://www.jach.hawaii.edu/UKIRT/publications/Newsletter/issue25/issue25.pdf

aug2000

Date	Day	Observers	Prog	Inst	SS	TSS	Comment	Moon	UT
1AUG	Tue	Kerr,Hirst	CGS4 ENG	CGS4	Kerr	Wold	Need to Chec	New	2-aug
2AUG	Wed	Kerr,Hirst	CGS4 ENG	CGS4	Kerr	*	before ORAC	-	3-aug
3AUG	Thu	Kerr,Hirst,Davis	ORAC ENG	UFTI,CGS4,FP,POL	Economou	*	Edge proj?	-	4-aug
4AUG	Fri	Kerr,Hirst,Davis,Leggett	ORAC ENG	UFTI,CGS4,FP,POL	Economou	Varricatt	Edge proj?	-	5-aug
5AUG	**Sat**	Kuhn	H29B	IRCAM+POL	Leggett	*	-	-	6-aug
6AUG	**Sun**	Kuhn	H29B	IRCAM+POL	(*)	*	-	-	7-aug
7AUG	Mon	Staff	QA	WFS,UFTI	Adamson	*	-	1st Q	8-aug
8AUG	Tue	Hirst,Davis,Economou	ORAC ENG	UFTI,CGS4,FP,POL	Davis	*	-	-	9-aug
9AUG	Wed	Crawford,Fabian,Gandhi	20 F +	UFTI	Hirst	Carroll	Flex 62 for	-	10-aug
10AUG	Thu	Crawford,Fabian,Gandhi	20 F	UFTI	(*)	*	Flex 62 for	-	11-aug
11AUG	Fri	Levine,Lada,Elston	62 F	UFTI	(*)	*	Flex 62 for	-	12-aug
12AUG	**Sat**	Levine,Lada,Elston	62 F	UFTI	(*)	*	Flex 62 for	-	13-aug
13AUG	**Sun**	Clark,Steele:staff	51 F :Serv	CGS4+150:UFTI	Kerr	*	Flex 60 for	-	14-aug
14AUG	Mon	Clark,Steele:Jones,Ellis,Fairly	51 F :26	CGS4+150:UFTI	(Kerr):Hirst	Wold	Flex 60 for	-	15-aug
15AUG	Tue	Clark,Steele:Jones,Ellis,Fairly	51 F :26	CGS4+150:UFTI	(*) :Hirst	*	Flex 60 for	Full	16-aug
16AUG	Wed	Rawlings,Adamson:Staff	60 F :DDT +	CGS4+40+UFTI+IRCA	Adamson:Adamson	*	Flex 60 for	-	17-aug
17AUG	Thu	Rawlings,Adamson:Staff	60 F :DDT	CGS4+40+UFTI+IRCA	(*) :Adamson	*	Flex 60 for	-	18-aug
18AUG	Fri	Rawlings,Adamson:Staff	60 F :DDT	CGS4+40+UFTI+IRCA	(*) :TBD	*	Flex 60 Admi	-	19-aug
19AUG	**Sat**	Rawlings,Adamson:Staff	60 F :DDT	CGS4+40+UFTI+IRCA	(*) :TBD	Kuhn	Flex 60 for	-	20-aug
20AUG	**Sun**	(Eyres),Geballe:Staff	LT46:Serv	CGS4+UFTI	Hirst	*	-	-	21-aug
21AUG	Mon	Staff	Serv +	CGS4+UFTI+IRCAM	Hirst	*	-	-	22-aug
22AUG	Tue	Bunker,Warren	25	CGS4+150	Kerr	*	-	3rd Q	23-aug
23AUG	Wed	TRISPEC Team	TRISPEC ENG	TRISPEC	Davis	*	-	-	24-aug
24AUG	Thu	Yamazaki,Sato,TRISPEC group,Hou	44	TRISPEC+IRPOL	(*)	Carroll	-	-	25-aug
25AUG	Fri	Chrysostomou, Hough,Gledhill,M	23	TRISPEC+IRPOL	(*)	*	-	-	26-aug
26AUG	**Sat**	Chrysostomou, Hough,Gledhill,M	23	TRISPEC+IRPOL	(*)	*	-	-	27-aug
27AUG	**Sun**	Harries,Cropper:Jameson,Dobbie	22:54	TRISPEC+IRPOL	Davis	*	-	-	28-aug
28AUG	Mon	Harries,Cropper:Jameson,Dobbie	22:54	TRISPEC	(*)	*	-	-	29-aug
29AUG	Tue	Harries,Cropper:Jameson,Dobbie	22:54	TRISPEC	(*)	Wold	-	New	30-aug
30AUG	Wed	Harries,Cropper:Jameson,Dobbie	22:54	TRISPEC	(*)	*	-	-	31-aug
31AUG	Thu	Kuhn	H29B	IRCAM+POL	Hawarden	*	-	-	1-sep

Fig. 7.6 The UKIRT schedule for August 2000, showing (9–20 August) a number of pair-flex combinations (indicated by an F in the Programme column)

were able to do other people's science, and the reactive scheduling experiment revealed the sorts of timescales on which data taking might proceed in a flex queue. The UKIRT Board was convinced and at its Spring 2002 meeting agreed to the implementation of a Flexible Schedule at UKIRT starting in Semester 2003A (February to July 2003). The unique points which marked UKIRT's model apart are:

1. All observing in the queue was carried out by visiting observers.
2. Considerable flexibility in the submission and alteration of science programmes in the database, as projects develop over the course of a semester.
3. PI involvement in projects even if they do not come to the telescope.

This model had as prerequisites the various innovations described previously:

1. DR pipeline for immediate feedback on data quality.
2. Distributed preparation tools.
3. Observation database.
4. Communications (automatic and manual) between the PI, the JAC staff support, the observers, and the database.

Over the period since flexible observing began, hundreds of observers, from seasoned veterans to new graduate students, have come to the telescope and executed both their own projects and those of others. Some have arrived on the island when their own project was already entirely or very nearly completed; those observers have in all cases faithfully stuck to the "deal" and executed the queue to the benefit of others (sometimes while writing up the results from their own queue observations). As an example, the final few nights of Cassegrain observing were given to a project of UIST spectroscopy of ionized winds in X-ray absorbed QSOs; but the observers also carried out observations of UKIDSS candidate T and Y dwarfs with UFTI, the H_3^+ molecule in translucent interstellar clouds, thermal-IR imaging of Nova V445 Pup, UIST long-slit spectroscopy of pulsating carbon stars, and photometry of a young embedded cluster with both UFTI and UIST.

eSTAR

Elsewhere in these proceedings, Terry Lee recalls the communications methods available to the observatory when it achieved first light. With no long-distance calling off-island, TELEX and airmail was the order of the day. Nothing demonstrates the advances in technology over the 30 years than the eSTAR system (e.g. Allan et al. 2007), which takes alerts from γ-ray satellites (e.g. SWIFT) and uses them to trigger rapid follow-up observations by a network of ground-based telescopes working in the optical and infrared. UKIRT is the largest telescope functionally connected to eSTAR; and as reported by N. Tanvir elsewhere was the first to observe the infrared afterglow of GRB 090423 – at the time of writing, the most distant object in the known Universe at a redshift of 8.2. In some ways this represents the culmination of many of the technical and operational innovations described in this paper, as the trigger caused the automatic insertion into the OMP database of an MSB tailored to the target, a siren[3] sounded in the control room, and the MSB was executed by an observer who had never seen or worked on the MSB. The next, and possibly most exciting step, is to enable eSTAR to trigger on neutrino signals, and possibly to catch the progenitor of a Galactic supernova before the light flash.

UKIRT's Current Science Mix

At the time of writing, UKIRT is operating exclusively in its wide-field mode, carrying out a science programme with a range of different components and scheduling styles:

- UKIDSS Sky Survey (*flexed*)
- UH, Japan, Korea PI projects (*mostly classical*)
- Campaigns (*mix of semi-classical, reactive, overrides*)
- Open Time (*flexed within defined schedule blocks*)
- Service Programme (*flexed within defined schedule blocks*)
- Monitoring projects (*UH, UKIDSS follow-up*)

---

[3]One with which viewers of the original series of Star Trek will be familiar!

All of the above is accomplished with the mix of tools and operational models developed as described in this paper, and it is a tribute to all those involved in those developments that the telescope is in one of its most productive phases ever.

References

Allan, A., et al.: Seamless Distributed Observing with eSTAR. ASP Conf. Ser. **376**, 81 (2007)
Bridger, A., et al.: ORAC: a modern observing system for UKIRT. SPIE **4009**, 227 (2000)
Cavanagh, B., et al.: The ORAC-DR data reduction pipeline. Astron. Nachr. **329**, 295 (2008)
Daly, P.N., et al.: Automated Observing at the UKIRT. ASP Conf. Ser. **61**, 457 (1994)
Davies, J.K.: Reactive Scheduling of The UK Infrared Telescope. ASP Conf. Ser. **87**, 76 (1996)
Davies, J.K.: Results of the UKIRT reactive scheduling experiment. SPIE **3349**, 76 (1998)
Economou, F., et al.: Flexible Software for Flexible Scheduling. ASP Conf. Ser. **281**, 488 (2002)

Chapter 8
The UKIRT Wide-Field Camera

Mark Casali

Abstract I present a history of the development and commissioning of the highly successful UKIRT wide-field camera, WFCAM, which has made huge advances in infrared surveys in terms of both depth, speed and quality.

Introduction

In the late 1990s, the worldwide development of 8-m class telescopes as new facilities for conducting ground-based optical and IR astronomy raised questions regarding the future use and scientific competitiveness of older existing 4-m class and smaller telescopes. Although 8 m telescopes will always outperform smaller telescopes for the study of single distinct objects, this is not true in the case of wide-area, imaging surveys. In this case the speed with which a survey covers a given area of sky to a given depth is, to first order, proportional to the product of aperture area and field-of-view (étendue). Since it is generally easier to achieve wide fields-of-view on small telescopes, their étendue may exceed that of larger facilities and so these telescopes are capable of cutting-edge scientific survey projects. The Schmidt telescopes are the best examples of this. The Palomar, UK and ESO Schmidt surveys have made major contributions to astronomy and provided countless targets for study by 2- and 4-m telescopes over the last 40 years, despite

M. Casali (✉)
European Southern Observatory, Karl-Schwarzschild-Straße 2, 85748
Garching bei München, Germany
e-mail: mcasali@eso.org

having only approximately 1-m aperture. More recently, the Sloan Digital Sky Survey has continued this work, while 2MASS has, for the first time, extended large-area moderately-deep surveys into the infrared. As general-purpose telescopes became larger, however, surveys need to go deeper to provide lists of suitable targets for study. So just as the 1.2 m Schmidt telescopes provided targets for study by 2- and 4-m telescopes, the development of survey capabilities on 2-m and 4-m class telescopes becomes essential to feed new 8-m projects.

The idea for using UKIRT as a wide-field survey facility grew out of UKIRT Board meetings in 1998. Tim Hawarden, then Head of Instrumentation at JAC, was asked to study various wide-field options, including both imaging and wide-field spectroscopy. At subsequent presentations Martin Ward, then chair of the UKIRT Board, seized on imaging as the priority. The UKATC was contacted and asked to initiate design studies and the WFCAM project was born. The intention, from the beginning, was to provide astronomers with unprecedented deep and large-scale infrared imaging capabilities.

For a more detailed description of the WFCAM camera one should refer to Casali et al. (2007, A&A 467,777).

Setting the Requirements

The key to understanding the reason for WFCAM's peculiar design is the set of specifications for the focal plane, and four aspects in particular. Early-on it was decided that given the cost of typical HgCdTe detectors, WFCAM would not be able to go beyond having four arrays in its focal plane. The next step up, for effective use of a circular focal field, would require $3 \times 3 = 9$ devices. This was beyond any possible funding budget that WFCAM could receive (though a bolder focal plane was later adopted for VISTA). Secondly, and crucially for the project, it was decided to make use of the first generation $2k \times 2k$ Hawaii-2 devices, rather than 2nd generation 2RG devices, because we felt that the construction time of the camera was better matched to the development time of the former. Thirdly, Hawaii-2 devices have substantial detector mechanical space around the active detector area due to the ceramic carrier and associated ZIF socket. When detectors cannot be close-butted they are best used at near 100 % spacing, and so we specified a detector spacing of 90 % to allow some field overlaps during survey offsets. Finally, a simple back of the envelope calculation, as well as realistic psf simulations showed that building an instrument for optimum survey speed, given a fixed number of pixels, requires a coarse under-sampled psf. At the time, given the median UKIRT seeing of 0.6 arcsec, and trying to not under-sample too much, we specified a pixel scale of 0.4 arcsec/pixel.

UK ATC	JAC Hilo	U Edinburgh	U. Cambridge	Subaru telescope	Imperial College
M. Casali, A. Pickup	A. Adamson	A. Lawrence	M. Irwin	K. Sekiguchi	S. Warren
D. Lunney, S. McLay	I. Robson	N. Hambly	J. Lewis		
D. Henry, D. Ives	P. Hirst	M. Read			
E.Atad, J. Elliot	T. Hawarden				
K. Burch, M. Folger	N. Rees				
M. Hastie, K. Laidlaw	T. Chuter				
D. Montgomery, D. Lee					
A. Vick, B. Woodward					

The Design

The ambitious focal plane requirements described in the previous section had far-reaching implications for the optical design. For a 3.8 m telescope, these requirements imply a final f-ratio of $f/2.4$, and a field-of-view of 0.9° in diameter. Approximately a year of investigation of various designs was done before the current solution was adopted as the baseline. The reasons for rejecting all the alternative possibilities are not discussed here. However, it is worth pointing out that two of the more obvious designs can be easily ruled out. Firstly, use of the prime focus, in principle ideal because the UKIRT primary is $f/2.5$, was not possible due to UKIRT's small dome and near-zero clearance for a prime focus instrument. Secondly, placing the instrument at a classical Cassegrain focus would have resulted in very large optics to achieve the final f-ratio required and would have required a complete dismantling of the normal Cassegrain instrument support unit each time WFCAM was installed. Figures 8.1 and 8.2 show the final forward-Cass configuration and the achieved focal plane coverage.

The WFCAM optical design is shown in two parts in Fig. 8.3. One may think of the design, broadly, as a large cryogenic quasi-Schmidt camera with finite conjugate foci. A new M2 for the telescope provides an $f/9$ intermediate focus 5.7 m above M1, UKIRT's parabolic primary mirror. A simple bi-convex field lens at this point (L1) re-images M1 at a cold, slightly-undersized 396 mm diameter pupil stop (S) inside the cryostat, necessary for thermal straylight control at K-band. With fused silica as the only practical IR optical material available in such large sizes, any significant optical power must be in reflective surfaces to avoid chromatic aberrations, and so the final f-conversion from 9 to 2.4 is achieved with M3, a Zerodur ellipsoidal mirror with double-arch, rear-shaping, mounted deep within the vacuum vessel (V). M3 is mounted on three flexures that allow differential contraction with respect to its aluminium mounting plate when cooled to cryogenic temperatures, while supporting the mirror to the required flexure tolerances as the telescope changes attitude during observing. M3 is one of the largest mirrors ever used cryogenically for astronomy; at 802 mm it is only slightly smaller than the Spitzer 850 mm primary mirror.

Fig. 8.1 WFCAM mounted above the UKIRT primary mirror

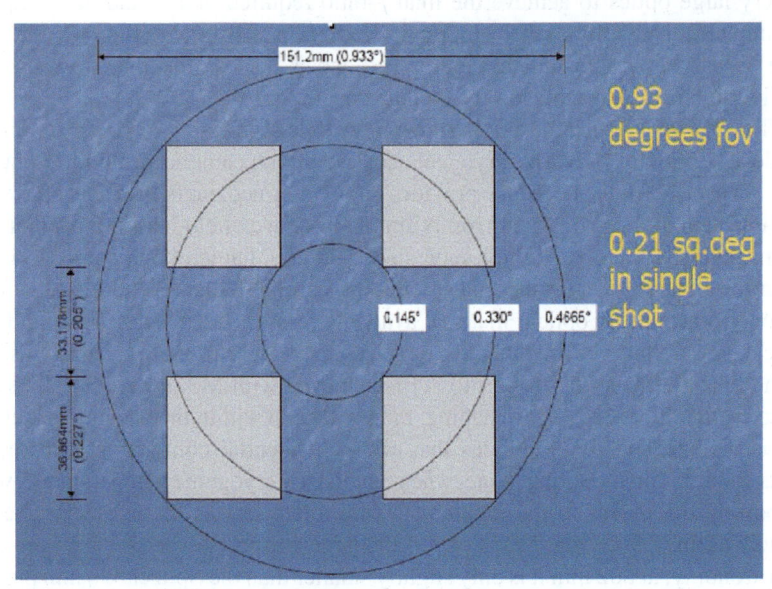

Fig. 8.2 The focal plane layout

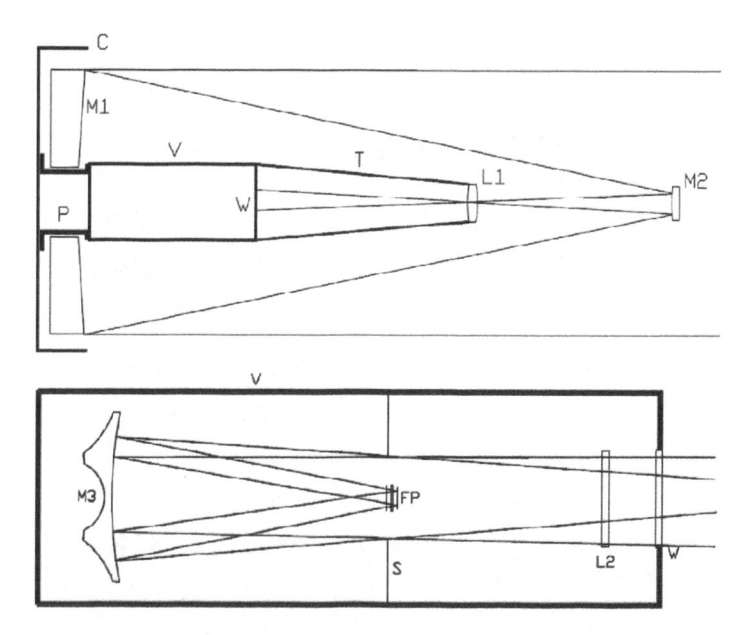

Fig. 8.3 The WFCAM optical layout

Since the new M2 is mismatched with M1, large spherical and off-axis aberrations result at the intermediate focus near L1. These are mainly corrected by the corrector plate L2, just inside the window (W). L2 has a single, aspheric surface and is a crucial component for the satisfactory optical performance of WFCAM.

The focal plane assembly (FP) involves three optical components. A filter plate contains four 60 mm square filters (one per detector) used for wavelength selection. A field flattening lens cut square, serves to give a flat-field over the detector plane. Thirdly, the square detector box itself holds the four detectors coplanar, incorporates their four PCBs, and encloses a shielded feed-through for the autoguider signals. Since the focal plane assembly directly vignettes the incoming beam, it was also designed to be as compact as possible, while allowing mechanical adjustments of detector position and coplanarity.

The entire optical design was optimised for best average image quality over the whole field from J to K bands, and system tolerancing was checked through Monte-Carlo analyses. The resulting design gives near diffraction limited performance at K-band over the whole field-of-view.

Project Pains

WFCAM represented a major step forward in capabilities for infrared imaging, involving significant leaps and hence risks in technology. Overall, the project went relatively smoothly. Areas of high risk were identified at the beginning of the project

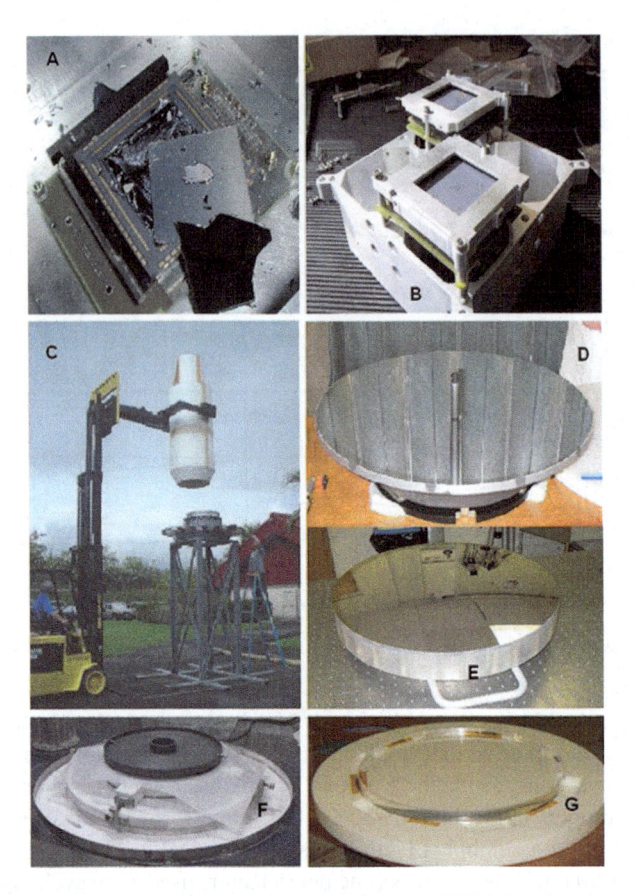

Fig. 8.4 Various components and areas of concern in the WFCAM project (see the following subsections for details)

and mitigated in various ways. For example, the cryogenic filter paddles were prototyped early. Concerns regarding the uniformity of contraction to cryogenic temperatures of M3 were addressed by specifying a special high homogeneity Zerodur. However, as any instrument builder knows, there is no way to protect a project from the "unknown unknowns", and these inevitably happened with WFCAM and a few major problems threatened the project at different times.

Delaminating Detectors (Fig. 8.4a)

Possibly the most serious and completely unexpected problem during instrument AIT was the sudden and explosive delamination of several Rockwell Hawaii-2 detectors. All IR detectors use hybrid materials and are designed to deal with the large internal stresses that result from differential contraction upon cooling to cryogenic temperatures. However, in general the detectors are very robust, so it was a great shock when one delaminated in the test cryostat. As a result the project was delayed for 6 months as we tested and studied our thermal design and cooling strategy. Finally, following consultation with Rockwell and after a long series of tests, we tentatively concluded that the problem was due to contamination of the detectors during manufacture. The faulty detectors were replaced by the manufacturer and no further problems of this nature were experienced.

Detector Coplanarity (Fig. 8.4b)

The Hawaii-2 detectors were first generation 2k × 2k PACE devices from Rockwell, glued onto ceramic holders by Rockwell with, unfortunately, limited accuracy. Not only were they rotated with respect to the holder principal axes, but they were also tilted. In addition the holder was designed to plug into a plastic ZIF socket, which ultimately meant there was no reference surface with which the detector orientation could be accurately held (other than the silicon multiplexer itself). WFCAM's fast final f-ratio (2.4) meant that there was very little tolerance for detector tilts. It thus became clearly that we had to design a fully adjustable focal plane, and this was achieved using shims and spring-loaded assemblies. This was a significant concern because of the contraction experienced by the assembly during cooldown. Would the adjustments move? Would the ZIF socket squirm slightly in its holder thus upsetting the alignment? In the end the design proved to be robust and stable and the detectors have been held with the required accuracy after final adjustment since 2004.

WFCAM Handling (Fig. 8.4c)

It was appreciated from the earliest days of the project that handling the 1 tonne instrument safely and with sufficient precision to lift it over the primary mirror and lower it in place on the central plinth would not be a straightforward process. Early concepts included assembling jib cranes on the telescope or using a system of winches to lift it into position. In the end, JAC took the lead in solving the problem and found that a specially modified lift truck from Hyster would meet our requirements. That the solution to our problem lay in asking one of the worlds largest lifting-equipment manufacturers to solve it for us should perhaps have been obvious earlier!

Cryogenic Tertiary Mirror (Fig. 8.4d)

A key element in WFCAMs optical design is the ellipsoidal tertiary mirror, which takes the place of the spherical mirror in a true Schmidt design. In principle, manufacturing of such a mirror (802 mm in diameter) is not a problem. However, its location deep in the camera cryostat meant the only practical operating temperature for it was very cold, around 100 K. Unfortunately, we could find no data on the use of such a large zerodur element at such a low temperature. Zerodur is designed for zero expansion at around ambient temperature, but at low temperatures significant contraction takes place. Would such a mirror contract homologously? To help ensure this we specified a special highly homogeneous zerodur mixture. No problems were experienced on cooling to operating temperature.

Lightweight Secondary Mirror (Fig. 8.4e)

The lightweighted new $f/9$ secondary for UKIRT was made from Zerodur and was 50 % lightweighted. It was designed to use the same fixation points as the standard $f/36$ mirror so that it could be mounted on the same piezo tip-tilt stages for fast guiding. It has performed well, despite initial concerns that the larger moment of inertia would give an unacceptably low bandwidth for the tip-tilt action.

Aspheric Corrector (Fig. 8.4f)

The aspheric corrector provides spherical and other aberration correction for WFCAM over its field-of-view. We always imagined this element would be difficult, given the large deviation from the nearest sphere of its single aspheric surface. Ultimately it proved to almost be a show-stopper as the manufacturer AMOS (Belgium) struggled to reach the required figure. The main problem was in the null-testing and in actually measuring the interferometric fringes. Finally, following a lot of determination, the final figure was reached, though well-behind schedule.

Cryostat Window (Fig. 8.4g)

The large cryostat window was a problem, not from the point of view of manu-facturing, but because the large distance from centre to edge, combined with the poor conductivity of fused silica meant that radiation into the cryostat would cool the centre to well below ambient and would certainly result in condensation during conditions of average humidity. So, a lot of care went into the thermal design - the key step eventually being to move the corrector plate into the cryostat (in the original design it was outside) immediately behind the window and allow it to thermally float by suitable isolation on G10 fixtures. With the corrector equilibrating at around 250 K this meant the window's radiative losses were greatly reduced and this 'double glazing' ensures the window stays condensation-free.

Fig. 8.5 Discussions on the WFCAM interface surface

Alignment on the Telescope (Fig. 8.5)

In a normal Cassegrain telescope there is a useful degeneracy between a decentre of the secondary mirror and its tilt. Small errors in one may be compensated in the other to obtain zero coma on-axis. Unfortunately, WFCAM with its long coaxial optical train, removes this degeneracy and all sub-systems had to be aligned to a high degree of accuracy with the telescope optical axis. But where was the UKIRT optical axis? WFCAM mounts on the UKIRT central machined steel plinth, but we had no information about its location relative to the optical axis. Measuring the optical axis (a parabolic primary, of course, has one unique axis of symmetry) is not trivial and essentially requires removing the top-end and observing stars.

Coma is zero exactly on axis, and symmetric about this point off axis. But removing the top-end, setting up a camera etc would be a major engineering task. So instead we proceeded step by step to understand the misalignment risk in simpler ways. First, we measured how tilted the WFCAM mounting interface on the plinth was to the local surface of the primary, with some simple precision jigs and micrometers. The result was reassuring – the errors in parallelism were much smaller than required. This still left the possibility of a simple decentre. Fortunately, an old Grubb-Parsons (UKIRT manufacturer) technical report was found which measured the on-axis coma (by eye) relative to the centre of the mirror cell, and indicated no observable coma in 0.5 arcsec images. This was very reassuring since the plinth is very accurately centred with respect to the cell, and suggested any error would be smaller than our error budget. Finally, it was found that a small amount of decentre relative to the optical axis could in fact be compensated by a small tilt of the whole cryostat. So a tilt adjustment was built into the base of WFCAM. Ultimately

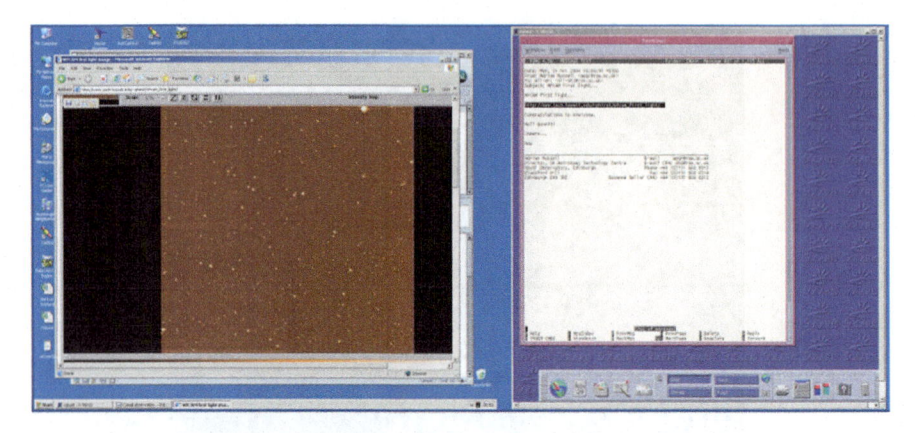

Fig. 8.6 First light with WFCAM. The stars are round!

a tilt adjustment of the whole cryostat was not required, and final alignment was done with the secondary mirror, and by one final internal adjustment of the focal plane tilt.

First Light

First light with the instrument was achieved on October 21 2004. The first image, shown in Fig. 8.6 was tremendously reassuring. The stars were approximately 1 arcsec in size, and were round and relatively uniform over the entire field. This indicated the optics were well aligned with no major residual astigmatism. Nevertheless, improving the image quality to specification took time and careful adjustments of the secondary mirror over a period of time by the JAC team. After this, WFCAM was finally ready to begin infrared surveys to depths and areas never before achieved in the infrared.

Acknowledgments I would like to firstly acknowledge the great work done by the engineers and technicians at the UKATC in making such a peculiar and challenging design a successful reality. Secondly, I would like to thank the staff at the Joint Astronomy Centre in Hawaii for doing all the hard work of final alignment, tweaking and adjustment of the instrument and software to achieve its current high observing efficiency. Thirdly, the staff of the Cambridge University Survey Unit and Edinburgh IfA widefield astronomy group deserve great praise for reducing the large amount of data and making it easily accessible to the public. Finally, I would like to thank our management and funding body at the time, PPARC, for being brave enough to fund such a risky instrument!

This article is dedicated to our old friend and colleague Tim Hawarden, who had a clever hand in the design of WFCAM and in so many other astronomical gizmos. His enthusiasm remains with us.

Reference

Casali, M., et al.: The UKIRT wide-field camera. Astron. Astrophys. **467**, 777 (2007)

Chapter 9
Polarimetry at UKIRT

J.H. Hough

Abstract Polarimetry has played an important role at UKIRT, both through visitor instruments and as an option on most facility instruments. A large range of science has been carried out covering galactic and extragalactic astronomy. To date, 93 refereed papers have been published, including the very first refereed UKIRT paper published in Nature in 1980, using observations taken in November 1979, soon after UKIRT opened. This paper outlines the development of polarimetry at UKIRT and also includes some of the many science programmes carried out over the last 30 years.

Introduction

In the early days of UKIRT, as with many telescopes at that stage of their development, there were few facility instruments and hence visitor instruments were welcomed. Groups that had made frequent use of the 1.5-m Flux Collector at Izaña, Tenerife, later known as the Carlos Sanchez Telescope, and the precursor to UKIRT, had been required to build their own instruments as the Flux Collector had no facility instruments. A number of early UKIRT observations used eye-pieces for guiding, rather than TV acquisition cameras and the author had the unnerving experience of guiding in this way when an earthquake caused UKIRT to be knocked off its shear pins but fortunately no major damage was found to the telescope (nor to the observer). One of the most significant changes came with the introduction of the first infrared arrays and thereafter most polarimetry was carried out with facility

---

J.H. Hough (✉)
Centre for Astrophysics Research, Science & Technology Research Institute,
University of Hertfordshire, College Lane, Hatfield AL10 9AB, UK
e-mail: j.h.hough@herts.ac.uk

instruments. Continued improvements to the stability of the telescope also led to far higher spatial resolution images, typically sub-arcsec, compared to apertures of several arcsecs used with the single-element detectors.

The Era of Single-Element Detectors

Hatfield Polarimeters

One of the first visitor instruments at UKIRT was one of the early Hatfield Polarimeters (known as HATPOLs) designed and built at the then Hatfield Polytechnic, later to become the University of Hertfordshire. This particular HATPOL used the facility f/9 photometer, incorporating a single element InSb detector and a B Halle super-achromatic waveplate designed for 0.31–1.1 μm but with sufficiently high efficiency to be useful in the near-infrared. In November 1979, J-band circular polarimetry observations were made of 2A0311-227 (Fig. 9.1), an X-ray source that had been recently identified as an AM Her binary (polar). Optical circular polarimetry had already been carried out at the 60-inch Flux Collector using HATPOL. The much lower degrees of near-infrared circular polarization (Fig. 9.1) were explained either by the emission at different wavelengths coming from different poles or a rapidly changing cyclotron optical opacity with wavelength (later observations showed a high degree of correlations between optical and infrared flickering, confirming the source of optical and near-infrared emission was from the same accretion column). The result, published in Nature (Bailey et al. 1980), was the first refereed paper from UKIRT.

Later versions of HATPOL included separate optical and near-infrared channels with the optical providing simultaneous U, B, V, R and I linear or circular polarimetry, with each channel switching between viewing the object and the sky. The optical channels used photomultiplier tubes and the infrared channel used the f/35 photometer, with a single element InSb detector. Although the advent of IR-arrays eventually made the HATPOLs redundant, the possibility of making simultaneous observations with five optical and one IR band (one of J, H, and K) was very important for objects that varied and it was also efficient for observations of many point sources, and the HATPOLs were used until 1990, often shuttling between the Anglo-Australian Telescope and UKIRT.

Programmes to take advantage of this large wavelength coverage included establishing a large number of polarized standards, covering the U through to the K-band, and were used by many groups at a number of telescopes (Whittet et al. 1992). Observations of 3C345 (1,641 + 399), a superluminal radio source and optically violently variable quasar, from the U through to the K band showed clear departures from a power-law spectrum with contributions from an unpolarized non-thermal component (the blue-bump) and possibly from hot dust (Fig. 9.2, Mead et al. 1988). It was becoming clear, however, that to make further progress in understanding the flux and polarization spectrum of such objects that spectropolarimetric observations, preferably in the optical and near-infrared, would be required.

Fig. 9.1 2A0311-227.
(**a**) K-band light curve;
(**b**) J-band light curve;
(**c**) broad-band optical light curve; (**d**) J-band circular polarization; (**e**) broad-band optical circular polarization. Infrared observations from UKIRT, November 1979

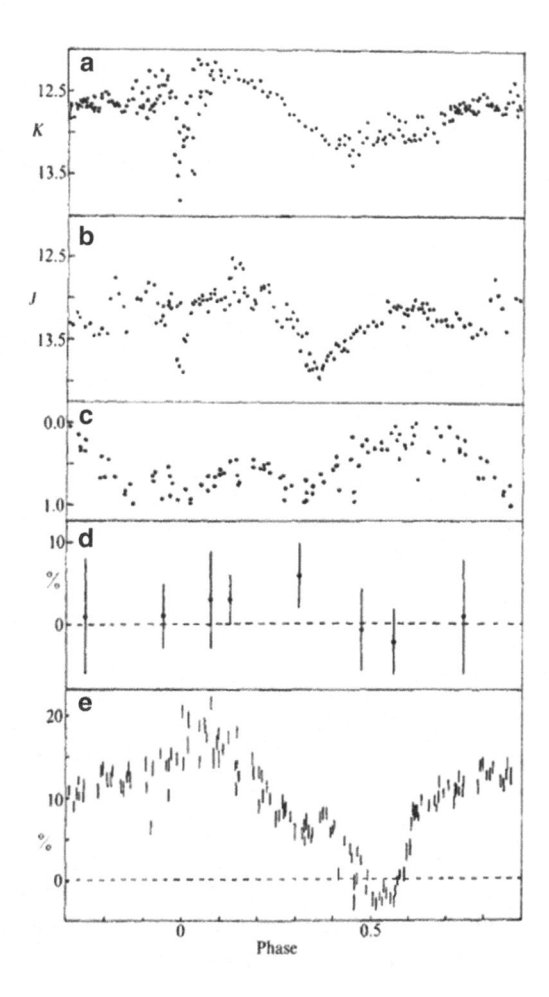

Kyoto Polarimeter

Another successful early visitor instrument was the Kyoto polarimeter, which consisted of a rotating waveplate mounted above the instrument support unit dichroic (see section "Introduction of IR Arrays"), and used between 1984 and 1989 with facility photometers UKT6 and UKT9. It was used, inter alia, to make large scale polarization maps using the 19.6 arcsec aperture (bucket mode) for which UKIRT, at that time, was well suited. Figure 9.3a and b show K-band and S(1) line polarization images of OMC-1 (Hough et al. 1986). The thicker line indicates the polarization close to BN/IRc2 (observations from the AAT).

The K-band observations clearly showed the centro-symmetric polarization pattern associated with reflection nebulosity centred around IRc2, although higher spatial resolution polarization observations with HATPOL at the AAT showed that polarization close to IRc2 and BN is along position angle (PA) 118°, the same PA

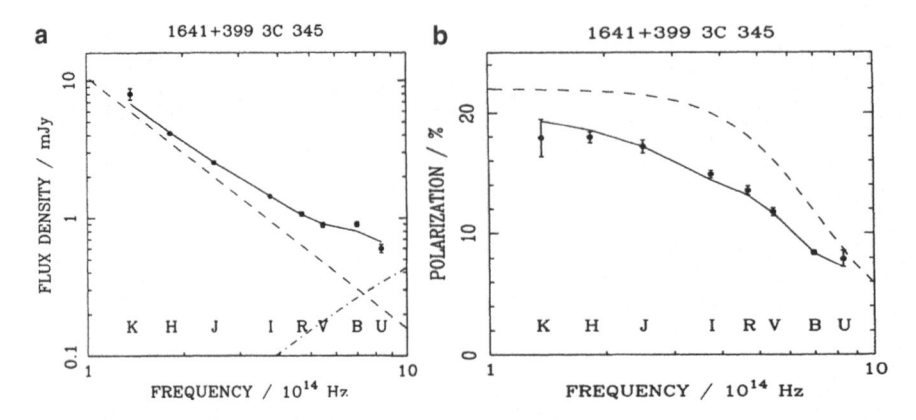

Fig. 9.2 (*Left*) Flux density and (*Right*) polarimetry for 3C345. The *solid line* connects model estimates at each observed frequency. The *dashed line* in (**a**) shows the flux density of the fitted power-law and blackbody components and in (**b**) shows the polarization predicted for the power-law and blackbody components combined

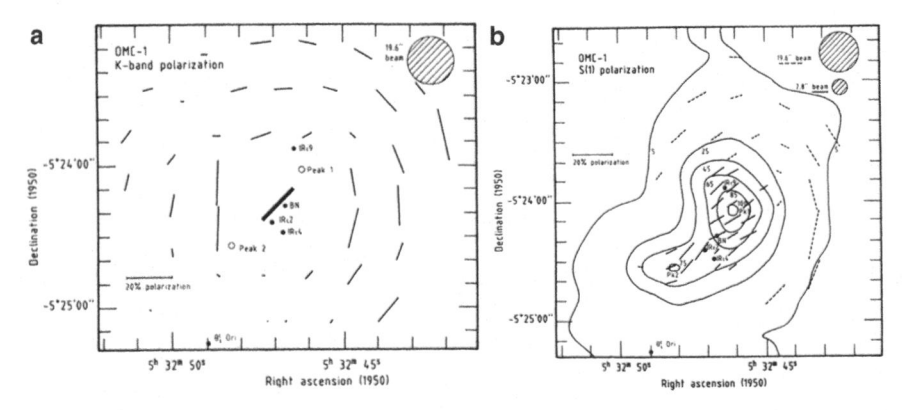

Fig. 9.3 (**a**) K-band polarization of OMC-1. The *thicker line* shows the polarization close to BN/IRc2 (AAT observations). (**b**) S(1) line polarization of OMC-1

as the outflow axis as defined by the redshifted and blueshifted CO gas, and by the two lobes of shocked molecular hydrogen emission. Polarimetry of the molecular hydrogen line emission showed aligned polarization vectors in the central ∼40 arcsec, again along PA 118°, surrounded by a reflection nebula, the first seen in line emission. The extent of the aligned vectors corresponds to the region of intrinsic line emission with the radiation passing through overlying aligned grains and being polarized by dichroic absorption. Knowing how grains align with respect to the local magnetic field showed that its PA was also 118°. More generally, polarization maps of regions of star formation, many carried out at UKIRT, were important in establishing the basic geometry of outflows associated with YSOs and more refined models were developed after the introduction of arrays.

Fig. 9.4 Signing of MoU for UK-Japanese Co-operation in Ground-Based Astronomy 1997. From *left* to *right*: J Hough, A Boksenberg, K Kodaira, N Kaifu

Joint polarimetry programmes with many Japanese astronomers, using the Kyoto polarimeter (Sato, Morimoto, Kaifu, Hasegawa, Tamura, Nagata and others), largely laid the foundations for the JCMT/Nobeyama collaborations, and the signing of the MoU for UK-Japanese Co-operation in Ground-Based Astronomy, which in turn led to the building of FMOS on Subaru (Fig. 9.4).

UCL Array Spectrometer

Another important early visitor instrument was the mid-IR UCL spectrometer, first used as a polarimeter on UKIRT in 1986. For mid-IR polarimetry on UKIRT see Pat Roche's article in these proceedings.

Introduction of IR Arrays

The introduction of IR arrays led to a reduction in the number of visitor instruments, certainly in the near-infrared, as most groups didn't have the resources or capacity to develop imagers and spectrometers incorporating the newly available arrays. All near-infrared instruments from this point on did include polarimetry options although these were, in some cases, added after the initial instrument design which led to some limitations in performance. The Instrument Support Unit (ISU) was

Fig. 9.5 Photograph of the IRPOL 2 unit

well suited to providing polarimetry for a range of instruments as it included a 45° dichroic that could be rotated to reflect the light to one of four ports where instruments were mounted. Importantly, with the UKIRT f/35 beam then in use, a polarimetry unit could be employed above the dichroic thereby minimizing any telescope-induced polarization (typical instrumental polarization is $\leq$0.2 %). The unit consisted of a rotating assembly which could hold various waveplates and the whole assembly could be lowered from a parked vertical position, out of the field-of-view of any instrument, to a horizontal position for polarimetry (Fig. 9.5).

The initial IRPOL, built at the ROE, could only accommodate waveplates of limited diameter. A second version, IRPOL2, was built at the University of Hertfordshire, funded by a PPARC grant, and this could hold waveplates of 95 mm diameter. The same grant provided an infrared $\lambda/2$ achromatic made from quartz and magnesium fluoride (1–2.5 μm) and zero-order magnesium fluoride $\lambda/2$ waveplates for the L and M bands; all manufactured by B Halle. A wiregrid polarizer is manually placed above the waveplate for calibrating the polarization efficiency.

The usual method of observing is to rotate the waveplate to 0, 45, 22.5 and 67.5° for linear polarimetry, and then to integrate the signal for several seconds in each position. This mode of operation, typical with array detectors, meant that modulation was slow. Polarimeters employing single elements detectors had normally used continuously rotating waveplates so that modulation rates were relatively high, a few Hz, reducing the effects of any changes in atmospheric transparency. In step-and-stare mode avoiding this problem is easily achieved by recording both the e- and o-images/spectra simultaneously. For extended sources a focal-plane mask is needed in each instrument to avoid the e- and o-images

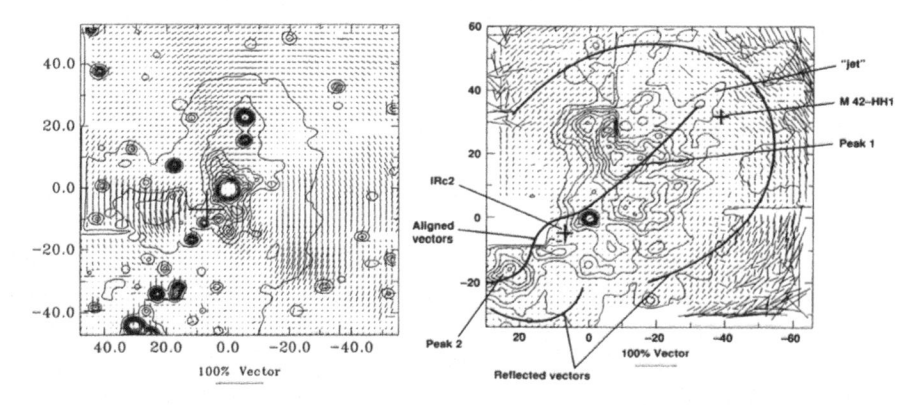

Fig. 9.6 *LH panel*, K-band imaging polarimetry of OMC-1 superimposed on the intensity contour plot; *RH panel*, S(1) line polarization image of OMC-1, superposed on the line intensity contour plot. The overall directions of the reflected and aligned vectors are shown by the *heavy lines*

overlapping. Although this means the telescope has to be moved in order to cover an extended image the gain in polarization accuracy in a given time is still considerably more and polarimetry can be carried out in poor weather condition.

UKIRT provided excellent data reduction software for imaging and spectropolarimetry based on *Gaia* and the STARLINK package *Polpack*. Having real-time data reduction is very important for polarimetry as it depends on the differences between similar images/spectra recorded at each waveplate position, and these will be very small for low degrees of polarization. For more details of polarimetry at UKIRT see www.jach.hawaii.edu/UKIRT/instruments/irpol/irpol.html.

IRCAM

IRCAM, the first InSb IR camera at UKIRT, with 58×62 pixels, was first used for imaging polarimetry in 1988. In January observations were made of the S(1) shocked molecular hydrogen line in OMC1 (Fig. 9.6, Burton et al. 1991) and broad band polarimetry of OMC1 (Fig. 9.6, Minchin et al. 1991). For these early observations the pixel scale was 0.62 arcsec/pixel and a cold wiregrid polarizer was used as the polarization analyser (dual-beam analysers were added to instruments at a later stage).

The vastly increased spatial resolution, compared to the results with the single element detectors (see above), didn't alter the general picture but did enable far more detailed models to be developed, including the shape of and inclination of the outflow cavity. The relative fluxes of the 1–0 S(1) line at 2.12 µm and the 1–0 Q branch at 2.4 µm, in the scattered nebulosity, were used to show the sizes of the scattering grains are comparable to those in the interstellar medium whereas Minchin et al. (1991) had found much larger grains in the core region of OMC1.

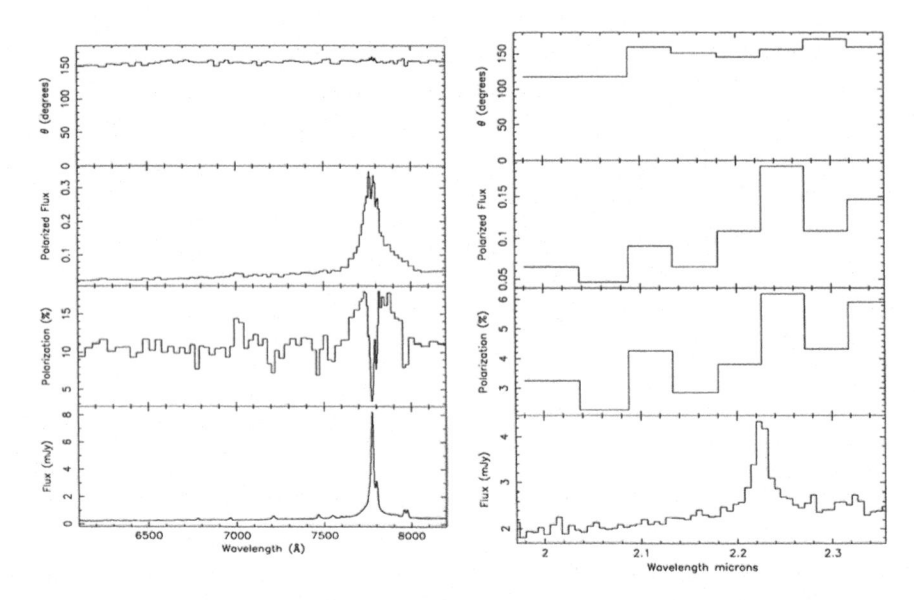

Fig. 9.7 3C234: *Left*, optical spectropolarimetry (WHT); *Right*, infrared spectropolarimetry

This combination of IRCAM and IRPOL was used to study a number of YSOs, and had real impact in our understanding of the environs of these objects.

CGS4

The second facility instrument to include polarimetry was CGS4, an array spectrometer. Although this instrument used a magnesium fluoride Wollaston prism as the analyzer, it was added after the basic instrument design was completed and the prism could only be placed above the slit, which meant that the slit had to be perfectly aligned with the line between the e- and the o-spectra. This was not always easy to accomplish, which meant that the advantages of dual-beam polarimetry were often lost. Young et al. (1998) observed the radio galaxy, 3C234 and were able to compare the data with optical spectropolarimetry from the William Herschel Telescope (WHT). Figure 9.7 shows the IR and optical data and although the former has a better signal to noise the importance of the IR is that many radio galaxies have extensive dust lanes and hence the scattered broad lines can suffer significant extinction in the optical. The IR shows an increasing polarization in the infrared caused by dichroic extinction in the AGN torus (see also section "TRISPEC").

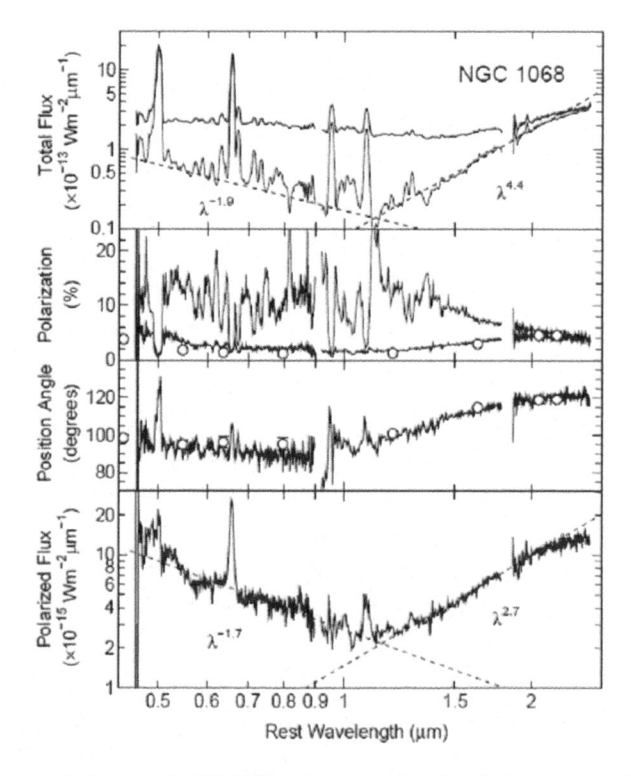

Fig. 9.8 Spectropolarimetry of NGC1068, after correction for Galactic interstellar reddening polarization. The *thin solid curves* in the *top two panels* show F_λ and P after correction for starlight dilution. The *dashed lines* show the power-law fit to the continuum in the optical and near-infrared. The *open circles* are the broadband polarimetry of Bailey et al. (1988)

TRISPEC

One visitor instrument making use of IR arrays was TRISPEC (Triple Range Imager and Spectrograph). This instrument had three channels: one optical covering 0.45–0.90 μm, and two infrared channels covering 0.90–1.85 and 1.85–2.5 μm. Simultaneous imaging and spectropolarimetry was possible in all three channels. A single custom-made half-wave plate, produced by B Halle, covered the full wavelength range of TRISPEC with β-BBO Wollaston prisms used in the optical and short-wavelength infrared channels, and a $LiNbO_3$ Wollaston in the second infrared channel. It was used in 2000 and 2001 to observe a number of Type 2 Seyfert galaxies (Watanabe et al. 2003). Figure 9.8 presents the results for NGC1068, showing the main polarization features arising from scattering at optical wavelengths, diluted by starlight and then increasing polarization in the near-infrared arising from dichroic extinction in the obscuring torus.

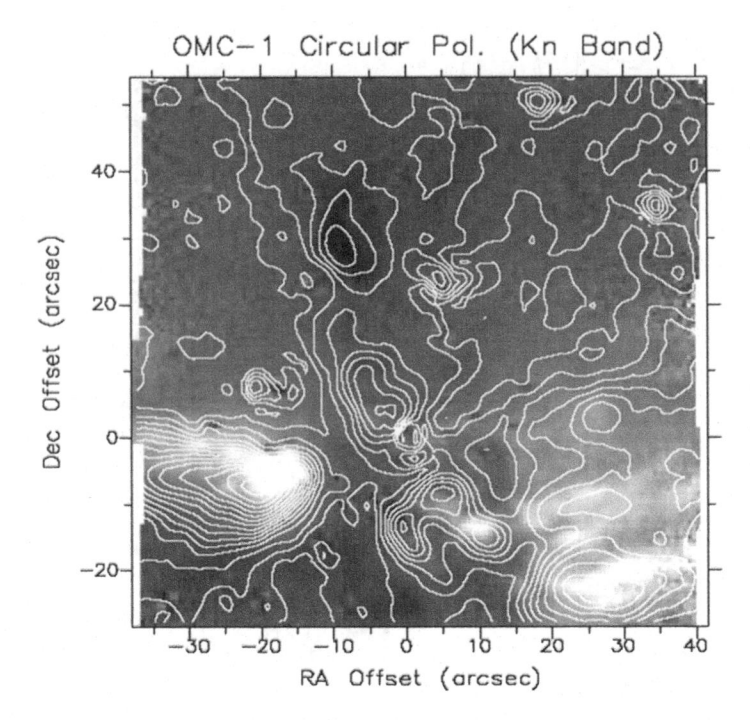

Fig. 9.9 K band circular polarization of OMC-1, overlaid on contours of linear polarization

Circular Polarimetry

Circular polarimetry was introduced by the University of Hertfordshire (UH) in 1998 and made available to the community through collaboration with UH. It consisted of a continuously rotating half-wave plate mounted above the IRPOL2 unit which then included a stepped quarter-wave plate. Calibration of polarization efficiency was carried out using a wire-grid polarizer and a second quarter-wave plate to produce 100 % circular polarization. Figure 9.9 presents circular polarimetry of OMC-1 (Chrysostomou et al. 2000) showing very high degrees of LH and RH circular polarization. Importantly, this confirmed earlier observations made at the AAT (Bailey et al. 1998), and later more extensive UKIRT observations with UIST (see section "UFTI and UIST") showed that the high degrees of CP are only associated with the region around IRc2 (Buschermöhle et al. 2005). The high degrees of polarization are either produced by scattering off aligned grains (Bailey et al. 1998), or by the passage of linearly polarized light through a medium of aligned grains (Lucas 2003). The Bailey et al. paper raised the interesting possibility that the high degrees of circularly polarized light, in regions where stars and planets are forming, could be responsible, through asymmetric photolysis, for initiating the homochirality found in all living organisms.

UFTI and UIST

Two later instruments, UFTI a 1–2.5 μm imager and UIST, a 1–5 μm imager spectrometer, both included polarimetry in the initial instrument design, thereby ensuring maximum performance. The polarising analyser used with UFTI and UIST was a beam-splitting Wollaston prism. With UIST, two prisms are installed, one in each grism wheel, so that spectropolarimetry is available with all grisms. UFTI uses β-BBO and UIST MgF_2 for the Wollaston prisms. Following are four examples of the diverse science programmes carried out with these instruments.

Post-AGB Stars

Figure 9.10 shows J-band polarimetry of the post-AGB object IRAS 06530-0213 (Gledhill 2005). Such objects are characterised by prodigious mass-loss, resulting in the formation of a circumstellar envelope of molecular gas and dust. Polarimetry is an excellent tool in imaging faint material close to a bright star. If the star is unpolarized, then a polarized flux image will have zero intensity for the star, effectively removing the star from the image. Figure 9.10 clearly shows the axisymmetric shell around the star with a more dense equatorial region.

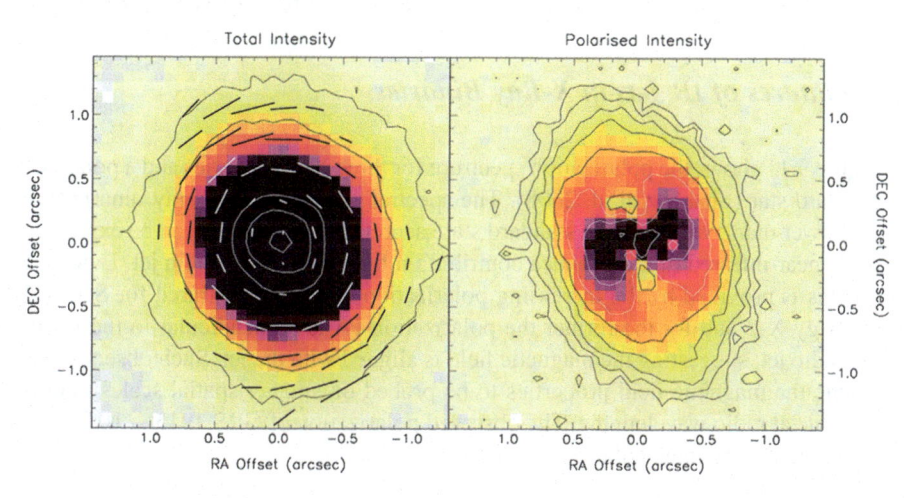

Fig. 9.10 IRAS 06530-0213: *LH panel*, J-band total intensity; *RH panel*, J-band polarized intensity

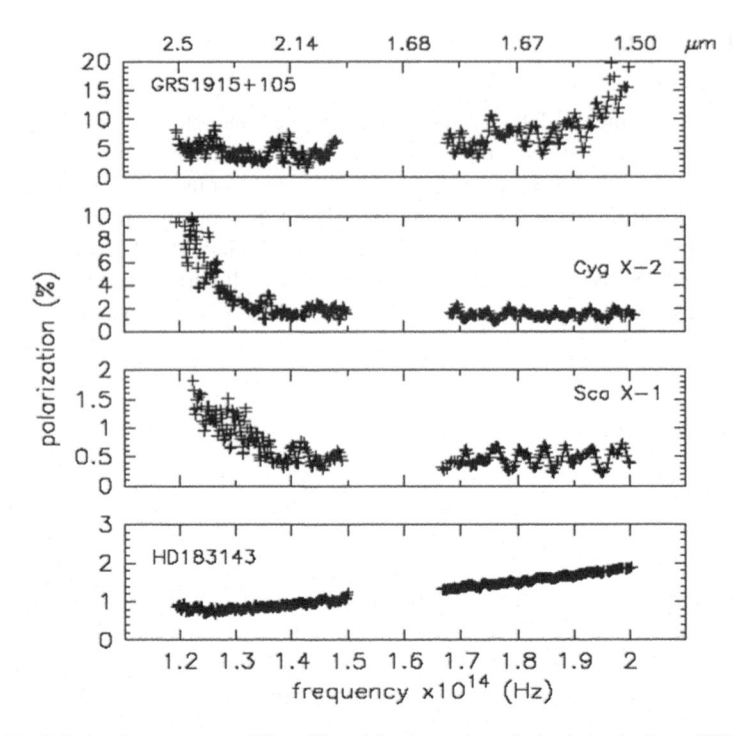

Fig. 9.11 Polarization spectrum of three X-ray binaries and a polarized standard star (HD183143)

Signatures of IR Jets in X-Ray Binaries

Figure 9.11 shows the polarization spectrum for three X-ray binaries and a polarized standard star (Shahbaz et al. 2008). The spectral slopes for the X-ray binaries are shallower than expected for a standard state accretion disk but this can be explained if the near-infrared flux contains a contribution from an optically thin jet. Evidence for this is provided by the increasing polarization in the near-infrared for Sco X-1 and Cgy X-2 and for the former the polarization PA is perpendicular to the PA of the radio jet, suggesting the magnetic field is aligned with the jet. Such observations enable the magnetic field properties to be probed on smaller spatial scales than is possible at radio wavelengths. The observed polarization of GRS 1915 + 105 could be dominated by interstellar polarization.

Grain Alignment

One of the main polarizing mechanisms in astronomy is produced by the passage of radiation through a medium of aligned grains (dichroic absorption). It is generally accepted that grains align with their short axis preferentially aligned along the

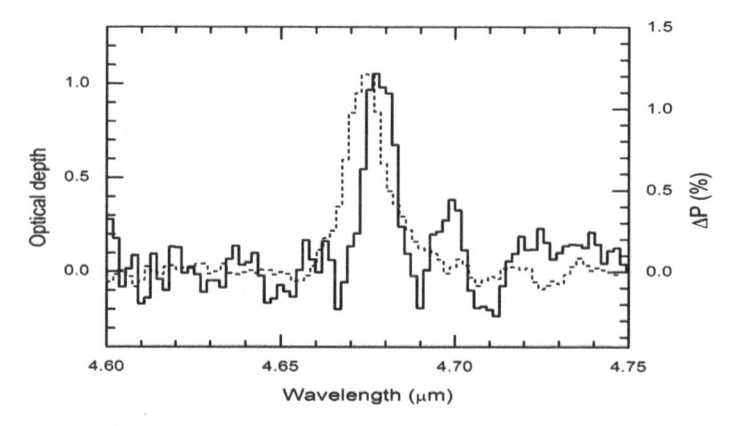

Fig. 9.12 Linear polarimetry of solid CO along the line of sight to Elias 16 (TDC)

direction of the local magnetic field and this has then been used to determine the structure of such fields. However, it was believed that grains should not align in dense cold clouds, where coupling of dust and gas temperatures should lead to poor alignment. To test this, spectropolarimetry of the 4.67 μm solid CO feature along the line-of-sight to Elias 16, background to the Elias Dark Cloud, was carried out using UIST (Hough et al. 2008). This clearly showed (Fig. 9.12) that the CO feature is polarized and the distinctive shift between the peak optical depth and the peak polarization is a clear signature of polarization produced by dichroic absorption. This confirms that grains do align in dense cold regions. The most likely explanation is that radiative torques are primarily responsible for producing an asymmetry in the distribution of grain spin axes.

Quasars

At the time of writing this article, the last published UKIRT polarimetry paper is by Kishimoto et al. (2008). They show that for a number of quasars the predicted blue spectrum for optically thick accretion disks does occur but that it is masked in total flux by thermal emission from hot dust at wavelengths longer than ∼1 μm, making the spectrum much flatter (Fig. 9.13). Scattering of disk radiation by electrons within the BLR reveals the intrinsic disk spectrum (electron scattering is wavelength independent).

Conclusions

Polarimetry has played a significant role in the excellent science that has been carried out at UKIRT over the last 30 years. Initially this was through visitor instruments but then all facility instruments included polarimetry as an option.

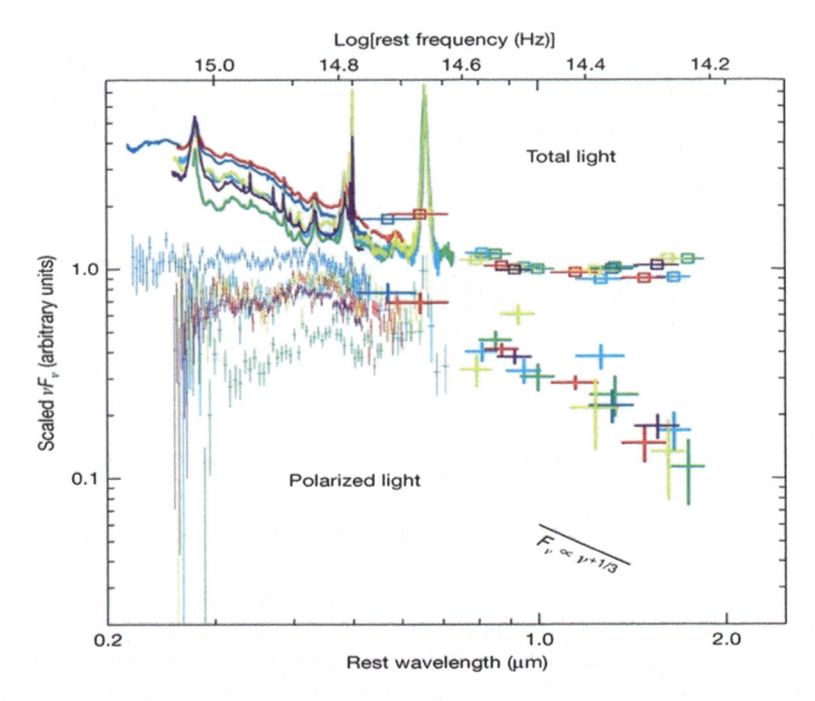

Fig. 9.13 Polarized and total light spectrum for six different quasars. See Kishimoto et al. (2008) for further details of the plots

Although the UKIRT instrument support unit made it relatively easy to include polarimetry for a range of instruments it did require top-level support from UKIRT management and from the Royal Observatory Edinburgh. It is very likely that these last 30 years will be seen as a golden age for UK polarimetry. Access is now more limited for visitor instruments as telescopes become increasingly dedicated to a small number of facility instruments carrying out major programmes and/or observatories, faced with limited resources, don't feel they have the capacity to support visitor instruments. A further problem arises, especially on the 8–10 m class telescopes, with limited access to the unfolded Cassegrain, or the ability to locate the polarization modulator above any inclined mirror. The additional instrumental polarization that results can be compensated for by the use of mirrors and retarders but this is not ideal.

The polarimetry community needs to make sure that the designers of future telescopes and instruments do include polarimetry in the base design. This will enable astronomers to build on the success of this technique, as amply demonstrated in 30 years of observing on UKIRT.

Acknowledgments I would like to thank the UKIRT staff who gave tremendous support during my many visits to UKIRT, particularly when bringing the HATPOL instruments which then required ISU2 and any facility instruments to be taken off the telescope. Not to be welcomed at 14,000 ft.

References

Bailey, J., Hough, J.H., Axon, D.J.: IR photometry and polarimetry of 2A0311-227. Nature **285**, 306 (1980)

Bailey, J., et al.: The polarization of NGC 1068. Mon. Not. R. Astron. Soc. **234**, 899 (1988)

Bailey, J., et al.: Circular polarization in star-formation regions: implications for biomolecular homochirality. Science **281**, 672 (1998)

Burton, M.G., et al.: Molecular hydrogen polarization images of OMC-1. Astrophys. J. **375**, 611 (1991)

Buschermöhle, M., et al.: An extended search for circularly polarized infrared radiation from the OMC-1 region of Orion. Astrophys. J. **624**, 821 (2005)

Chrysostomou, A., et al.: Polarimetry of young stellar objects – III. Circular polarimetry of OMC-1. Mon. Not. R. Astron. Soc. **312**, 103 (2000)

Gledhill, T.M.: Axisymmetry in protoplanetary nebulae – II. A near-infrared imaging polarimetric survey. Mon. Not. R. Astron. Soc. **356**, 883 (2005)

Hough, J.H., et al.: Infrared polarization in OMC-1 – discovery of a molecular hydrogen reflection nebula. Mon. Not. R. Astron. Soc. **222**, 629 (1986)

Hough, J.H., et al.: Grain alignment in dense interstellar environments: spectropolarimetry of the 4.67-μm CO-ice feature in the field star Elias 16 (Taurus dark cloud). Mon. Not. R. Astron. Soc. **387**, 797 (2008)

Kishimoto, M., et al.: The characteristic blue spectra of accretion disks in quasars as uncovered in the infrared. Nature **454**, 492 (2008)

Lucas, P.W.: Computations of light scattering in young stellar objects. J. Quant. Spectrosc. Radiat. Transf. **79**, 921 (2003)

Mead, A.R.G., Brand, P.W.J.L., Hough, J.H., Bailey, J.A.: Polarimetric observations of the quasar 3C 345. Mon. Not. R. Astron. Soc. **233**, 503 (1988)

Minchin, N.R., et al.: Near-infrared imaging polarimetry of bipolar nebulae. I – the BN-KL region of OMC-1. Mon. Not. R. Astron. Soc. **248**, 715 (1991)

Shahbaz, T., Fender, R.P., Watson, C.A., O'Brien, K.: The first polarimetric signatures of infrared jets in X-Ray binaries. Astrophys. J. **672**, 510 (2008)

Watanabe, M., et al.: Simultaneous optical and near-infrared spectropolarimetry of type 2 Seyfert galaxies. Astrophys. J. **591**, 714 (2003)

Whittet, D.C.B., et al.: Systematic variations in the wavelength dependence of interstellar linear polarization. Astrophys. J. **386**, 562 (1992)

Young, S., et al.: The obscured BLR in the radio galaxy 3C 234. Mon. Not. R. Astron. Soc. **294**, 478 (1998)

Chapter 10
UKIRT in the Mid-Infrared

P.F. Roche

Abstract The historical development of and scientific results from the mid-infrared instruments used on the UK Infrared Telescope from 1980 to 2005 are described. That near- and mid-infrared instruments (and submm photometers) were mounted on the telescope and easily available for use was a particular strength of UKIRT, which encouraged multi-wavelength and long-term monitoring programmes. The mid-infrared spectrometers UCLS and CGS3 were the most sensitive available in the world in the 1980s and 1990s and they established the spectral properties of many classes of astronomical objects. Papers describing mid-infrared measurements accounted for over 10 % of the total output from UKIRT in this period.

Introduction

From the beginning, UKIRT was intended to provide a comprehensive astronomical capability across the near- and mid-infrared windows accessible from the ground. Programmes with mid-infrared photometers were included in the first telescope time allocations; allocations for the first years of operation (1979–1982) are published in QJRAS. Indeed, mid infrared observations have played a substantial role in the scientific output of the telescope, accounting for over 10 % of the published papers. I will describe the mid-infrared instrumentation available at UKIRT over the last 30 years and summarise some of the scientific results produced.

P.F. Roche (✉)
Department of Astrophysics, Oxford University, Denys Wilkinson Building,
Keble Road, Oxford OX1 3RH, UK
e-mail: p.roche1@physics.ox.ac.uk

UKIRT as a Mid-Infrared Telescope

The first mid-infrared instrument was the single bolometer instrument UKT7. The initial telescope configuration employed the f/9 secondary and a focal-plane chopper was used until the f/35 chopping-secondary mirror was installed in mid-1980. The substantial thermal imbalance resulting from the motion of the instrument beam across the telescope mirrors with the focal plane chopper gave large and unstable differential signals that compromised sensitivity in the early days of operation. At that time, the television guide camera could only detect objects significantly brighter than ~ 12 magnitude, requiring the observers to guide the telescope using an eyepiece on all but the brightest targets. This made observing at UKIRT inefficient – and it must be said a frustrating experience. However, the installation of the chopping secondary mirror, and especially the replacement of the ill-fated hydraulically-operated mechanism by a more conventional electro-mechanical system brought UKIRT's mid-infrared performance closer to a level comparable with that achieved at other leading infrared telescopes.

Even when the chopping secondary was installed, the telescope collimation remained problematic. The secondary mirror position relative to the optical axis of the primary was usually set to give reasonable image quality in the near infrared, but this resulted in the instrument beam falling beyond the edge of the primary mirror. In the mid-infrared this would lead to large thermal gradients, and so a compromise position had to be used, accepting some image distortions (coma) to give acceptable thermal performance. Because the mid-IR instruments throughout the 1980s were aperture photometers or spectrometers, this was not too much of a problem, but the advent of array instruments in the second half of the 1980s, starting with IRCAM and including the NASA/GSFC mid-IR camera in 1989 required further improvements to the mirror alignment. When the new top end with its 5-axis secondary mirror motion was installed in the mid-1990s as part of the JAC/ROE/MPIA upgrades project, the alignment problems were finally solved and UKIRT's image quality and thermal stability became truly excellent.

From inauspicious beginnings, UKIRT's guiding and nodding or beam-switching performance improved rapidly. Whilst the telescope today has a conventional guide-camera, the mid-1980s saw a very different and innovative approach that provided a pragmatic and surprisingly effective guidance system. It consisted of a quad-cell guider stuck to a TV monitor displaying the output from a camera mounted on the cross-head below the Instrument Support Unit (ISU). This arrangement mitigated a number of deficiencies in the telescope performance and provided guiding and beamswitching to quite a high accuracy. I believe that this is just one of Ian Gatley's contributions to UKIRT operation, and certainly one that had a big impact. It was particularly useful on one of my observing runs in May 1983 when comet IRAS-Araki-Alcock made its closest approach to the Earth at a distance of less than five million km. The comet whizzed across the sky, covering over $30°$ in a day, at a rate that would have taxed most guiding systems but UKIRT's autoguider coped just fine. The only requirement was for the observers to pay close attention to the TV screen as occasionally a star brighter than the comet nucleus would pass by the

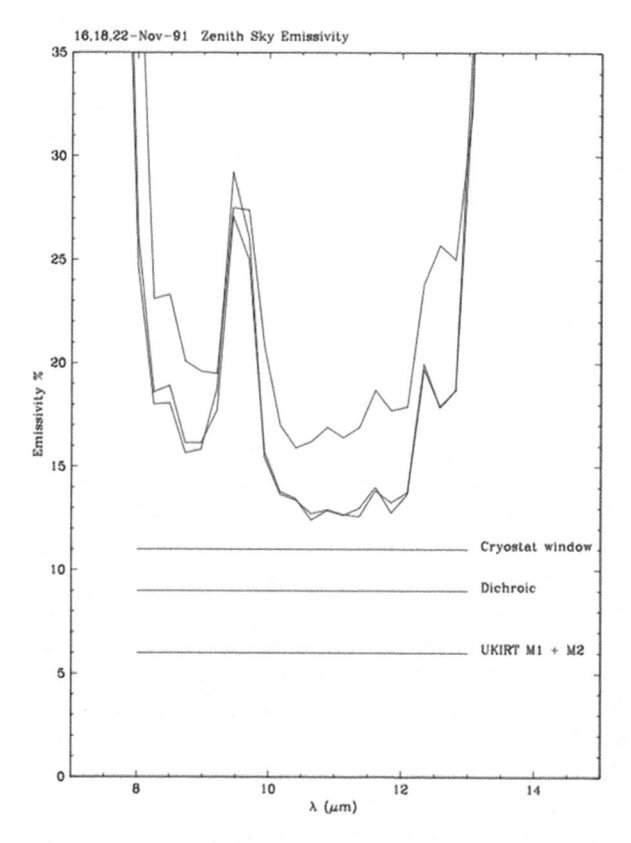

Fig. 10.1 Emissivity measurements at 8–13 μm made in November 1991 with the UCL spectrometer. The PWV column as measured with the CSO radiometer was high on Nov 16 ($\tau_{225GHz} \sim 0.2$), and the sky emissivity is significantly higher than the other two nights with $\tau_{225GHz} \sim 0.095$ and 0.08 respectively

quad-cell detector, which could then lock onto the bright intruder rather than the comet. UKIRT's current guiding and tracking arrangements are excellent and were installed as part of the upgrades programme.

One of the components that made UKIRT successful was the ability to mount several instruments simultaneously on the telescope, and switch between them by rotating a dichroic mirror that reflected the IR light to the elected instrument and passed the visible light to the guide camera below. The extra reflection from the dichroic necessarily increased the system emissivity seen by thermal infrared instruments, compared to the emissivities seen by up-looking instruments on other IR telescopes (such as the IRTF). It undoubtedly did compromise the thermal background of UKIRT but the advantages in stability of the instruments, guiding capabilities and ease of operation outweighed this disadvantage in my opinion.

The telescope emissivity as been monitored over the years, and has achieved a best figure of $\sim$10 %, with enhanced values due to dust on the mirrors and deterioration of the dichroic coatings. Figure 10.1 shows measurements of the UKIRT system emissivity made in November 1991. The contributions from the telescope, dichroic and cryostat window are separated from the sky through measurements made off the telescope and by tilting the dichroic so that the instrument beam bypassed the secondary mirror and are assumed to be wavelength-independent. The sky emissivity near 9 μm is higher than predicted from models, and may arise from particles injected into the atmosphere by the Mt Pinatubo eruption in June 1991.

The UKIRT Mid-Infrared Photometers

The initial complement included two single-channel bolometric photometers, UKT7 and UKT8. UKT8 had a remarkably long life from its commissioning in July 1980 until it was retired in November 1994. In addition two multi-channel photometers were constructed. IRASFU was an array of 2×8 Si:As photoconductor detectors (from the same batch as those used in the UCLS and CGS3 spectrometers) designed for efficient mid-IR mapping and particularly for the follow-up of sources identified by the IRAS satellite. It was followed by UKT16 or the 8-banger, an array of 2×4 germanium bolometers, built to increase the mapping speed in the mid-IR. Despite the significant potential multiplex advantage, the array instruments were not popular, because the individual channels could not match the sensitivity achieved by the single channel instruments and the calibration overheads were significant, requiring standard stars to be scanned across all of the pixels. I have been able to locate only one paper presenting data with IRASFU, wherein Richardson et al. (1986) present a 20 μm map of DR21 and one with UKT16 where Greidanus and Strom (1991) present a map of the 20 μm emission in part of Cass A.

In the early days of UKIRT, a major uncertainty was the nature of the mid-infrared emission from the nuclei from different classes of active galaxies. Observations with the bolometers contributed significantly to investigations by several teams, often combining data with photometers on the IRTF or other telescopes. (e.g. Willner et al. 1985; Lawrence et al. 1985). Close collaboration between the telescopes on Mauna Kea was an early feature of many programmes at UKIRT, but especially those conducting photometry and spectroscopy in the mid-IR. A second feature exploited UKIRT's wide range of capabilities. Observations across the near- and mid-infrared regions were combined with submillimetre measurements to investigate the variable emission from quasars and blazars. Indeed some of the earliest mid-IR results reflect this mode of operation, providing key information on the frequency-dependence of flares in OV236 and 3C273 (Gear et al. 1983; Robson et al. 1983), tracking the changes from millimeter wavelengths into the infrared (Fig. 10.2). Throughout their lives, the bolometers often provided important mid-infrared data in multi-wavelength observations of astronomical objects. The longest running programme was monitoring of dust formation in Wolf-Rayet binary systems in which Williams et al. were able to demonstrate that dust

10 UKIRT in the Mid-Infrared

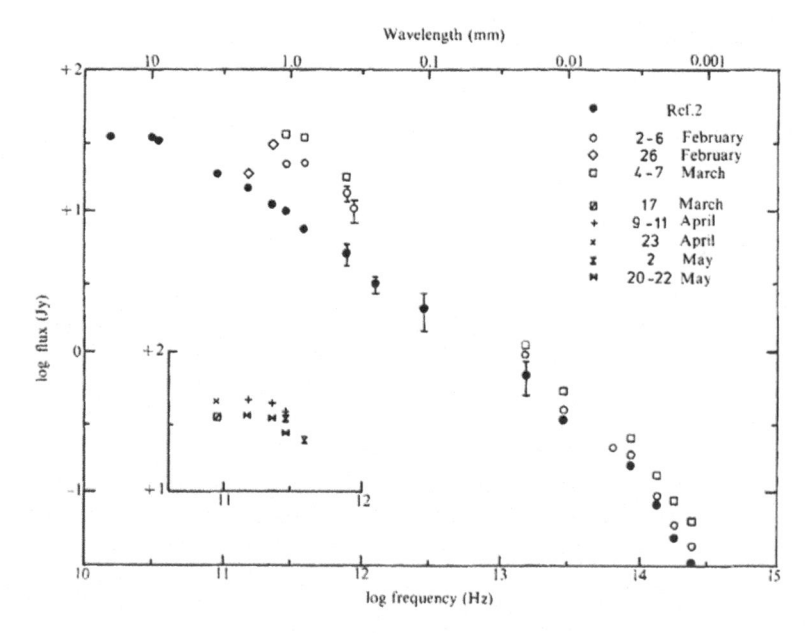

Fig. 10.2 The evolution of a flare in the mm to IR emission of the QSO 3C 273 traced by monitoring with the UKIRT photometers (Robson et al. 1983)

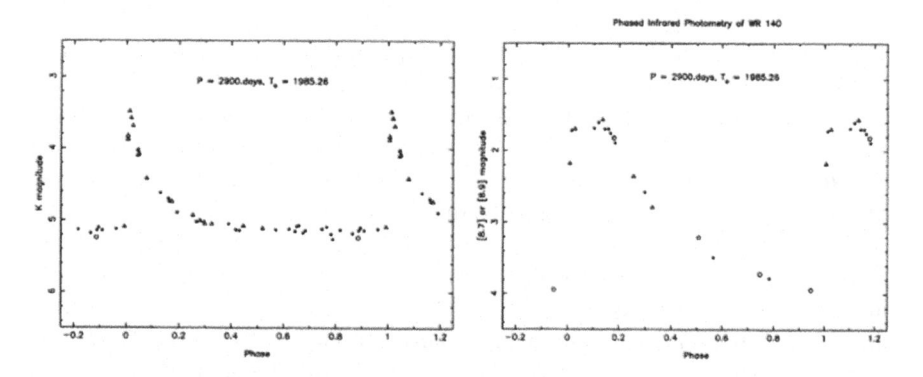

Fig. 10.3 K- and N-band photometry of WR140, showing the increase in infrared emission associated with dust condensation (Williams et al. 1990)

formation in the carbon-rich envelopes of late-type WC stars was triggered by the close approach of the WR and O star (e.g. Williams et al. 1990) (Fig. 10.3).

Observations of Solar System objects, and especially asteroids and comets featured prominently, sometimes as a result of coordinated campaigns (e.g. measurements of comet p/Halley (Green et al. 1986)) and the trial run measurements of p/Crommelin (Zarnecki et al. 1984). In the latter case, observations were curtailed not only by telescope computer problems, but also by an eruption of

Mauna Loa. Other comets were observed as targets of opportunity, often carried out in service mode. The mid-infrared observations, usually combined with near-infrared observations, provide estimates of dust production rates and temperatures as well as albedos and other properties of grains.

The UKIRT Mid-IR Spectrometers

At first sight, it is perhaps surprising that the spectrometers account for the bulk of the mid-IR publications from UKIRT, but they provided truly unique capabilities and were by far the most sensitive instruments of their kind for observations of compact objects throughout the 1980s and 1990s. The arrays of 30 discrete detectors provided a substantial multiplex advantage over their competitor instruments, which employed single detectors and circular variable filters. After 2000, instruments on 8-m telescopes became available, but the UKIRT capabilities were only completely eclipsed in terms of N-band spectroscopy of compact objects by the IRS on the Spitzer Space Telescope.

UCLS: The UCL Mid-Infrared Spectrometer (1980–1992)

The first spectroscopic run was with the UCL spectrometer as a visiting instrument, in March 1980, with the last occurring more than a decade later. In its initial configuration it used an array of 5 Si:As detectors at a spectral resolution of 0.09 μm in the N-band and a focal-plane chopper. The instrument was upgraded the following year to an array of 30 detectors, sampling the 8–13 μm spectrum with 3 grating positions. In 1982, a lower resolution grating was procured, providing full coverage of the 8–13 μm window and maximum sensitivity, but at a lower spectral resolution of 0.22 μm. In conjunction with observations at the AAT and the IRTF, this instrument defined the spectral properties of many classes of astronomical object in the 10 and 20 μm windows. It made important contributions to our understanding of PAHs. The good correlation between the dominant mid-IR dust emission properties, in terms of silicate, silicon carbide or PAH band emission, and gas-phase Carbon-to-Oxygen ratio demonstrated that the PNe displaying the PAH emission bands have the highest C/O ratios, consistent with them arising from carbon-rich species (Barlow 1983).

The mid-infrared extinction properties in the solar neighbourhood and towards the Galactic Centre were measured, showing enhanced silicate absorption in the central parts of the Galaxy (Roche and Aitken 1984, 1985). A highlight of the UKIRT observations is the differentiation of the spectral properties in star-forming and active galaxies, presented in a series of papers between 1981 and 1991 (e.g. Roche et al. 1991) (Fig. 10.4). It was proposed that the high energy flux in AGN preferentially destroyed small grains and in particular PAHs (Aitken and

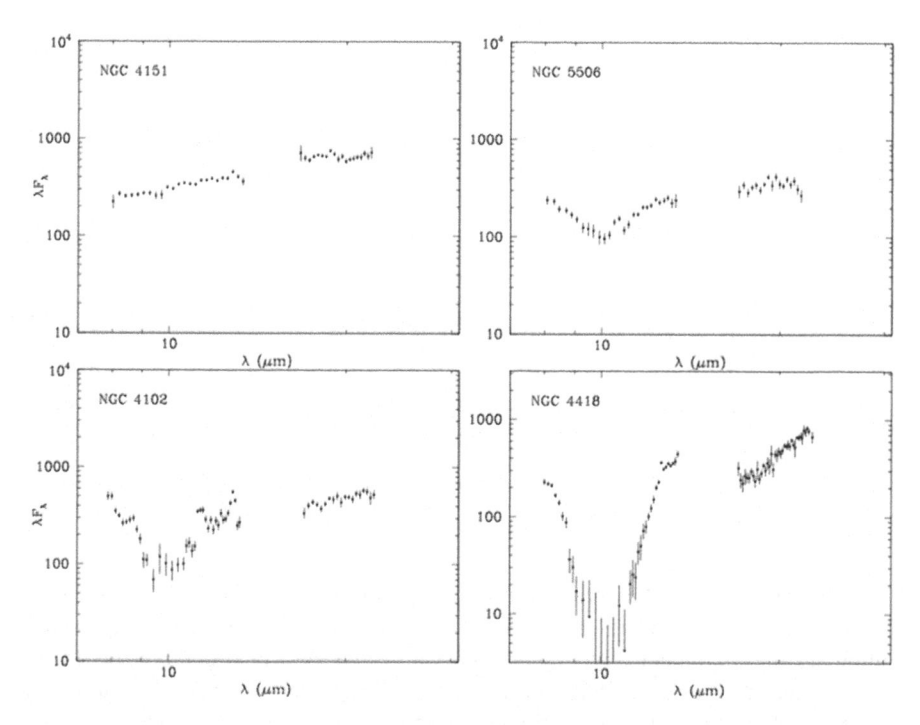

Fig. 10.4 N and Q-band spectra of four galaxies obtained with the UCL spectrometer between 1982 and 1985. The PAH emission bands dominate the starburst galaxy NGC 4102 together with the 12.8 μm [NeII] line, contrasting with the featureless spectrum of the Seyfert 1.5 galaxy NGC 4151 and the shallow silicate absorption in NGC 5506. The highly obscured nucleus of NGC 4418 still has the deepest known extragalactic silicate absorption feature (Roche et al. 1986, 1991)

Roche 1985) explaining the weakness of the PAH emission bands in active nuclei. This naturally led to the establishment of PAH emission as a diagnostic of a galaxy nucleus where star-formation dominates; this has been developed into a key diagnostic for high-z galaxies from Spitzer data.

A spectropolarimetric mode was added in 1983, using an internal wire-grid analyser and warm rotating CdS and CdSe waveplates for the 10 and 20 μm regions respectively. In this mode, the UCLS was again used on the AAT and IRTF as well as on UKIRT, but the importance of Mauna Kea as a site is particularly clear in the 20 μm spectropolarimetry results. Results from spectropolarimetry include the detection of ordered magnetic fields in the central parsec of our Galaxy and the first evidence for crystalline silicates through the identification of the increased band strength of sharp 11.2 μm structure in the polarized spectrum of AFGL 2591 (Aitken et al. 1988) (Fig. 10.5).

Even today, apart from some exploratory results obtained by Capps et al. (1978); Dyck & Lonsdale (1981) and a few objects measured with TIMMI-2, all of the mid-infrared spectropolarimetric measurements available are those obtained with the UCL spectrometer. Half of the objects listed in the Mid-Infrared Spectropolarimetry

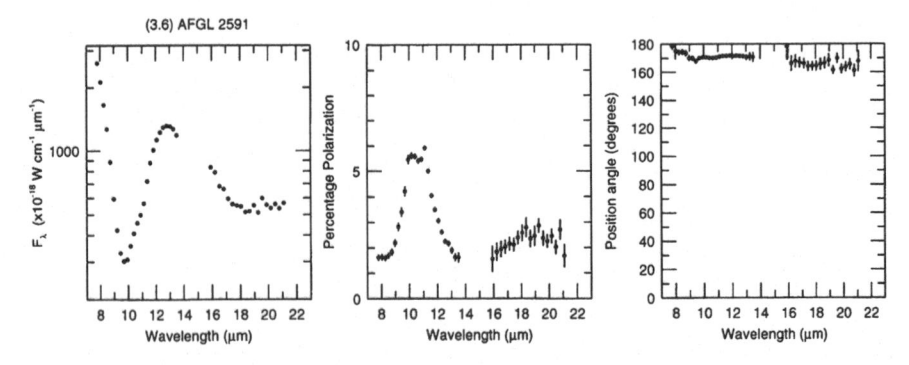

Fig. 10.5 N and Q-band spectropolarimetry of the obscured YSO, AFGL 2591, obtained at UKIRT in 1986. The main polarization peaks near 10 and 20 μm are at slightly longer wavelengths than the absorption minima in the intensity spectra, and are attributed to absorption by aligned silicate grains. The secondary peak at 11.2 μm provided some of the first evidence of crystalline silicates, with higher band strength than the more common amorphous grains (Aitken et al. 1988)

Atlas (Smith et al. 2000) were measured at UKIRT. Spectropolarimetry in the midinfrared allows the separation of emissive and absorptive polarization components, allowing the alignment directions of the warm and cool dust components to be determined. The results include constraints on silicate band strengths and variations in the 10–20 μm silicate polarization ratios. In massive YSOs, the polarization spectral data indicate that the magnetic fields in the YSO envelopes are predominantly toroidal, perhaps indicating that they have been wound up in the YSO disks.

CGS3, the UKIRT Facility Spectrometer 1990–1998

The facility mid-infrared spectrometer, CGS3 was built at UCL and commissioned on UKIRT in July 1990, where it remained in operation until 1998. A description of the instrument, and its contributions to the mid-infrared calibration project, for which it was crucial, is provided by Cohen and Davies (1995). The instrument had a similar performance to the UCL Spectrometer with an array of 32 Si:As detectors and 3 interchangeable gratings providing medium ($\Delta\lambda \sim 0.06$ μm) and low-resolution ($\Delta\lambda \sim 0.2$ μm) 8–13 μm and low resolution 20 μm spectroscopy. It was anticipated that CGS3 would be largely replaced by visiting instruments with 2-D detector array (e.g. MICS) and eventually by a UKIRT facility mid-IR spectrometer.

CGS3 followed the successful IRAS mission and overlapped with ISO. Most of the programmes conducted with CGS3 investigated the spectral properties of dust in late type stars (Fig. 10.6) and in Vega disks, although it also continued programmes of solar system spectroscopy and some galaxy spectra. Sylvester et al. (1996) presented CGS3 spectra of a dozen Vega-disk candidates along with optical, near-IR and sub-mm photometry. They detected silicate or PAH emission bands in

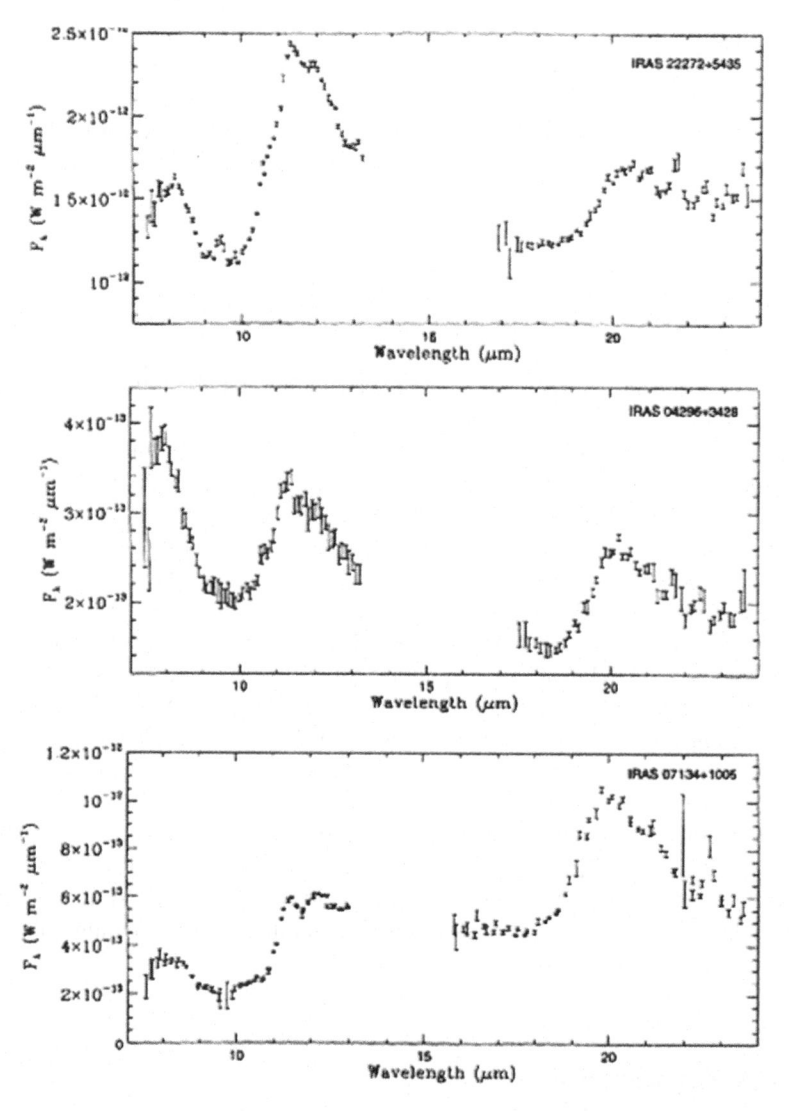

Fig. 10.6 10 and 20 μm spectra of four post-AGB objects believed to be in transition to the planetary nebula phase, showing complex hydrocarbon emission spectra (Justtanont et al. 1996)

most of their targets, concluding that the dust in Vega disk systems can be either oxygen- or carbon-rich, with no clear dependence upon the spectral type of the star.

The nature of circumstellar dust around evolved stars has been probed in a number of programmes. Groenewegen et al. (1998) used radiative-transfer models to interpret the SEDs and mid-infrared spectra of carbon stars obtained with CGS3 or UCLS. They estimated mass-loss rates and dust-to-gas ratios. Speck et al. (2000) investigated the profiles of the silicon carbide emission bands in carbon star envelopes, identifying a number of objects with SiC in absorption and large inferred mass loss rates, and proposed that self-absorption could affect many carbon stars and thus the SiC production rates. They pointed out that the detailed circumstellar profiles are consistent with the alpha-SiC structures, unlike meteorites in which beta-SiC dominates, though their later work pointed to spectral shifts in laboratory data from SiC embedded in KBr.

Mid-IR Array Spectrometers

An array spectrometer for UKIRT was seen as the natural follow-up to CGS4, providing long-slit spectroscopy at 10 and 20 μm and a range of spectral resolving powers. This instrument was initially conceived as CGS5, but with the decision to join the Gemini project, it was seen that an instrument that could be used on both UKIRT and Gemini-N would be a real benefit, spending 50 % of its time on each telescope. After a long gestation period, Michelle was commissioned on UKIRT in late 2001, with science programmes conducted in 2002. Michelle was transferred to Gemini in 2003 and after a second short period on UKIRT it was delivered to Gemini North on long-term loan in the Spring of 2004. The relatively short periods of availability on UKIRT meant that quite a lot of time was spent in commissioning the various modes of Michelle. Several papers describe Solar System observations, and especially the determinations of asteroid albedos. For example, Fernandez et al. (2005) found that most objects are best fitted by low thermal inertia models and that many near earth asteroids have comet-like albedos.

Some other array instruments had spectroscopic modes. MICS was a forerunner for COMICS on Subaru and had spectroscopic capabilities in the N band (e.g. Miyata et al. (2000) measured the dust profiles of 18 Mira system systems, concluding that both silicates and amorphous alumina grain are required to match the spectra).

The Array Cameras

A number of mid-infrared array cameras were used on UKIRT. The first of these instruments was the Goddard camera developed by Gezari et al. It used a Si:Ga array with 58×62 pixels and was used on UKIRT in 1989 to map out the 12 μm polarization in the central 10 arcsec of our Galaxy (Aitken et al. 1991), demonstrating strong fairly uniform polarization (and hence grain alignment) in the northern arm of the Galactic centre mini-spiral. The MIRACLE camera with a 58×62 pixel Si:As array was developed at MPE and ROE and used in 1991 to map-out the dust emission in the nucleus of NGC 1068 at 10 and 20 μm, showing that

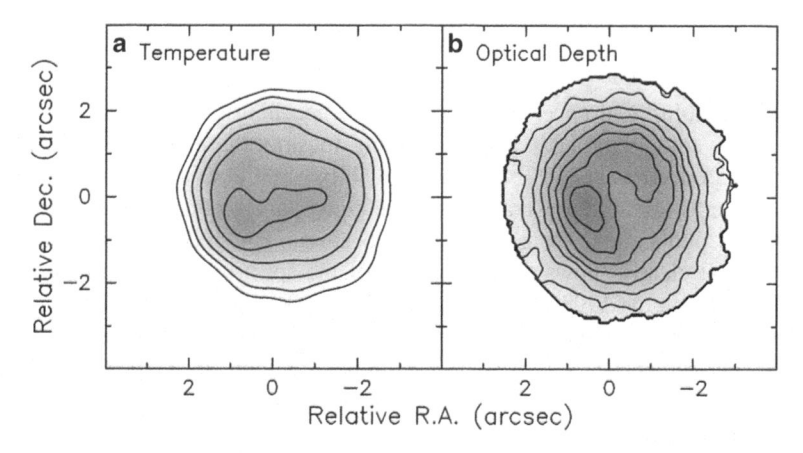

Fig. 10.7 Colour temperature (*left*) and 20.6 μm optical depth (*right*) maps of the dust in the circumstellar envelope of 07134 + 1005. The temperatures range from 160 to 185 K and are derived from images at 8.8 and 17.4 μm obtained with MIRAC2 on UKIRT in 1994 (Dayal et al. 1998)

most of the mid-infrared flux emerges from extended regions in the ionization cones rather than from a compact nuclear torus in this galaxy (Cameron et al. 1993). The Berkeley camera had a smaller format detector (10 × 64 Si:Ga pixels) but higher sensitivity and was used to image a number of compact planetary nebulae and evolved stars from 1992 (e.g. Meixner et al. 1997), revealing asymmetric structures in some of them and providing information on the mass loss history.

As detector technologies improved, larger format arrays became available, and the instruments used in the 2nd half of the 1990s were typically 128 × 128 pixel devices. Imaging polarization measurements of the BNKL complex in Orion were extended to 17 μm from measurements with NIMPOL at UKIRT in 1995, showing two distinct polarized regions at almost orthogonal position angles associated with BN and IRc2 (Aitken et al. 1997). The complex of polarized structures over small spatial scales may account for the relatively low submm polarization measured in the core of Orion with large beams with the JCMT. The MAX mid-IR camera was developed at MPIA to take full advantage of the improved thermal and imaging performance delivered to UKIRT by the upgrades programme. It was used from 1995 to resolve disks around some TT Tauri stars and with careful image processing, produced wide-field ∼ (5 × 3.5 arcmin) maps of the Orion nebula at 10 and 20 μm (Robberto et al. 2005) revealing extended filamentary structures together with compact sources identified as disks around young objects in the cluster. The Berkeley camera was replaced by MIRAC2, which extended the imaging surveys of proto-planetary nebulae and other dusty stellar objects (Meixner et al. 1999), revealing a variety of dust shell structures and profiles (Fig. 10.7).

Mid-Infrared Impact

The first mid-infrared paper appeared in 1980, reaching a peak of ten papers per year in the mid-1990s and slowly declining towards the end of that decade. A resurgence occurred in 1992 with the availability of CGS3 and subsequently the array cameras, giving an average of ten refereed papers per year from then until 2000. The publications from 1981 to 2006 are shown, coded by instrument, in Fig. 10.8. It clearly shows the dominance of the spectrometers in producing the peaks (UCLS in the 19980 s and CGS3 in the 1990s) and the steadier output from the photometers and imagers. There were over 150 papers published from mid-infrared instruments in this period, accounting for more than 10 % of the total output from UKIRT, though some of these publications also included data from other instruments.

In my view, the major mid-infrared legacies from UKIRT were:

- Establishing the nature of mid-IR emission from different classes of galaxy through near- and mid-infrared photometry and spectroscopy.
- Demonstrating the remarkable differences in spectral characteristics of AGN and HII region nuclei, with the latter dominated by emission from PAH grains that are destroyed by the hard photons from the active nuclei.
- Helping clarify the properties of the PAH bands – leading to their identification as C-rich particles that exist in the regions surrounding HII regions

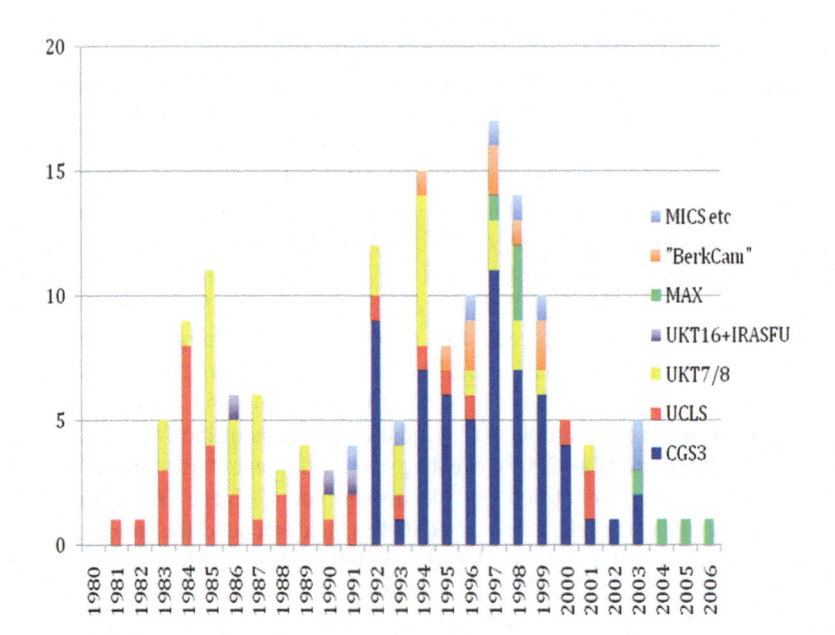

Fig. 10.8 UKIRT publications from the first 25 years, coded by instrument

- Measuring the thermal properties of asteroids and comets and finding links between different families of solar system objects
- Elucidating the properties of Vega excess disks and ProtoPlanetary nebulae
- Identifying the polarimetric signatures of aligned amorphous and crystalline silicate grains and demonstrating their use as tracers of magnetic fields at arcsec resolution

The combination of mid-IR with near-infrared and submm instruments offered real advantages. Long-term and multi-wavelength programmes, often using the UKIRT Service observing capability were critical to many projects such as monitoring episodic dust formation in Wolf Rayet Stars and the development of flares in blazars.

Acknowledgements The world-leading infrared astronomy at UKIRT has been developed through the dedication of numerous people in Hawaii and Edinburgh. They are too many to mention, but I am extremely grateful for the help and support received over the years from the early days in Leilani Street to the present day at the JAC in Komohana Street. UKIRT is a wonderful telescope with a fantastic staff. It still has some foibles but the scientific results and the enthusiastic atmosphere are a real testament to the excitement of our field. Thank you all.

References

Aitken, D.K., Gezari, D.Y., McCaughrean, M.J., Smith, C.H., Roche, P.F.: Polarimetric imaging of the Galactic center at 12.4 microns – the detailed magnetic field structure in the northern arm and the east-west bar. Astrophys. J. **380**, 419 (1991)

Aitken, D.K., Roche, P.F.: 8–13 micron spectrophotometry of galaxies. IV – six more Seyferts and 3C345. V – the nuclei of five spiral galaxies. Mon. Not. R. Astron. Soc. **213**, 777 (1985)

Aitken, D.K., Smith, C.H., James, S.D., Roche, P.F., Hough, J.H.: Infrared spectropolarimetry of AFGL 2591 – evidence for an annealed grain component. Mon. Not. R. Astron. Soc. **230**, 629 (1988)

Aitken, D.K., Smith, C.H., Moore, T.J.T., Roche, P.F., Fujiyoshi, T., Wright, C.M.: Mid- and far-infrared polarimetric studies of the core of OMC-1: the inner field configuration. Mon. Not. R. Astron. Soc. **286**, 85 (1997)

Barlow, M.J.: Observations of dust in planetary nebulae. In: Flower, D.R. (ed.) IAU symposium 103, 'Planetary Nebulae', p. 105. Reidel, Dordrecht (1983)

Cameron, M., Storey, J.W.V., Rotaciuc, V., Genzel, R., Verstraete, L., Drapatz, S.: Subarcsecond mid-infrared imaging of warm dust in the narrow-line region of NGC 1068. Astrophys. J. **419**, 136 (1993)

Capps, R.W., Gillett, F.C., Knacke, R.F.: Infrared observations of the OH source W33A. Astrophys. J. **226**, 863 (1978)

Cohen, M., Davies, J.K.: Spectral irradiance calibration in the infrared – V. The role of UKIRT and the CGS3 spectrometer. Mon. Not. R. Astron. Soc. **276**, 715 (1995)

Dayal, A., Hoffmann, W.F., Bieging, J.H., Hora, J.L., Deutsch, L.K., Fazio, G.G.: Mid-infrared (8–21 micron) imaging of proto – planetary Nebulae. Astrophys. J. **492**, 603 (1998)

Dyck, H.M., Lonsdale, C.J.: Polarimetry of infrared sources. In: Wynn-William, C.G. (ed.) IAU symposium 96 'Infrared Astronomy', p. 223. Reidel, Dordrecht (1981)

Fernandez, Y.R., Jewitt, D.C., Sheppard, S.S.: Albedos of asteroids in Comet-like orbits. Astrophys. J. **130**, 308 (2005)

Gear, W.K., Robson, E.I., Ade, P.A.R., Griffin, M.J., Smith, M.G., Nolt, I.G.: Multifrequency observations of OV236 (1921–293) reveal an unusual spectrum. Nature **303**, 46 (1983)

Green, S.F., McDonnell, J.A.M., Pankiewicz, G.S.A., Zarnecki, J.C.: The UKIRT infrared observational programme. ESA SP250 **II**, 81 (1986)

Greidanus, H., Strom, R.G.: 20-micron observations of Cassiopeia A. Astron. Astrophys. **249**, 521 (1991)

Groenewegen, M.A.T., Whitelock, P.A., Smith, C.H., Kerschbaum, F.: Dust shells around carbon Mira variables. Mon. Not. R. Astron. Soc. **293**, 18 (1998)

Justtanont, K., Barlow, M.J., Skinner, C.J., Roche, P.F., Aitken, D.K., Smith, C.H.: Mid-infrared spectroscopy of carbon-rich post-AGB objects and detection of the PAH molecule chrysene. Astron. Astrophys. **309**, 612 (1996)

Lawrence, A., Ward, M., Elvis, M., Fabbiano, G., Willner, S.P., Carleton, N.P., Longmore, A.: Observations from 1 to 20 microns of low-luminosity active galaxies. Astrophys. J. **291**, 117 (1985)

Meixner, M., Skinner, C.J., Graham, J.R., Keto, E., Jernigan, J.G., Arens, J.F.: Axially symmetric superwinds of proto–planetary Nebulae with 21 micron dust features. Astrophys. J. **482**, 897 (1997)

Meixner, M., Ueta, T., Dayal, A., Hora, J.L., Fazio, G., Hrivnak, B.J., Skinner, C.J., Hoffmann, W.F., Deutsch, L.K.: A Mid-infrared imaging survey of proto-planetary Nebula candidates. Astrophys. J. Suppl. **122**, 221 (1999)

Miyata, T., Kataza, H., Okamoto, Y., Onaka, T., Yamashita, T.: A spectroscopic study of dust around 18 oxygen-rich mira variables in the N band. I. Dust profiles. Astrophys. J. **531**, 917 (2000)

Richardson, K.J., White, G.J., Phillips, J.P., Avery, L.W.: The structure and kinematics of the DR21 region. Mon. Not. R. Astron. Soc. **219**, 167 (1986)

Robberto, M., Beckwith, S.V.W., Panagia, N., Patel, S.G., Herbst, T.M., Ligori, S., Custo, A., Boccacci, P., Bertero, M.: The Orion Nebula in the mid-infrared. Astrophys. J. **129**, 1534 (2005)

Robson, E.I., Gear, W.K., Clegg, P.E., Ade, P.A.R., Smith, M.G., Griffin, M.J., Nolt, I.G., Radostitz, J.V., Howard, R.J.: A flare in the millimetre to IR spectrum of 3C273. Nature **305**, 194 (1983)

Roche, P.F., Aitken, D.K.: An investigation of the interstellar extinction. I – towards dusty WC Wolf-Rayet stars. Mon. Not. R. Astron. Soc. **208**, 481 (1984)

Roche, P.F., Aitken, D.K.: An investigation of the interstellar extinction. II – towards the mid-infrared sources in the Galactic Centre. Mon. Not. R. Astron. Soc. **215**, 425 (1985)

Roche, P.F., Aitken, D.K., Smith, C.H., James, S.D.: NGC 4418 – a very extinguished galaxy. Mon. Not. R. Astron. Soc. **218**, 19P (1986)

Roche, P.F., Aitken, D.K., Smith, C.H., Ward, M.J.: An atlas of mid-infrared spectra of galaxy nuclei. Mon. Not. R. Astron. Soc. **248**, 606 (1991)

Speck, A.K., Barlow, M.J., Sylvester, R.J., Hofmeister, A.M.: Dust features in the 10-micron infrared spectra of oxygen-rich evolved stars. Astron. Astrophys. **146**, 437 (2000)

Smith, C.H., Wright, C.M., Aitken, D.K., Roche, P.F., Hough, J.H.: Studies in mid-infrared spectropolarimetry – II. An atlas of spectra. Mon. Not. R. Astron. Soc. **312**, 327 (2000)

Speck, A.K., Barlow, M.J., Skinner, C.J.: The nature of the silicon carbide in carbon star outflows. Mon. Not. R. Astron. Soc. **288**, 431 (1997)

Speck, A.K., Barlow, M.J., Sylvester, R.J., Hofmeister, A.M.: Astron. Astrophys. **146**, 437 (2000)

Sylvester, R.J., Skinner, C.J., Barlow, M.J., Mannings, V.: Optical, infrared and millimetre-wave properties of Vega-like systems. Mon. Not. R. Astron. Soc. **279**, 915 (1996)

Williams, P.M., van der Hucht, K.A., Pollock, A.M.T., Florkowski, D.R., van der Woerd, H., Wamsteker, W.M.: Multi-frequency variations of the Wolf-Rayet system HD 193793. I – infrared, X-ray and radio observations. Mon. Not. R. Astron. Soc. **243**, 662 (1990)

Willner, S.P., Elvis, M., Fabbiano, G., Lawrence, A., Ward, M.J.: Infrared observations of LINER galactic nuclei. Astrophys. J. **299**, 443 (1985)

Zarnecki, J.C., McDonnell, J.A.M., Carey, W.C., MacDonald, G.H., Eaton, N., Meadows, A.J.: Recent infrared observations of comets with UKIRT and IRAS. Adv. Space Res. **4**(203) (1984)

Part II
UKIRT Science

Chapter 11
Thirty Years of Star Formation at UKIRT

Chris Davis

Abstract It's safe to say that UKIRT's contribution to star formation at near-infrared (near-IR), mid-infrared (mid-IR) and even sub-millimetre (sub-mm) wavelengths has been considerable. From the early days of single-detector photometers, through the development of 2-D arrays and complex multi-mode imager-spectrometers, to the present-day large-format imager WFCAM, UKIRT has offered the international community access to some of the world's most innovative, competitive, and versatile instrumentation possible. Suffice to say, UKIRT users have made the most of these instruments! Below I try to give a taste of the variety of star formation research that has come to pass at UKIRT (with apologies to those whose important work I fail to mention).

Photometry of Young Stars – There's No Hiding from UKIRT!

Some of the earliest observations at UKIRT were conducted at mid-IR and sub-mm wavelengths, using single-channel bolometers such as UKT7 and UKT8 (affectionately known as Little Bertha and Big Bertha!), the popular and prolific UKT14, and, for spectral line work, sub-mm receivers from Queen Mary College and the Rutherford Appleton Lab (e.g. Padman et al. 1985; White et al. 1986; Gear et al. 1988; Ward-Thompson et al. 1989). Moreover, almost from day 1, UKIRT was open to observers who wished to bring their own instruments. For example, in one of the first papers to present mid-IR and far-IR photometry of embedded young stars, Davidson and Jaffe (1984) presented 400 μm photometry, obtained at UKIRT with the University of Chicago f/35 SMM photometer. The UKIRT data

---

C. Davis (✉)
Astrophysics Research Institute, Liverpool John Moores University, Liverpool Science Park,
IC2 Building 146 Brownlow Hill Liverpool, L3 5RF United Kingdom
e-mail: c.j.davis@ljmu.ac.uk

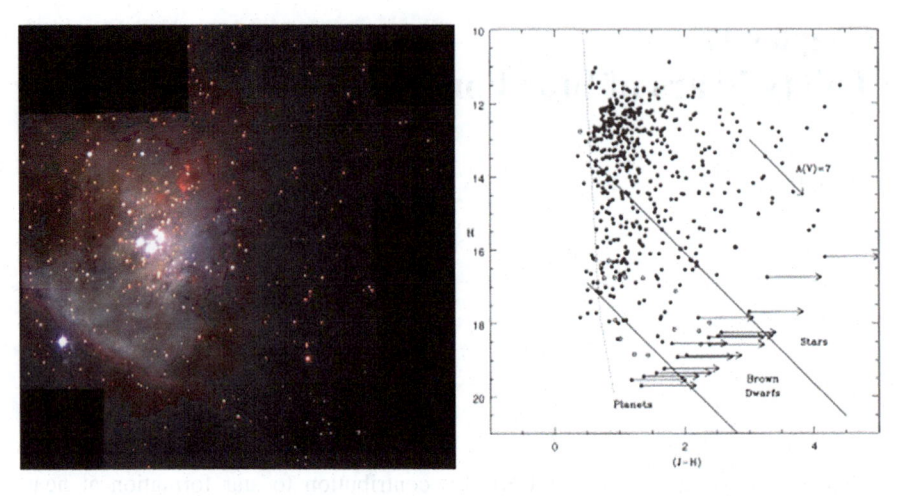

Fig. 11.1 *Left* – IJH false-colour image of the Trapezium cluster in Orion. *Right* – Colour-magnitude diagram extracted from these UFTI data, from which brown dwarfs and "free-floating planets" were identified (Data from Lucas and Roche (2000))

were used in conjunction with *Kuiper Airborne Observatory* photometry at shorter wavelengths to demonstrate the excess associated with cold circumstellar dust. At that time, similar observations existed for only one other object, B 335.

At shorter wavelengths, mapping the pre-main-sequence population really took off with the commissioning of UKIRT's first 2-D imaging array, IRCAM. Aspin et al. (1994) and Aspin and Sandell (1997) presented early photometry of dozens of young stars in NGC 1333 in Perseus, stressing (almost a decade before the launch of Spitzer) the importance of thermal imaging in their JHKL colour-colour and colour-magnitude diagrams, the longer-wavelength data being crucial for distinguishing the youngest sources from reddened background stars and weak-line T Tauri stars. Similar broad-band photometry was presented by Eiroa and Casali (1992) and Aspin and Barsony (1994), who used IRCAM to study the Serpens cluster and the red sources in LkHα 101, respectively. A few years later, Carpenter et al. (1997) analysed JHKL photometry of the Mon R2 cluster, complementing their IRCAM3 data of a $15' \times 15'$ region with CGS4 spectroscopy of 16 stars. Carpenter et al. found that two-thirds of the sources in the cluster exhibited infrared excess in the K and L-bands, and that the ratio of high to low mass stars was consistent with a Miller-Scalo Initial Mass Function (IMF).

In 1998 the UFTI commissioning team of Lucas and Roche used the new wide field and high spatial resolution afforded by this instrument to search for young stars and, particularly, for cool, low mass objects in and around the busy Trapezium cluster in Orion (Fig. 11.1). They used IJH photometry to identify ~165 brown dwarf candidates, several of which were found to be young "free-floating planets" with masses below the deuterium burning limit (Lucas and Roche 2000). They followed up on these discoveries with deep spectroscopy at Gemini, identifying six sources with spectral types later than M9, consistent with planetary-mass objects (Lucas et al. 2006).

More recently, with the launch of the *Spitzer Space Telescope* and the availability of large-format near-IR cameras like WFCAM, there has been a surge in research on young stellar clusters. The UKIDSS Galactic Plane and Galactic Clusters surveys (described elsewhere in this volume) are yielding data that complement marvelously well the longer-wavelength *Spitzer* observations. These data will in the near future supercede 2MASS as the near-IR photometry of choice, because of their depth and superior resolution. In the meantime, PIs of non-UKIDSS projects are making excellent use of combined WFCAM and *Spitzer* datasets: for example, Kumar et al. (2007) have used photometry of some 60,000 stars to map the distributions of young, deeply embedded "Class I" sources and their more evolved contemporaries, the "Class II" T Tauri stars, across the massive star-forming region DR21/W75. They find the more abundant Class II sources to be spread widely throughout the region, the Class I proto-stars being tightly confined to regions of high extinction. Luhman et al. (2008), working closer to home, have combined WFCAM and Spitzer data to search for discs around young brown dwarfs, while Wright and Drake (2008) have used WFCAM to search for near-IR counterparts to Chandra X-ray sources in the massive star-forming region CyG OB2. They identify counterparts for some 1,500 sources.

WFCAM (and UKIRT as a whole) has also been popular with those wishing to conduct variability studies. Young stars – FU Ori, EXor, T Tauri and Herbig Ae/Be stars – are known to be variable on periods of months, weeks, or even days, due largely to bursts in accretion or to photospheric activity. Recently, Alves de Oliveira and Casali (2008) analysed WFCAM data, collected on 14 separate nights, to search for variable sources in the spectacular ρ Oph region (shown in Fig. 11.2). They found 41 % of the stellar population to be variable on periods of days and weeks; they associate this behavior with star spots and varying extinction. Lately, UKIRT has been monitoring a number of other regions on a nightly basis; currently (winter of 2009), monitoring of Orion is being conducted *in sync* with the *Spitzer Space Telescope*.

Excitation in the Interstellar Medium – Some Like It Hot

With its suite of imagers and spectrometers, UKIRT has played a pivotal role in advancing our understanding of the physical conditions associated with star forming regions and the Inter-Stellar Medium (ISM). Early observations of infrared nebula by Gatley et al. (1987), Hasegawa et al. (1987) and Tanaka et al. (1989 – see Fig. 11.3) showed that the excitation of molecular gas in star forming regions was far from simple, being at best a combination of shock excitation and fluorescence.

Fig. 11.2 Deep WFCAM imaging of the low-mass star forming region Ophiuchus (Alves de Oliveira and Casali 2008)

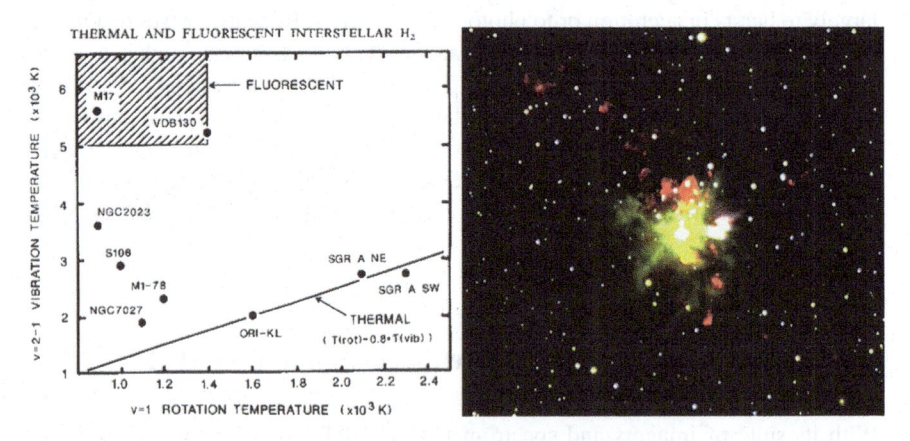

Fig. 11.3 *Left* – Tanaka et al. (1989) used line ratios to identify the H_2 excitation mechanisms in a number of well-known galactic sources. *Right* – A more recent WFCAM JHH$_2$ image of the intermediate-mass star-forming cluster AFGL 961, showing the complexity of the region

With the benefit of modern high-resolution imagers and spectrometers, this seems hardly surprising, given the complexity of star forming clusters (see e.g. the WFCAM image of the intermediate-mass star-forming region, AFGL 961, in Fig. 11.3). But back in the day, with only a single-element detector to hand (UKT9 in this case), a Circular Variable Filter (CVF) for order sorting, and a Fabry-Perot (FP) etalon for scanning in frequency to build up each spectrum pixel-by-pixel, these results were hard-won and certainly new and exciting.

In the 1980s and early 1990s, UKIRT very much led the way in studies of galactic nebula (Jourdain de Muizon et al. 1986; Gatley et al. 1987; Brand et al. 1988; Burton et al. 1989; Geballe et al. 1989; Chrysostomou et al. 1993). For example, Burton et al. (1990) observed H_2 1-0S(1) line profiles at high spectral resolution in a number of galactic nebulae, finding the lines to be narrow and peaked at ambient velocities – a result they interpreted in terms of non-thermal excitation (fluorescence). At about the same time, Geballe and Garden (1990), building on earlier UKIRT observations (Geballe and Garden 1987) were mapping OMC-1 (in Orion) in pure-rotation H_2 0-0S(9) and CO 1-0P(8) emission at 4.7 μm; in this region at least, much broader lines were observed, which were symptomatic of shock excitation. As we shall see in the next section, work pushed on in this area largely via studies of individual low-mass objects, as spatial resolution improved and observers gained access to instruments equipped with 2-D arrays of pixels.

Accretion and Outflow – What Goes Down, Must Come Up!

Access to a wonderful new cooled grating 1–5 μm spectrometer, CGS4, drove observational studies of perhaps the most fundamental processes associated with star formation – accretion and outflow. As part of her thesis work in Edinburgh, Chandler, together with her collaborators, obtained some of the first high-resolution spectroscopic observations of ro-vibrational CO band-head emission at 2.3 μm (Chandler et al. 1990 – see also Fig. 11.4). In a sample of a half-dozen young stars they observed a variety of band-head profile shapes, which they modeled in terms of excitation in the inner regions of a rotating Keplarian disc (see also Chandler et al. 1995). Similar observations were later obtained by Casali and Matthews (1992), who observed CO in absorption; by Aspin (1994), who detected CO band-head emission scattered by the dense gas that envelopes the massive young star GGD-27; and by Reipurth and Aspin (1997), who surveyed a number of outflow sources in the K-band at low spectral resolution, including some well-known FU Ori variables (all of which possessed strong CO in absorption).

Focusing on more evolved sources, Folha and Emerson (2001) used CGS4 in an ambitious survey of 50 T Tauri stars. Observing with the echelle, they used Paβ and Brγ hydrogen recombination lines as probes of accretion and outflow processes. They observed a variety of line shapes including P Cygni profiles which they interpreted in terms of magnetic accretion with velocities of hundreds of kilometers per second.

A few years later, Sheret et al. (2003), inspired by claims in the literature of detections of pure-rotational H_2 emission from the discs of pre-main sequence stars – discs viewed as perhaps the precursors to proto-planets – tried to confirm these observations with Michelle (during its brief stay at UKIRT). They obtained a

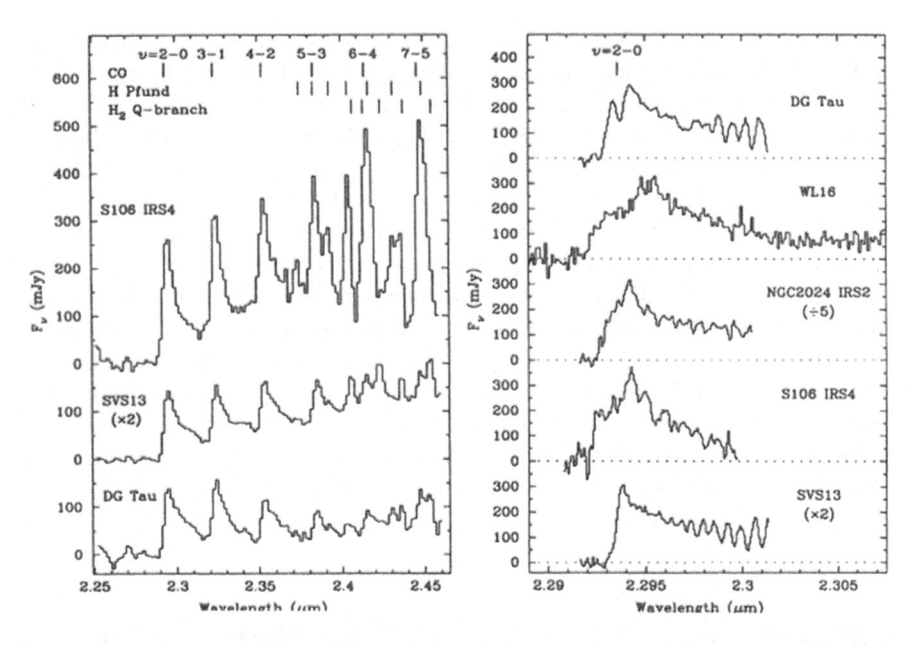

Fig. 11.4 CGS4 spectroscopy of accreting proto-stars, at low spectral resolution (*left*) and at high spectral resolution using the echelle (*right*). The CO band-heads in these spectra were thought to be associated with the warm, dense inner regions of accretion discs (Chandler et al. 1990)

marginal detection of the 4-2 emission line a 12.2 μm in one of their two targets (AB Aur), though failed to detect any emission from the other.

Studies of outflows at UKIRT were very much inspired by the early work of Zealey, Williams and collaborators (Zealey et al. 1984, 1992), who used IRCAM to image a number of well-known Herbig-Haro (HH) objects in H_2 1-0S(1) line emission (Fig. 11.5), and by Zinnecker et al. (1989), who used IRCAM with the FP to resolve H_2 line profiles in a number of HH objects. Carr (1993) likewise used IRCAM with the FP, though rather than observe multiple sources, he mapped H_2 line profiles across one of the brightest HH objects known at that time, HH 7 (Fig. 11.6); similar observations, of the explosive outflow activity in OMC-1 in Orion, were conducted by Chrysostomou et al. (1997). CGS4 was later used to map H_2 profiles across the L 1448 and DR 21 outflows (now known as MHO 539 and MHO 898/899, respectively), and along two of the spectacular "bullets" that emanate from the Orion nebula (Davis and Smith 1996a, b; Tedds et al. 1999). At about the same time Fernandez and co-workers used CGS4 with its low-resolution grating to examine gas excitation in a number of flows (Fig. 11.6); they found that in HH 7 and DR 21 a combination of shocks and fluorescence was needed to account for their excitation diagrams (Fernandes and Brand 1995; Fernandes et al. 1997).

More recently, Davis et al. (2003) and Whelan et al. (2004) have used CGS4, its echelle grating, and a technique known as spectro-astrometry to measure the relative positions of H_2, [FeII] and Paβ emission-line peaks to within a few tens of AU

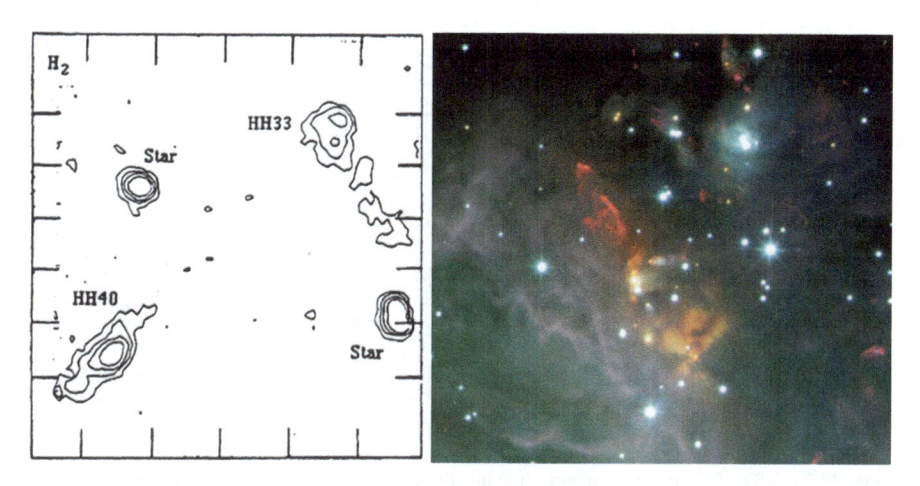

Fig. 11.5 *Left* – Early H_2 imaging of outflows from young stars, in this case the Herbig-Haro objects HH 33/40 (Zealey et al. 1992). *Right* – WFCAM JKH$_2$ false-colour image of jets and embedded young stars in the OMC-2 region of Orion (Davis et al. 2009)

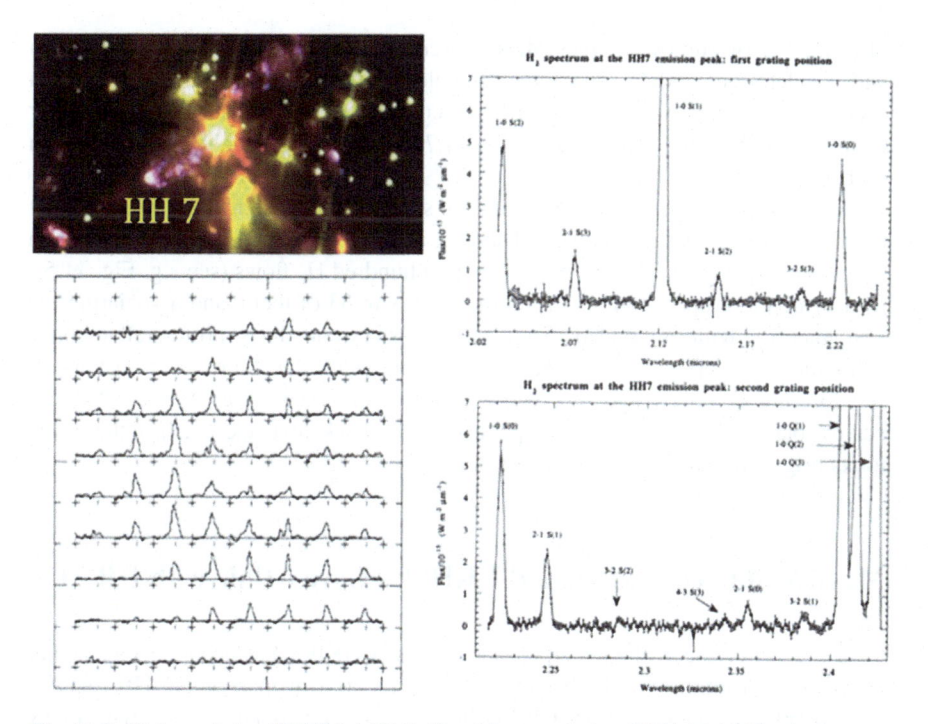

Fig. 11.6 High-resolution IRCAM + FP (*left*) and low-resolution CGS4 (*right*) H_2 spectroscopy of the HH 7 bow shock (Carr 1993; Fernandes and Brand 1995). HH 7 is evident to the south-east of the bright young star SVS13 in the false-colour WFCAM/Spitzer image shown *top-left*

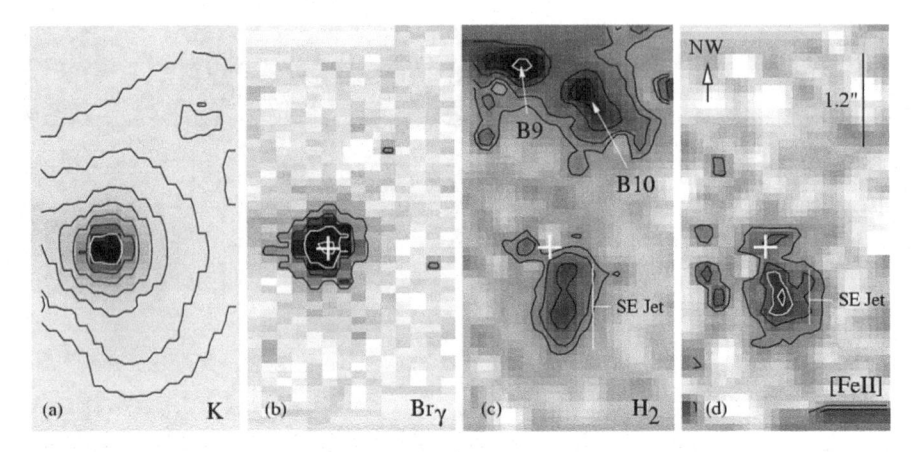

Fig. 11.7 UIST IFU spectral images showing the collimated jet associated with the luminous young star IRAS 18151 + 1208 (Davis et al. 2004). The jet is clearly seen in continuum-subtracted H$_2$ and [FeII] images; the Brg coincident with the source is thought to be associated with magnetospheric accretion flows

of a number of outflow sources; these observations help constrain jet collimation and acceleration models. The Integral Field Unit (IFU) in UIST has also been used to image, in various emission lines, collimated "micro-jets" from a number of intermediate-mass young stars (Fig. 11.7; Davis et al. 2004, Varricatt et al. in prep.)

We've come a long way since the days of single-object imaging and spectroscopy in the mid-1990s: earlier this year, Davis et al. (2009) mapped 8 square degrees in Orion with WFCAM, identifying well over a hundred H$_2$ flows (see e.g. Fig. 11.5), measuring the proper motions of multiple knots in 33 of them, and associating the vast majority with embedded (Spitzer) proto-stars and dusty proto-stellar cores. With the completion of the UKIDSS H$_2$ survey of the Taurus-Auriga-Perseus star-forming complex, and the recent approval of PATT time for the UWISH2 narrow-band survey of the galactic plane (http://astro.kent.ac.uk/uwish2), the future looks bright for outflow studies at UKIRT.

Massive Star Formation and UKIRT's Love Affair with DR 21

Over the years, one of the most popular targets at UKIRT has been the spectacular massive star forming region, bright radio source, nest of masers and HII regions, and complex of molecular outflows collectively known as DR 21. As with many areas in astronomy at UKIRT, the earliest observations were conducted at sub-mm wavelengths. For example, Richardson et al. (1986) produced a very JCMT-*esque* study of DR 21 using receivers from Queen Mary College. They published a remarkably detailed work that included CO, CS, HCN, HCO+, H^{13}CO + and CS

11 Thirty Years of Star Formation at UKIRT

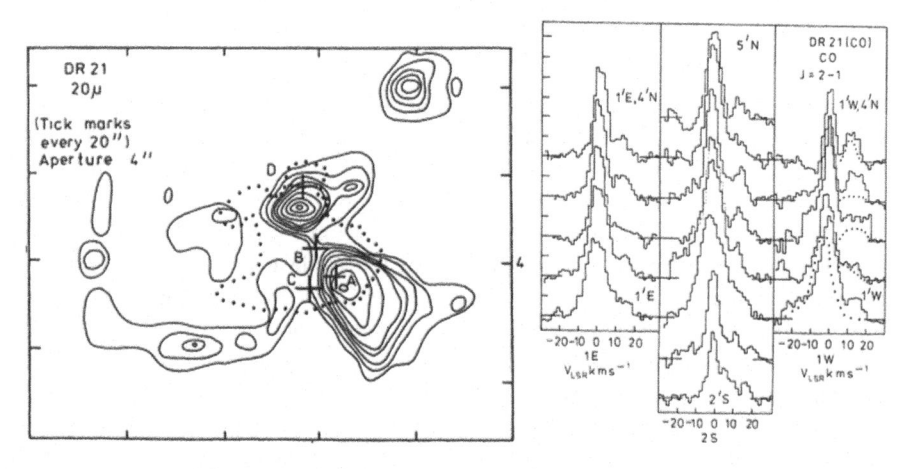

Fig. 11.8 Mid-IR and sub-mm observations of the massive star forming region/cluster of compact HII regions known as DR 21; *left* – a 20 μm map taken with a 4″ aperture, *right* – CO J = 2-1 spectra (All data are from Richardson et al. (1986))

spectroscopy, as well as continuum observations at 20 and 300 μm (Fig. 11.8). With these data they were able to map the distribution of high velocity molecular gas around the central HII regions, and model the overall ambient gas distribution.

At shorter wavelengths, Roelfsema et al. (1989) used CGS2 and the FP to probe the ionized gas in DR 21 via Brα observations at high resolution, mapping the emission and adjacent 4 μm continuum across the central region. Meanwhile, Garden and co-workers were focusing on the bright bipolar outflow associated with DR 21, first mapping the H_2 emission, one velocity-resolved spectrum at a time with a CVF and the FP (Garden et al. 1986). These were painstaking observations that involved switching the FP between line and line-free frequencies every 3 s, and sampling blank sky every 10 pointings. The final map of DR 21 comprised observations at 400 positions. Garden et al. followed up on this work with IRCAM imaging and high resolution H_2 line profile mapping (Garden et al. 1990, 1991).

These data (and similar studies of Orion, e.g. Brand et al. 1988) subsequently inspired a slew of theoretical papers aimed at interpreting line shapes and excitation diagrams in terms of planar and curved (bow-shaped) "Jump" shocks and magnetically-cushioned "Continuous" shocks (Smith and Brand 1990a, b, c; Smith et al. 1991a, b; Smith 1994). These tried-and-tested models still represent some of the most comprehensive work done on molecular shock physics.

DR 21 was of course not the only high mass star-forming region targeted by UKIRT observers. Bunn et al. (1995) used Brα, Brγ and Pfγ as tracers of high velocity winds in a number of massive young stars (see also Drew et al. 1993; Lumsden and Hoare 1996, 1999) and, in a very ambitious new project, Varricatt et al. (2010) have recently used UFTI to search for collimated molecular outflows in 50 massive star forming regions. In regions where collimated jets have been observed, follow-up CGS4 echelle and UIST IFU observations have also been secured.

Polarimetry of Star Forming Regions

Polarimetric observations have for many years been a mainstay at UKIRT (see the article by Jim Hough in this volume), and a review of star formation would not be complete without mentioning some of the innovative observations and ground-breaking results obtained at UKIRT. Polarimetry has been possible largely because of the continued support from the University of Hertfordshire; users have also benefited from improved instrument design (putting the Wollaston prism downstream of the slit rather than upstream, as was the case with CGS4), enhanced data acquisition, and reliable imaging and spectro-polarimetry data reduction software (as part of the ORAC-DR program). In fact, in the last few years polarimetry has become so straight-forward, particularly with UIST, that during recent Cassegrain blocks it was incorporated into queue scheduling and the service observing program.

Linear spectro-polarimetric observations of scattered light from dust grains around T Tauri stars were made as early as November 1979 using the Hatfield Polarimeter (Hough et al. 1981). A few years later H_2 and K-band imaging polarimetry of the Orion BN-KL nebula were made with the Kyoto-UKIRT infrared polarimeter and UKT9 (Hough et al. 1986); these observations revealed for the first time a molecular hydrogen reflection nebula around the flow from BN, and represented one of a number of early collaborations between UK and Japanese astronomers at UKIRT (see also Nagata et al. 1986; Yamashita et al. 1987, etc.; note also that Chrysostomou et al. 1994 subsequently used H_2 line polarisation measurements to map a twist in the magnetic field about IRc2, the most luminous young star in the region). Sato et al. (1988) observed 20 sources in Ophiuchus in polarized light, thereby mapping the degree of polarisation and changes in the polarisation position angle across the densest regions of the cloud, while Aitken et al. (1988) used the UCL array spectrometer with a wire grid to observe polarisation at 10 and 20 μm in AFGL 2591, thereby showing how spectro-polarimetry could be used as a sensitive indicator of grain chemistry.

In the early 1990s polarimetrists were not slow to take advantage of the 2-D array cameras arriving at UKIRT. In a series of papers, Minchin et al. (1991a, b, c) mapped and modeled the polarised reflection nebulae associated with a number of young IR sources, while Whittet et al. (1992) used UKIRT and the Anglo-Australian Telescope to examine the wavelength dependence of polarisation, developing at the same time a comprehensive catalogue of polarised standards that is still used today.

In recent years the full suite of Cassegrain instruments at UKIRT has been used for polarimetry: Chrysostomou et al. (2000) used IRCAM3 to map the circularly polarised emission across OMC-1 (Buschermohle et al. 2005 conducted similar observations with UFTI); Kuhn et al. (2001) instead used IRCAM3 to search for circumstellar discs around young stars; Holloway et al. (2002) measured the polarisation of the 3 μm water-ice feature towards a number of young stars; Oudmaijer et al. (2005) observed polarised line emission from massive young

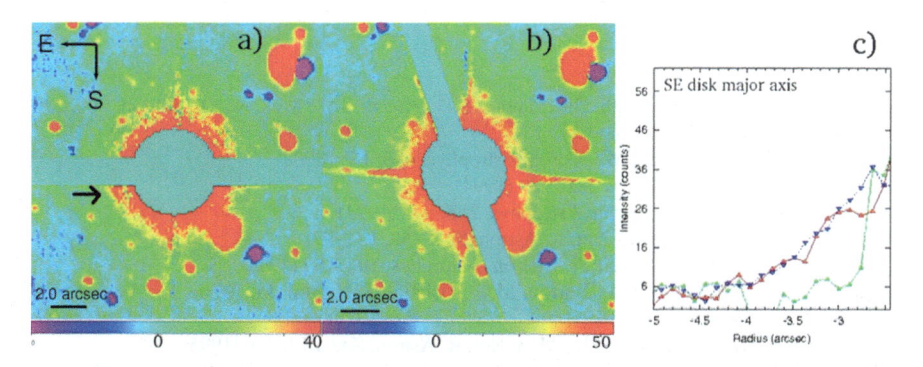

Fig. 11.9 Coronagraphic-imaging-polarimetry of the Herbig Ae proto-planetary disc system HD 163296 (John Wisniewski et al. private communication). *Panels* (**a**) and (**b**) show the bright star at two UIST position angles after subtraction of a PSF star. Note that the features associated with the scattered light disc remain at both position angles. *Panel* (**c**) presents radial profile cross-sections taken across the south-east quadrant at each UIST position angle (*red and blue lines*). The *green line* shows a similar profile for a source with no known disc. Note that data were taken through a Wollaston prism and waveplate; analysis of the polarised light associated with the disc is pending

stars with UIST in spectroscopy mode, while Hales et al. (2006) used UIST in imaging mode to map the polarised light around a number of dusty low-mass young stars. Hough et al. (2008) recently used UIST to examine grain alignment through spectro-polarimetry of the 4.7 μm CO ice feature, and very recently Wisniewski and co-workers have been using UIST in its newest mode, coronographic-imaging-polarimetry, to trace the scattered light discs around very bright Herbig Ae stars (Fig. 11.9).

The Future

Clearly, UKIRT has made a major contribution to the field of star formation, through observations at near-IR, but also mid-IR and sub-mm wavelengths. British astronomers and their collaborators have always been quick to take advantage of the variety of modes made available to them, and continue to do so in the era of wide-field astronomy and large, multi-national collaborations on legacy surveys like UKIDSS. WFCAM observations are hugely complementary to *Spitzer* mid-IR imaging and photometry, JCMT/HARP sub-mm molecular line observations of large-scale cloud structure and dynamics, and JCMT/SCUBA-2 mapping of the cold dust emission in dense pre-stellar and proto-stellar cores. It is in these areas that UKIRT, as she passes her 30th birthday, is now making an impact on star-formation research, and will continue to do so for a number of years to come.

References

Aitken, D.K., Roche, P.F., Smith, C.H., James, S.D., Hough, J.H.: Mon. Not. R. Astron. Soc. **230**, 629 (1988)
Alves de Oliveira, C., Casali, M.: Astron. Astrophys. **485**, 155 (2008)
Aspin, C.A.: Astron. Astrophys. **281**, L29 (1994)
Aspin, C.A., Barsony, M.: Astron. Astrophys. **288**, 849 (1994)
Aspin, C.A., Sandell, G.: Mon. Not. R. Astron. Soc. **289**, 1 (1997)
Aspin, C.A., Sandell, G., Russell, A.: Astron. Astrophys. **106**, 165 (1994)
Brand, P.W.J.L., Moorhouse, A., Burton, M.G., Geballe, T.R., Bird, M., Wade, R.: Astrophys. J. **334**, L103 (1988)
Bunn, J.C., Hoare, M.G., Drew, J.E.: Mon. Not. R. Astron. Soc. **272**, 346 (1995)
Burton, M.G., Brand, P.W.J.L., Geballe, T.R., Webster, A.S.: Mon. Not. R. Astron. Soc. **236**, 409 (1989)
Burton, M.G., Geballe, T.R., Brand, P.W.J.L., Moorhouse, A.: Astrophys. J. **352**, 625 (1990)
Buschermohle, M., Whittet, D.C., Chrysostomou, A., Hough, J.H., Lucas, P.W., Adamson, A.J., Whitney, B.A., Wolff, M.J.: Astrophys. J. **624**, 821 (2005)
Carpenter, J.M., Meyer, M.R., Dougados, C., Strom, S.E., Hillenbrand, L.A.: Astron. J. **114**, 198 (1997)
Carr, J.: Astrophys. J. **406**, 553 (1993)
Casali, M., Matthews, H.E.: Mon. Not. R. Astron. Soc. **258**, 399 (1992)
Chandler, C.J., Carlstrom, J.E., Scoville, N.Z., Dent, W.R.F., Geballe, T.R.: Astrophys. J. **412**, L71 (1990)
Chandler, C.J., Carlstrom, J.E., Scoville, N.Z.: Astrophys. J. **446**, 793 (1995)
Chrysostomou, A., Brand, P.W.J.L., Burton, M., Moorhouse, A.: Mon. Not. R. Astron. Soc. **265**, 329 (1993)
Chrysostomou, A., Hough, J.H., Burton, M.G., Tamura, M.: Mon. Not. R. Astron. Soc. **268**, 325 (1994)
Chrysostomou, A., Burton, M.G., Axon, D.J., Brand, P.W.J.L., Hough, J.H., Bland-Hawthorn, J., Geballe, T.R.: Mon. Not. R. Astron. Soc. **289**, 605 (1997)
Chrysostomou, A., Gledhill, T.M., Menard, F., Hough, J.H., Tamura, M., Bailey, J.: Mon. Not. R. Astron. Soc. **312**, 103 (2000)
Davidson, J.A., Jaffe, D.T.: Astrophys. J. **277**, L13 (1984)
Davis, C.J., Smith, M.D.: Astron. Astrophys. **309**, 929 (1996a)
Davis, C.J., Smith, M.D.: Astron. Astrophys. **310**, 961 (1996b)
Davis, C.J., Whelan, E., Ray, T.P., Chrysotomou, A.C.: Astron. Astrophys. **397**, 693 (2003)
Davis, C.J., Varricatt, W.P., Todd, S., Ramsay Howat, S.K.: Astron. Astrophys. **425**, 981 (2004)
Davis, C.J., Froebrich, D., Stanke T., et al. (12 authors): Astron. Astrophys. **496**, 153 (2009)
Drew, J.E., Bunn, J.C., Hoare, M.G.: Mon. Not. R. Astron. Soc. **265**, 12 (1993)
Eiroa, C., Casali, M.: Astron. Astrophys. **262**, 468 (1992)
Fernandes, A., Brand, P.W.J.L.: Mon. Not. R. Astron. Soc. **274**, 639 (1995)
Fernandes, A., Brand, P.W.J.L., Burton, M.G.: Mon. Not. R. Astron. Soc. **290**, 216 (1997)
Folha, D., Emerson, J.: Asron. Astrophys. **365**, 90 (2001)
Garden, R., Geballe, T.R., Gatley, I., Nadeau, D.: Mon. Not. R. Astron. Soc. **203**, 221 (1986)
Garden, R., Russell, A., Burton, M.G.: Astrophys. J. **354**, 232 (1990)
Garden, R., Geballe, T.R., Gatley, I., Nadeau, D.: Astrophys. J. **366**, 474 (1991)
Gatley, I., Hasegawa, T., et al. (9 authors): Astrophys. J., **318**, L73 (1987)
Gear, W.K., Chandler, C.J., Moore, T.J.T., Cunningham, C.T., Duncan, W.D.: Mon. Not. R. Astron. Soc. **231**, p47 (1988)
Geballe, T.R., Garden, R.: Astrophys. J. **365**, 602 (1987)
Geballe, T.R., Garden, R.: Astrophys. J. **317**, L107 (1990)

Geballe, T.R., Tielens, A.G.G.M., Allamandola, L.J., Moorhouse, A., Brand, P.W.J.L.: Astrophys. J. **341**, 278 (1989)
Hales, A.S., Gledhill, T.M., Barlow, M.J., Lowe, K.T.E.: Mon. Not. R. Astron. Soc. **365**, 1348 (2006)
Hasegawa, T., Gatley, I., Garden, R., Brand, P.W.J.L., Ohishi, M., Hayashi, M., Kaifu, N.: Astrophys. J. **318**, L77 (1987)
Holloway, R.P., Chrysostomou, A., Aitken, D.K., Hough, J.H., McCall, A.: Mon. Not. R. Astron. Soc. **36**, 425 (2002)
Hough, J.H., Bailey, J., Cunningham, E.C., McCall, A., Axon, D.J.: Mon. Not. R. Astron. Soc. **195**, 429 (1981)
Hough, J.H., Aitken, D.K., Whittet, D.C.B., Adamson, A.J., Chrysotomou, A.: Mon. Not. R. Astron. Soc. **387**, 797 (2008)
Hough, J.H., et al. (16 authors): Mon. Not. R. Astron. Soc. **22**, 629 (1986)
Jourdain de Muizon, J., Geballe, T.R., D'Hendecourt, L.B., Baas, F.: Astrophys. J. **306**, L105 (1986)
Kuhn, J.R., Potter, D., Parise, B.: Astrophys. J. **553**, 189 (2001)
Kumar, M.S.N., Davis, C.J., Grave, M.J., Ferreira, B., Froebrich, D.: Mon. Not. R. Astron. Soc. **374**, 54 (2007)
Lucas, P.W., Roche, P.F.: Mon. Not. R. Astron. Soc. **314**, 858 (2000)
Lucas, P.W., Weights, D.J., Roche, P.F., Riddick, F.C.: Mon. Not. R. Astron. Soc. **373**, L60 (2006)
Luhman, K.L., Hernandez, J., Downes, J.J., Hartman, L., Briceno, C.: Astrophys. J. **688**, 362 (2008)
Lumsden, S., Hoare, M.G.: Astrophys. J. **464**, 272 (1996)
Lumsden, S., Hoare, M.G.: Mon. Not. R. Astron. Soc. **305**, 701 (1999)
Minchin, N.R. (9 authors): Mon. Not. R. Astron. Soc. **248**, 715 (1991a)
Minchin, N.R., Hough, J.H., McCall, A., Aspin, C., Yamashita, T., Burton, M.G.: Mon. Not. R. Astron. Soc. **249**, 707 (1991b)
Minchin, N.R., Hough, J.H., Burton, M.G., Yamashita, T.: Mon. Not. R. Astron. Soc. **251**, 522 (1991c)
Nagata, T., Yamashita, T., Sato, S., Suzuki, H., Hough, J.H., Garden, R., Gatley, I.: Mon. Not. R. Astron. Soc. **223**, 7 (1986)
Oudmaijer, R.D., Drew, J.E., Vink, J.S.: Mon. Not. R. Astron. Soc. **364**, 725 (2005)
Padman, R., Scott, P.F., Vizard, D.R., Webster, A.S.: Mon. Not. R. Astron. Soc. **214**, 251 (1985)
Reipurth, B., Aspin, C.A.: Astron. J. **114**, 6 (1997)
Richardson, K.R., White, G.J., Phillips, J., Avery, L.W.: Mon. Not. R. Astron. Soc. **219**, 167 (1986)
Roelfsema, P.R., Goss, W.M., Geballe, T.R.: Astron. Astrophys. **222**, 247 (1989)
Sato, S., Tamura, M., Nagata, T., Kaifu, N., Hough, J.H., McLean, I.S., Garden, R., Gatley, I.: Mon. Not. R. Astron. Soc. **230**, 321 (1988)
Sheret, I., Ramsay Howat, S.K., Dent, W.R.F.: Mon. Not. R. Astron. Soc. **343**, L65 (2003)
Smith, M.D.: Mon. Not. R. Astron. Soc. **266**, 238 (1994)
Smith, M.D., Brand, P.W.J.L.: Mon. Not. R. Astron. Soc. **203**, 221 (1990a)
Smith, M.D., Brand, P.W.J.L.: Mon. Not. R. Astron. Soc. **242**, 495 (1990b)
Smith, M.D., Brand, P.W.J.L.: Mon. Not. R. Astron. Soc. **245**, 108 (1990c)
Smith, M.D., Brand, P.W.J.L., Moorhouse, A.: Mon. Not. R. Astron. Soc. **248**, 451 (1991a)
Smith, M.D., Brand, P.W.J.L., Moorhouse, A.: Mon. Not. R. Astron. Soc. **248**, 730 (1991b)
Tanaka, M., Hasegawa, T., Hayashi, S.S., Brand, P.W.J.L., Gatley, I.: Astrophys. J. **336**, 207 (1989)
Tedds, J.A., Brand, P.W.J.L., Burton, M.G.: Mon. Not. R. Astron. Soc. **307**, 37 (1999)
Varricatt, W.P., Davis, C.J., Ramsay, S.K., Todd, S.P.: Mon. Not. R. Astron. Soc. **404**, 661 (2010)
Ward-Thompson, D., Robson, E.I., Whittet, D.C.B., Gordon, M.A., Walther, D.M., Duncan, W.D.: Mon. Not. R. Astron. Soc. **241**, 119 (1989)
Whelan, E., Ray, T.P., Davis, C.J.: Astron. Astrophys. **417**, 247 (2004)
White, G.J., Richardson, K.J., Avery, L.W., Lesurf, J.C.G.: Astrophys. J. **302**, 701 (1986)

Whittet, D.C.B., Martin, P.G., Hough, J.H., Rouse, M.F., Bailey, J.A., Axon, D.J.: Astrophys. J. **386**, 562 (1992)
Wright, N.J., Drake, J.J.: Astrophys. J. **184**, 84 (2008)
Yamashita, T. et al. (10 authors): Publ. Astron. Soc. Jpn. **39**, 809 (1987)
Zealey, W.J., Williams, P.M., Sandell, G.: Astron. Astrophys. **140**, L31 (1984)
Zealey, W.J., Williams, P.M., Sandell, G., Taylor, K.N.R., Ray, T.P.: Astron. Astrophys. **262**, 570 (1992)
Zinnecker, H., Mundt, R., Geballe, T.R., Zealey, W.J.: Astrophys. J. **342**, 337 (1989)

Chapter 12
Comets and Asteroids from UKIRT

John K. Davies

Abstract I present a personal view of my experiences using UKIRT to observe small solar system bodies between 1984 and 2002. This period saw a huge increase in capability from single element bolometers and photometers like UKT8 and UKT9 to imaging devices like IRCAM and UIST. Spectroscopic capability improved vastly from the linear instruments CGS2 and CGS3 to array spectrometers like CGS4 and Michelle. At the same time the discovery of the trans-Neptunian population greatly expanded the number of solar system objects available for observation. A small number of other important observations of planetary atmospheres and satellites are also highlighted.

Near Earth Objects (NEO)

3200 Phaethon

Initially designated 1993 TB, Earth approaching asteroid 3200 Phaethon was discovered during the IRAS fast moving object search (Davies et al. 1984). It was realised in a matter of days (Whipple 1983) that the object's orbit was a close match to that of the Geminid meteor stream, the only major stream for which no parent comet had ever been identified. This raised the question of whether 3200 Phaethon was the first detected example of the long postulated, but never observed, extinct cometary nucleus. With only optical and near infrared photometry available it is extremely difficult to determine if an object is reflective, like a stony asteroid or dark like a comet nucleus covered with a refractory organic crust. Simultaneous

---

J.K. Davies (✉)
UK Astronomy Technology Centre, Royal Observatory,
Edinburgh, EH9 3HJ, UK
e-mail: jdavies@roe.ac.uk

reflected and emitted (thermal infrared) observations can break this dichotomy by making it possible to solve simultaneously for size and albedo. Using a combination of UKT8 and UKT9 Green et al. (1985) were able to show that 3200 had a thermal inertia more typical of bare rock than a thermally insulating mantle over an icy body. If so, and this conclusion has not been overturned, then 3200 probably originated as a main belt asteroid and its associated meteor stream cannot result from cometary activity.

The NEATM

The thermal models used by Green et al. (1985) were improved versions of models developed for main belt asteroids. However such simple models tend to make unrealistic assumptions and this makes them difficult to apply to Earth approaching objects. NEOs tend to be highly irregular and may be observed over a wide range of phase (Earth-Asteroid-Sun) angles never seen when observing main belt asteroids. Harris (1998) developed a modified asteroid thermal model which includes a beaming parameter to account for phase effects and can be fitted to spectrophotometric data across the 5–20 µm range. This model was applied to CGS3 data for asteroids 2100 Ra-Shalom and 1991 EE (Harris et al. 1998) and later to 433 Eros, 1980 Tezcatlipoca and 3691 Dionysus (Harris and Davies 1999). They found that the NEATM produced consistent results for size and albedo over a wide range of phase angles. The availability of Michelle, which could take both long slit spectra and images for photometric calibration increased UKIRT's capability for this kind of work and Michelle was used for several such studies (e.g. Wolters et al. 2005, 2008) before it was transferred to Gemini.

Asteroid Mineralogy with CGS4

Broad-band optical photometry has been used since the 1970s to characterise asteroids into various taxonomic schemes, but linking these taxonomies to specific minerals requires spectroscopic observations which can be used to search for meteoritic analogues and so develop links between the meteorite and asteroid populations. Leader in the field is undoubtedly the work of R. Binzel and collaborators who have obtained spectra of many hundreds of NEOs (e.g. Binzel et al. 2004) but UKIRT has also contributed by adding CGS4 1–2.5 µm spectroscopy to the mix. CGS4 is not the ideal instrument for this task since its high spectral resolution is rather overspecified for the detection of broad mineralogical features and the need to observe each of the J, H and K windows separately and then splice the results together makes the observations complicated and usually overhead dominated. Nonetheless some useful results have been obtained, for example the set of 12 NEO spectra obtained between 1998 and 2003 by Davies et al. (2007).

By pushing CGS4 to longer wavelengths, Rivkin et al. (2003) detected a hydrated mineral absorption feature in 11 of 16 main belt asteroids observed between 1996 and 2000. These are quite challenging observations since the 2.9 μm feature lies close to the edge of the window separating the H and K bands, so a good understanding of the atmospheric subtraction is quite critical when reducing and interpreting these observations.

The Centaurs

The first outer planet crossing asteroid 2060 Chiron was discovered by Kowal in 1977 but it was not until 1992 that the second such object 1992 AD/5145 Pholus was located. Mueller et al. (1992) quickly reported that 5145 Pholus was, unlike 2060 Chiron, very red in the UBVRI range. Davies et al. (1993b) confirmed that this very red slope continued into the near-IR with JHK observations of Pholus taken in UKIRT service observing time soon after its discovery. They also presented a 1.4–2.5 μm spectrum (taken by Gillian Wright during CGS4 engineering time) showing significant absorption features at 1.9 and 2.25 μm never before seen in asteroidal spectra. Cruikshank et al. (1998) combined UKIRT UKT9 photometry and CGS4 spectroscopy with optical data to produce a detailed model of the surface composition involving a mixture of methanol and water ice plus olivine and Titan Tholin darkened by spectrally neutral amorphous carbon.

The discovery rate of Centaurs picked up slowly in the mid 1990s and UKIRT was often the first telescope to obtain IR data on newly discovered objects. In most cases this was restricted to JHK photometry since the objects were too faint for IR spectroscopy. However these data, often combined with optical information from either the INT or the UH 88 in. telescope, confirmed the rapidly developing suspicion that the Centaur population was highly diverse, having a range of V–J colours which spanned the spectrally neutral 2060 Chiron to the extremely red 5145 Pholus. Several examples were found with colours representative of the D class of dark and reddened main belt objects (e.g. Davies et al. 1998a, b; Farnham and Davies 2003).

A few other Centaurs were targeted spectroscopically, (e.g. McBride et al. 1999) but only the brightest ones were accessible to UKIRT/CGS4 and the approval of a VLT large programme on outer solar system objects led by Antonella Barruci of the Paris observatory made UKIRT uncompetitive in this field after about 2000.

Other Strange Asteroids

As well as the Centaurs, a number of other unusual asteroids were discovered in the mid-late 1990s. Whenever possible these were targeted by UKIRT in order to compare their properties with the, at that time, slowing growing dataset of Centaurs and trans-Neptunian objects. 1998WU$_{24}$ had the orbital characteristics of a Halley type comet but showed no signs of cometary activity. UKIRT and UH 88 in. observations (Davies et al. 2001) were able to determine the rotation period and show that it had colours typical of D type asteroids. No infrared spectral features were detected.

1996PW was another extinct comet candidate, this time with orbital characteristics of a long period comet. Once again VRIJHK data revealed a D type asteroid spectrum and no spectral features. Minor planet (20461) 1999 LD$_{31}$ is one of only a few objects in retrograde orbits listed in the Minor Planet catalogue as an asteroid rather than a comet. UKIRT provided JHK photometry to support UH 88 in. optical and Keck thermal infrared observations which, once again, matched D type asteroids leading to the conclusion that this too was a dormant or extinct comet nucleus (Harris et al. 2001).

Comet Spectroscopy

The first detections of 3 μm spectral features, usually attributed to C–H stretches in unspecified organic molecules, in comets came with detections in Comet 1P/Halley in 1985/1986. By the early 1990s, UKIRT was able to contribute to this field with first CGS2, and later CGS4, taking a number of cometary spectra in the near IR. A significant result was the detection in 1990 of a strong 3.28 μm feature in Comet Levy (Davies et al. 1991). The feature was much stronger, relative to the by then classical 3.4 μm feature, than in any comet to date. However, the relatively low resolving power and sensitivity of CGS2 made precise identification of the various features, seen at 3.28, around 3.4 and 3.52 μm difficult.

The arrival of CGS4 provided a significant increase in capability and excellent spectra were obtained of comet P/Swift-Tuttle in 1992 (Davies et al. 1993a). In this case a combination of higher spectral resolution plus improved models of the emission from cometary organics showed that much of the 3.4 μm feature, plus the weaker feature at 3.52 μm, could be ascribed to methanol. Subtracting the methanol contribution from the broad 3.4 μm feature revealed a residual feature centred at 3.43 μm and eliminated the need to add formaldehyde to the mix in order to explain the 3.52 μm feature. Disanti et al. (1995) carried out detailed modelling of this and other spectra of Comet P/Swift-Tuttle and reported that the unidentified feature accounted for one-half of the flux contained in the 3.4 μm feature and exhibited a heliocentric dependence consistent with a volatile parent species. They attributed this to be representative of asymmetric CH$_2$ (or CH–X) vibrational stretching in some as yet unidentified organic compound.

The bright comet Hale-Bopp presented a mixed picture at UKIRT. The comet was discovered far outside the region, usually inside 3 AU, when comets become active due to the sublimation of water from the nucleus. Although it was long considered to be present, the detection of water ice in comets had proved elusive. Direct observations of nuclei when near the Sun are impossible since the nucleus resides inside an optically thick coma, but outside 3 AU typical comets are too faint for high signal-to-noise spectroscopy with 4 m telescopes. Hale-Bopp was exceptionally bright and so was observable at much greater distances than usually possible. UKIRT spectra in September 1995, when the comet was 7 AU from the Sun, (Davies et al. 1997c) showed broad absorption features at 1.5 and 2.05 µm. These could be attributed to water ice, either on the nucleus or in icy grains surrounding it. Significantly the absences of a spectral feature at 1.65 µm due to crystalline ice suggests that the cometary ice was probably in an amorphous state when the observations were made.

As the comet approached the Sun attempts were made to study the 3.4 µm feature as had been done with earlier comets. However in this case no significant results were obtained. Observing conditions were difficult due to the low altitude as seen from Mauna Kea (although some daytime observations were possible) and any molecular emission was completely swamped by the huge background of thermal emission from dust in the coma. The dust was however detected with CGS3 (Davies et al. 1997a) and revealed a complex 10 µm silicate absorption feature and excess flux at 20 µm. These spectra were in generally good agreement with the higher signal to noise and greater coverage of spectra taken with the ISO SWS instrument from space (Crovisier et al. 1997).

Trans-Neptunian Objects

As well as Centaur 5145 Pholus, 1992 saw another important discovery in the outer solar system, the new trans-Neptunian object 1992 QB$_1$ (Jewitt and Luu 1991). This object was followed by first a trickle then a flood of new trans-Neptunian objects such that a whole new field of research opened up very rapidly. With V magnitudes of around 22–24, and hence JHK magnitudes of about 20–22 depending on colour, this new population was at the limits of UKIRT capabilities. None the less J, and occasionally H and K observations were possible both to compare the objects' colours with those of Centaurs (since the Centaurs are almost certainly trans-Neptunians injected into planet crossing orbits) and to probe the then controversial subject that the trans-Neptunian population might have a bi-modal colour distribution. Some successes were had with 1993SC (Davies et al. 1997b) and then samples of brighter objects observed from both UKIRT and INT (Davies et al. 2000; McBride et al. 2003). However, as with the Centaur programme, UKIRT could not compete with the power of the VLT large programme of photometry, nor with spectroscopy of the brighter objects possible from Keck and VLT so observations of these distant asteroids had drawn to a close by about 2002.

Other Highlights

UKIRT, particularly in combination with CGS4, made a number of important observations of outer planet atmospheres and satellites.

Trafton et al. (1993) detected 11 emission features of the H_3^+ fundamental vibration-rotation band between 3.89 and 4.09 μm in the atmosphere of Uranus. These features allowed the rotational temperature (740 ± 25 K), ortho-H_3^+ fraction (0.51 ± 0.03) and disk-averaged $H_3(+)$ column abundance (6.5×10^{10} (± 10 %) molecules/cm^2) to be calculated. The same year, Geballe et al. (1993) reported three emission lines of the nu-2 fundamental vibration-rotation band of H_3^+ in the ionosphere of Saturn near and at its poles. The peak observed column density was more than two orders of magnitude lower than the column density at the south pole of Jupiter, and less than that detected from Uranus.

Ices in the outer solar system were also observed with CGS4, often leading to order of magnitude improvements over data taken with filter photometry or using CVFs. Neptune's satellite Triton was observed by Cruikshank et al. (1993), as was Pluto by Owen et al. (1993).

A final highlight to be reported here are UKIRT observations of the impact of Comet Shoemaker-Levy-9 with Jupiter. A number of observatories took excellent optical and infrared images of the post-impact disturbances in Jupiter's atmosphere. UKIRT was one of only a handful of observatories that attempted spectroscopy of the event and its aftermath. Two teams shared the observing window which lasted over a few nights as the fragments of the disrupted comet ploughed in to Jupiter. Their results were presented in Miller et al. (1995) and Knacke et al. (1997).

Acknowledgements None of the remarkable results described here would have been possible without the dedicated UKIRT staff who maintained, developed and operated this telescope. In particular I thank the long suffering Telescope Systems Specialists (Thor Wold, Tim Carroll, Dolores Walther and Joel Aycock) who had to find and guide on targets with sometimes uncertain positions and which were often faint but *always* moving. Tom Geballe reminded me of some key results outside my own, rather narrow, focus

References

Binzel, R.P., Rivkin, A.S., Stuart, J.S., Harris, A.W., Bus, S.J., Burbine, J.H.: Observed spectral properties of near Earth objects, results for population distribution, source regions and space weathering. Icarus **170**, 259–294 (2004)

Crovisier, J., Leech, K., Bockelee-Morvan, D., Brooke, T.Y., Hanner, M.S., Altieri, B., Keller, H.U., Lellouch, E.: The spectrum of comet Hale-Bopp (C/1995 01) observed with the infrared space observatory at 2.9 Au from the sun. Science **275**, 1904 (1997)

Cruikshank, D.P., Roush, T.L., Owen, T.C., Geballe, T.R., De Bergh, C., Schmitt, B., Brown, R.H., Bartholomew, M.J.: Ices on the surface of Triton. Science **261**, 742–745 (1993)

Cruikshank, D.P., Roush, T.L., Moroz, L.V., Geballe, T.R., Pendleton, Y.J., White, S.M., Bell, J.F., Davies, J.K., Owen, T.C., de Bergh, C., Tholen, D.J., Bernstein, M.P., Brown, R.H., Tryka, K.A., Dalle Ore, C.M.: The composition of centaur 5145 Pholus. Icarus **135**, 389–407 (1998)

Davies, J.K., Green, S., Stewart, B., Meadows, A.J., Aumann, H.H.: IRAS and the search for fast moving objects. Nature **309**, 315–319 (1984). 24 May 1984

Davies, J.K., Green, S.F., Geballe, T.R.: The detection of a strong 3.28 m emission feature in Comet Levy. Mon. Not. R. Astron. Soc. **251**, 148–151 (1991)

Davies, J.K., Mumma, M., Rueter, D., Hoban, S., Weaver, H., Puxley, P., Lumsden, S.: The infrared spectrum of comet P/Swift Tuttle. Mon. Not. R. Astron. Soc. **265**, 1022–1026 (1993a)

Davies, J.K., Sykes, M.V., Cruikshank, D.P.: Near infrared spectroscopy and photometry of the unusual minor planet 5145 Pholus. Icarus **102**, 166–169 (1993b)

Davies, J.K., Geballe, T.R., Hanner, M.S., Weaver, H.A., Crovisier, J., Bockele e-Morvan, D.: Thermal infrared spectra of Comet Hale-Bopp at heliocentric distances of 4 and 2.9 AU. Earth Moon Planet **78**, 293–298 (1997a)

Davies, J.K., McBride, N., Green, S.F.: Optical and infrared photometry of Kuiper belt object 1993SC. Icarus **125**, 61–66 (1997b)

Davies, J.K., Roush, T.L., Cruikshank, D.P., Bartholomew, M.J., Geballe, T.R., Owen, T., De Bergh, C.: The detection of water ice in Comet Hale-Bopp. Icarus **127**, 238–245 (1997c)

Davies, J.K., McBride, N., Ellison, S.E., Green, S.F., Ballantyne, D.: Optical and infrared observations of six centaur objects. Icarus **134**, 213–227 (1998a)

Davies, J.K., McBride, N., Green, S.F., Mottola, S., Carsenty, U., Basran, D., Hudson, K.A., Foster, M.J.: The lightcurve and colours of unusual minor planet 1996PW. Icarus **132**, 418–430 (1998b)

Davies, J.K., Green, S.F., McBride, N., Muzerall, E., Tholen, D.J., Whiteley, R.J., Foster, M.J., Hillier, J.K.: Visible and infrared photometry of fourteen Kuiper belt objects. Icarus **146**, 253–262 (2000)

Davies, J.K., Tholen, D.J., Whiteley, R.J., Green, S.F., Hillier, J.K., Foster, M.J., McBride, N., Kerr, T.H., Muzerall, E.: The lightcurve and colors of unusual minor planet 1998 WU24. Icarus **150**, 69–77 (2001)

Davies, J.K., Harris, A.W., Rivkin, A.S., Wolters, S.D., Green, S.F., McBride, N., Mann, R.K.: Near infrared spectra of 12 near Earth objects. Icarus **186**, 111–125 (2007)

Disanti, M., Mumma, M., Geballe, T.R., Davies, J.K.: Systematic observations of methanol and other organics in Comet Swift-Tuttle. Icarus **116**, p1–p17 (1995)

Farnham, T.L., Davies, J.K.: The rotational and physical properties of the Centaur (32532) 2001 PT13. Icarus **164**(2), 418–427 (2003)

Geballe, T.R., Jagod, M.-F., Oka, T.: Detection of H_3^+ infrared emission in Saturn. Astrophsy. J. Lett. **408**, L109–L112 (1993)

Green, S.F., Meadows, A.J., Davies, J.K.: Infrared observations of the extinct Cometary Candidate Minor Planet (3200) 1983TB. Mon. Not. R. Astron. Soc. **214**, 29–36 (1985). Short Communication

Harris, A.W.: A thermal model for near Earth asteroids. Icarus **131**, 291–301 (1998)

Harris, A.W., Davies, J.K.: Physical characteristics of near Earth asteroids from thermal infrared spectrophotometry. Icarus **142**, 464–475 (1999)

Harris, A.W., Davies, J.K., Green, S.F.: Thermal infrared spectrophotometry of the near Earth asteroids 2100 Ra-Shalom and 1991 EE. Icarus **135**(441–450), 1998 (1998)

Harris, A.W., Delbo, M., Binzel, R.P., Davies, J.K., Roberts, J., Tholen, D.J., Whiteley, R.J.: Visible to thermal-infrared spectrophotometry of a possible inactive cometary nucleus. Icarus **153**, 322–337 (2001)

Jewitt, D.J., Luu, J.X.: QB1 IAUC 5611, 14 September 1992 (1991)

Knacke, R.F., Fajardo-Acosta, S.B., Geballe, T.R., Noll, K.S.: Infrared spectra of the R impact of Comet Shoemaker-Levy 9. Icarus **125**, 340–347 (1997)

McBride, N., Davies, J.K., Green, S.F., Foster, M.J.: Optical and infrared observations of the Centaur 1997CU26. Mon. Not. R. Astron. Soc. **306**, 799–805 (1999)

McBride, N., Green, S.F., Davies, J.K., Tholen, D.J., Shepperd, S.S., Whiteley, R.J., Hillier, J.K.: Optical-infrared colors of KBO's- searching for trends. Icarus **161**, 501–510 (2003)

Miller, S., Achilleos, N., Dinelli, B.M., Laml, H.A., Tennyson, J., Jagod, M.-F., Oka, T., Geballe, T.R., Trafton, L.M., Joseph, R.D., Ballester, G.E., Baines, K., Brooke, T.Y., Orton, G.: The effect of the impact of Comet Shoemaker-Levy 9 on Jupiter's aurorae. Geophys. Res. Lett. **22**, 1629–1632 (1995)

Mueller, B.E.A., Tholen, D.J., Hartmann, W.K., Cruickshank, D.J.: Extraordinary colours of asteroidal object (5145) 1992AD. Icarus **97**, 150–154 (1992)

Owen, T., Roush, T.L., Cruikshank, D.P., Elliott, J.L., Young, L.A., De Bergh, C., Schmitt, B., Geballe, T.R., Brown, R.H., Bartholomew, M.J.: Surface ices and atmospheric composition of Pluto. Science **261**, 745–748 (1993)

Rivkin, A.S., Davies, J.K., Johnson, J.R., Ellison, S.L., Trilling, D.E., Brown, R.H., Lebofsky, L.A.: Hydrogen concentration on C-Class asteroids from remote sensing. Meteor. Planet. Sci. **38**, 1383–1398 (2003)

Trafton, L.M., Geballe, T.R., Miller, S., Tennyson, J., Ballester, G.E.: Detection of H_3^+ from Uranus. Astrophys. J. **405**, 761–766 (1993)

Whipple, F.: TB and the Geminid Meteors IAUC 3881, 25 October 1993 (1983)

Wolters, S.D., Green, S.F., McBride, N., Davies, J.K.: Optical and thermal infrared observations of six near-Earth asteroids in 2002. Icarus **175**, 92–110 (2005)

Wolters, S.D., Green, S.F., McBride, N., Davies, J.K.: Thermal infrared and optical observations of four near-Earth asteroids. Icarus **193**, 535–552 (2008)

Chapter 13
Spectroscopic Tomography of a Wind-Collision Region

Peredur Williams, Watson Varricatt, and Andy Adamson

Abstract Changes in the P Cygni profile of the 1.083-μm He I line in the spectrum of the colliding-wind Wolf-Rayet system WR140 as it passes through periastron passage allow us to map the region where the fast winds of the WC7 and O5 stars collide.

Introduction

The first spectrum recorded by UKIRT was observed on 1979 January 28 during the commissioning of UKT2, a single-element InSb photometer incorporating a 1 %-resolution CVF (circular variable filter) covering 2.3–4.6 μm. It was controlled by an LSI-11 microprocessor. The target was the Wolf-Rayet (WR) star EZ CMa (HD 50896) and the spectrum showed a strong emission line at 3.09 μm, which was identified with the He II (7–6) transition array. The broad emission lines of WR stars were well matched to the low resolutions of the CVFs and more were observed using UKT1, which incorporated a 1.4–2.8 μm CVF, by Williams et al. (1980). By the end of 1979 we had the instrument computer (PDP-11), graphics display (Tektronix) and hardcopy (Polaroid camera) at the telescope.

The broad emission lines characteristic of WR stars are formed in fast ($\sim$2,000 km/s), radiatively driven winds carrying $\sim 10^{-5}$ solar masses/year mass loss with $\sim 10^4$ solar luminosities of kinetic power (KP). The winds have significant influence on the dynamics and composition of the ISM, forming conspicuous wind-blown bubbles. When WR stars are in binary systems ("colliding-wind binaries",

---

P. Williams (✉)
Institute for astronomy, University of Edinburgh, Edinburgh, UK
e-mail: pmw@roe.ac.uk

W. Varricatt • A. Adamson
Joint Astronomy Centre, Hilo, USA

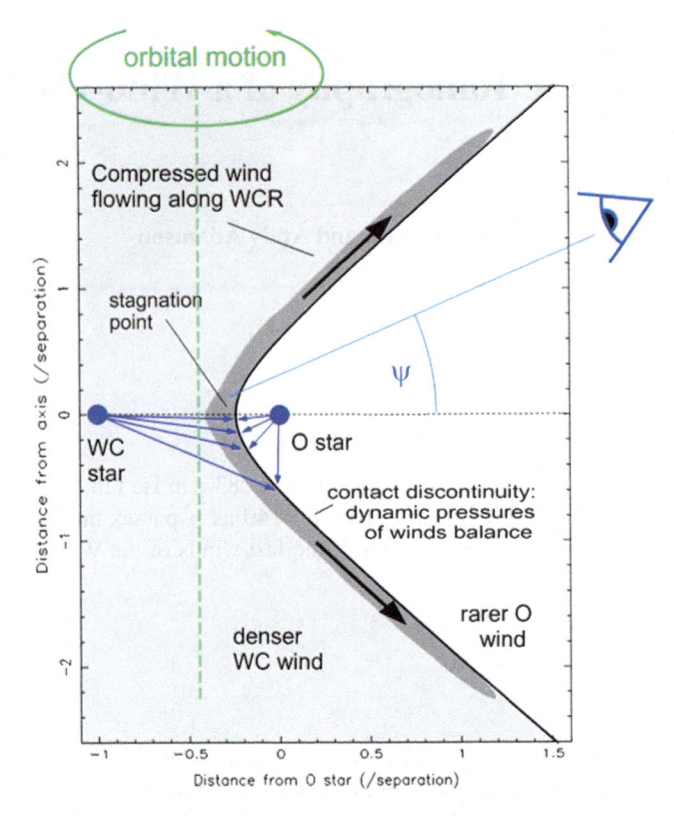

Fig. 13.1 Sketch of the WCR perpendicular to the orbital plane showing the contact discontinuity between the winds of the two stars and the shock-compressed WC wind flowing along it (there is shocked O star wind on the other side). The angle ψ between the axis of the WCR and our line of sight varies round the orbit

CWBs) with companions luminous enough to have their own fast winds, some of the KP is dissipated where the winds of the two stars collide, causing shock heating to $\sim 10^7$ K, the production of X-rays, acceleration of electrons and synchrotron emission. The winds collide where their dynamic pressures balance, and the shape of the wind-collision region (WCR) is determined by the ratio of the momenta of the two winds. In a WR + O CWB, the wind velocities are comparable but the mass-loss rate of the WR star is an order of magnitude greater than that of the O star, so that the WCR is much closer to, and folds around, the O star with a form approximating a cone at large distances from the stars (Fig. 13.1, cf. Eichler and Usov 1993). The stellar winds are shocked where they reach the WCR. Excess line emission from the compressed shocked gas flowing in the WCR can be observed on some line profiles, and this allows us to map the WCR as it moves with the stars in their binary orbit. If the shocked wind can cool efficiently (Usov 1991), dust can condense from the compressed gas and be observed from its infrared emission. We know of a few dozen CWBs showing one or more of these observable phenomena.

A CWB that shows all of these phenomena is the WC7 + O5 binary WR 140. Its dust formation and non-thermal radio emission were first observed 30 years ago. These, and the X-ray emission, have been observed to vary dramatically with the 7.94-year binary period; the demonstration that WR140 was a binary followed from the repeatability of its dust-formation episodes (Williams et al. 1990). Observed flux variations arise from the movement of the WCR and associated X-ray and non-thermal radio sources in the stellar winds as the orbit progresses, causing periodic changes in the extinction towards us. The intrinsic strengths of the CW effects also vary if the binary orbit is eccentric and variation of the separation of the components causes the WCR to occur at a range of local wind densities. This is a strong effect in WR140, whose high orbital eccentricity (e = 0.88, Marchenko et al. 2003) causes some CW effects, including dust formation and changes to line profiles from compressed wind, to occur for small fractions of the period close to periastron, when the stellar separation is favourable and the shocks may be radiative instead of adiabatic. These effects have made WR140 a superb laboratory for studying high-energy phenomena (e.g. Pittard and Dougherty 2006) and a template for understanding systems like η Carinae, recently shown to be a CWB. Both η Car and WR140 went through periastron passages in their orbits in early 2009 and there was a Joint Discussion (JD13) at the 2009 IAU General Assembly to discuss preliminary results from intensive observing campaigns directed at both systems.

As part of the multi-frequency observing campaign on WR140, we observed the 1.083-μm He I line, which has a broad P Cygni profile. The absorption component gives the terminal velocity of the wind (−2,860 km/s for the WC7 star), while the emission component is about 5,500 km/s wide and (most of the time) has a flat top indicative of formation in an expanding optically thin wind. This is an appropriate place to recall that the first interpretation of such a line profile in terms of stellar outflow was made at the ROE, by Jacob Halm (1904), Lecturer in Astronomy at the University of Edinburgh and Assistant Astronomer at the ROE.

Early observations of the 1.083-μm He I line in WR140 with CGS2 and CGS4 had shown a classic P Cygni profile at orbital phases 0.66 and 0.83, but further observations by Varricatt et al. (2004, Paper I) with CGS4 and UIST starting at orbital phase 0.96, before the 2001 periastron passage, showed the appearance of a conspicuous sub-peak on top of the previously flat emission profile. This indicated increasing interaction of the WC7 and O5 winds as the stars approached each other in their orbit: between phases 0.83 and 0.96, the separation would have fallen from 1.28a to 0.5a and the pre-shock wind density at the interaction region would have risen by a factor of 6.5. Varricatt et al. tracked the evolution of the emission sub-peak as WR140 went through periastron and interpreted the changes in terms of the changing orientation of the WCR along which the compressed wind flowed using a simple geometric model (e.g. Lührs 1997). At the same time, the changes in the depth of the absorption component set constraints on the size of the WCR depending on the orbital inclination.

Observations

In 2008 June-August and December, we observed a further 11 spectra of WR140 covering the 1.083-μm He I line using UIST with the short-J grism, giving a resolution of R = 1,500. Their orbital phases (0.93–0.95 and 0.99) extend the study in Paper I, providing more detail at critical phases. In particular, they cover the partial eclipse of the stars as the WCR crosses our line of sight (Fig. 13.2). These observations constrain the CWB geometry more tightly than before, giving a cone opening angle $\theta = 50°$. This implies a wind-momentum ratio $\eta = 0.10$, rather larger than that derived from the wind velocities and mass-loss rates of the two stars determined in earlier multi-frequency studies (e.g. Pittard and Dougherty 2006). The wind terminal velocities are directly measured, but determination of the mass-loss rate of the WR star from radio or X-ray observations depends on the composition and clumping in the wind and this direct measurement of the geometry will help constrain the models.

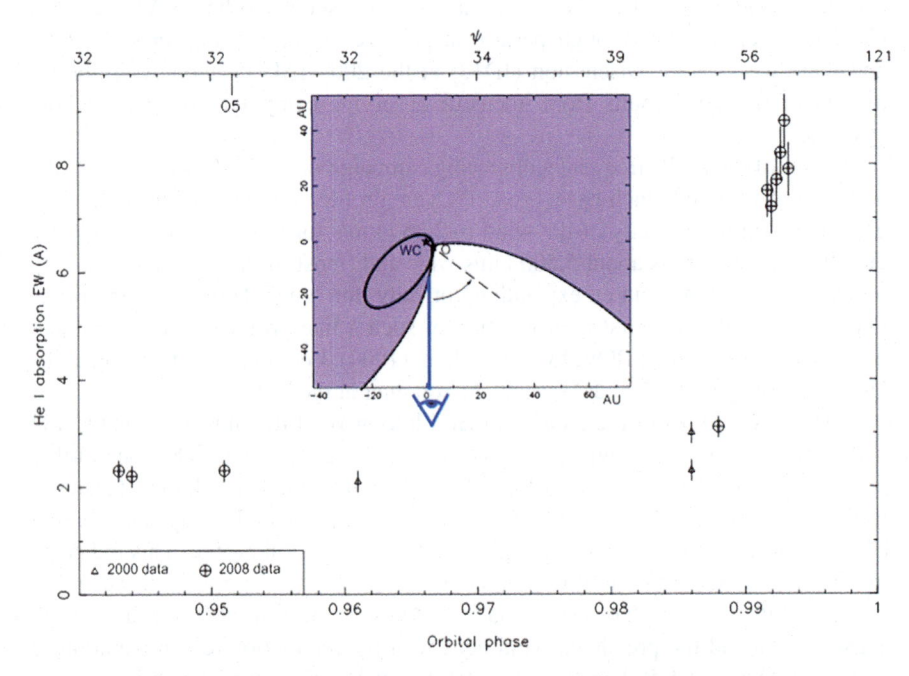

Fig. 13.2 Absorption component EW plotted against phase showing rise near $\varphi = 0.99$ when both stars suffer extinction in the dense WC wind. The phase range covers conjunction (O5 star in front) when extinction is minimal. The angle ψ between sightline and WCR axis is marked along *the top*; the highly non-linear variation arises from the eccentricity of the orbit. Inset: sketch in the plane of the orbit for $\varphi = 0.99$ showing orbit of O star relative to the WC star, our sightline and the WCR axis (*broken line*) as the O star begins to suffer extinction in the WC stellar wind

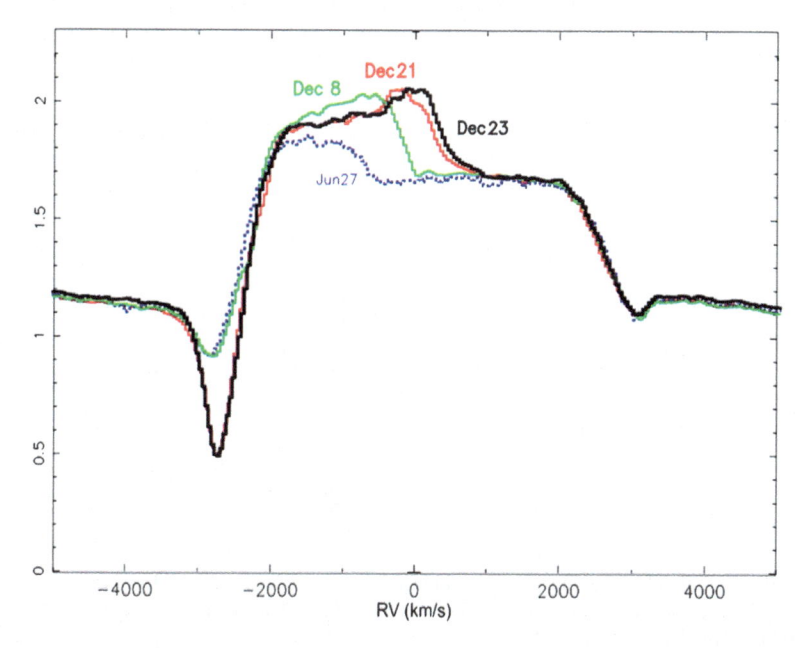

Fig. 13.3 Four line profiles from the 2008 observation set showing the change of absorption and also the evolution of the emission sub-peak on top of the profile from 2008 Jun 27 ($\varphi = 0.932$, *dotted line*) to Dec 23 ($\varphi = 0.993$, *heavy line*) in the run-up to the 2009 periastron. Further away from periastron, the profile has a flat top

The emission component of the profile showed re-appearance of a sub-peak at the 'blue' end with a central velocity $-1,320$ km/s (Fig. 13.3). The July and three August spectra showed a similar feature at the same velocity but with gradually increasing strength consonant with the increase of CW effects as the stars approach each other. The December profiles show a stronger, asymmetrical sub-peak initially similar to those observed in the previous cycle, and latterly a more complex structure that evolved to longer wavelengths. These profiles can, in principle, be modelled with knowledge of the shape of the WCR and the emissivity and velocity of the compressed wind material flowing along it. For a first approximation, we consider compressed wind flowing along the asymptotic, conical region of the WCR: that is, where the angle relative to the WCR axis is constant and equal to the opening angle θ determined from the absorption profile and to have reached a constant velocity. The latter can be calculated from the winds of the two stars in the thin shell approximation (Cantó et al. 1996) to be $v_{\text{flow}} = 2,370$ km/s. Using these values, and the orbital elements from Marchenko et al. (2003) and Dougherty et al. (2005), we can calculate the variation of the RV as a function of orbital phase.

The model RV variation is compared with measured RVs in Fig. 13.4 and it can be seen that the simple model reproduces the variation of central RV and width well. This is not a fit to the observations: the stellar and orbital parameters can be

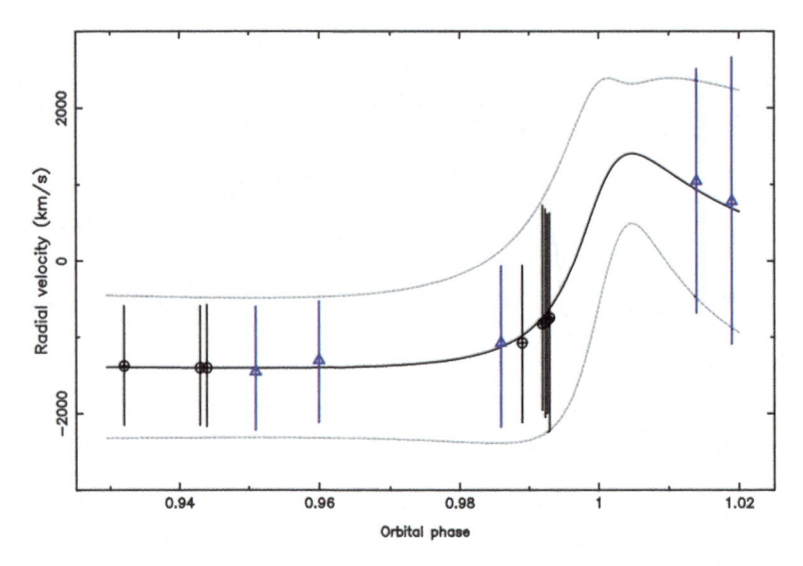

Fig. 13.4 Radial velocities and velocity widths of the sub-peaks measured from the line profiles (*circles* 2008 observations, *triangles* from Paper I) compared with simple model (*solid line* the central RV, *grey lines* the RV range)

adjusted to fit the simple model. More realistic models aiming to fits to the detailed line profiles indicate that we are seeing emission from the accelerating compressed wind closer to the stagnation point (Fig. 13.1) and this is borne out by the close similarity of the He I profile to that of the Si XIV 6.18Å line observed at $\varphi = 0.997$ by Pollock et al. (2005).

References

Cantó, J., Raga, A.C., Wilkin, F.P.: Exact algebraic solutions for the thin-shell two-wind interaction problem. Astrophys. J. **469**, 729 (1996)

Dougherty, S.M., Beasley, A.J., Claussen, M.J., Zauderer, A., Bolingbroke, N.J.: High-resolution radio observations of the colliding-wind binary WR 140. Astrophys. J. **623**, 447 (2005)

Eichler, D., Usov, V.V.: Particle acceleration and non-thermal radio emission in binaries of early-type stars. Astrophys. J. **402**, 271 (1993)

Halm, J.: On Professor Seeliger's theory of temporary stars. Proc. R. Soc. Edinb. **25**, 513 (1904)

Lührs, S.: A colliding-wind model for the Wolf-Rayet System HD 152270 (WR 79). Publ. Astron. Soc. Pac. **109**, 504 (1997)

Marchenko, S.V., Moffat, A.F.J., Ballereau, D., et al.: The unusual 2001 periastron passage in the "clockwork" colliding-wind binary WR 140. Astrophys. J. **596**, 1295 (2003)

Pittard, J.M., Dougherty, S.M.: Radio, X-ray and gamma-ray emission models of the colliding-wind binary WR 140. Mon. Not. R. Astron. Soc. **372**, 801 (2006)

Pollock, A.M.T., Corcoran, M.F., Stevens, I.R., Williams, P.M.: Bulk velocities, chemical composition, and ionization structure of the X-ray shocks in WR 140 near periastron as revealed by the CHANDRA gratings. Astrophys. J. **629**, 482 (2005)

Usov, V.V.: Stellar wind collision and dust formation in long-period, heavily interacting Wolf-Rayet binaries. Mon. Not. R. Astron. Soc. **252**, 49 (1991)

Varricatt, W.P., Williams, P.M., Ashok, N.M.: Near-infrared spectroscopic monitoring of WR 140 during the 2001 periastron passage. Mon. Not. R. Astron. Soc. **351**, 1307 (2004). Paper I

Williams, P.M., Adams, D.J., Arakaki, S., Beattie, D.H., Born, J., Lee, T.J., Robertson, D.J., Stewart, J.M.: Near infrared spectroscopy of WC stars. Mon. Not. R. Astron. Soc. **192**, 25p (1980)

Williams, P.M., van der Hucht, K.A., Pollock, A.M.T., Florkowski, D.R., van der Woerd, H., Wamsteker, W.M.: Multi-frequency variations of the Wolf-Rayet system HD 193793 – I. Infrared, X-ray and radio observations. Mon. Not. R. Astron. Soc. **243**, 662 (1990)

Chapter 14
Highlights of Infrared Spectroscopy of the Interstellar Medium at UKIRT

Thomas R. Geballe

Abstract This paper gives a review of UKIRT's major contributions to our understanding of the interstellar medium. I focus on the studies of several phenomena in which UKIRT has made particularly key contributions. These are (1) the physics of shock waves, (2) fluorescent molecular hydrogen, (3) grain mantles in dense clouds, (4) hydrocarbons in diffuse clouds, and (5) properties of dense and diffuse clouds revealed by H_3^+.

Introduction

UKIRT has made major contributions to our understanding of the interstellar medium (ISM) since its very early days. During the first few years of its existence, the majority of UKIRT's refereed publications on the ISM were sub-millimetre studies of CO in molecular clouds. However, the very first refereed paper on the ISM is a Pink Pages Monthly Notices paper by Doug Whittet and colleagues (Whittet et al. 1981), containing a crude 3-μm spectrum of a star in Taurus (Fig. 14.1). This paper showed that the weakness of the ultraviolet dust absorption feature at 2,200 Å observed toward that star is not due to its carrier being masked by water ice on the dust particles in front of the star, as had been suggested, because the characteristic 3.0-μm absorption by water ice is weak or absent. In 1983 Doug and colleagues published a more significant paper in Nature (Whittet et al. 1983) relating the optical depths of dust and water ice in the Taurus Molecular Cloud, based on data obtained in 1981. That discovery, as well as the journal of publication, is an early indication of UKIRT's impact on the infrared astronomy world.

T.R. Geballe (✉)
Gemini Observatory, 670 N A'ohoku Place, Hilo, HI 96720, USA
e-mail: tgeballe@gemini.edu

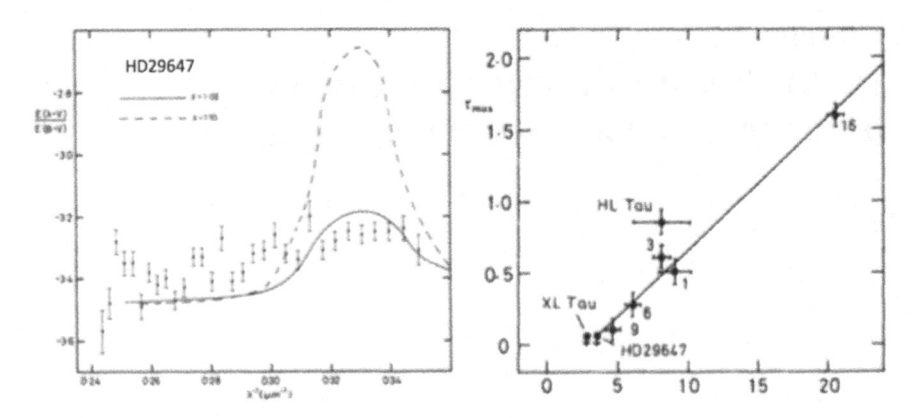

Fig. 14.1 *Left*: The first published infrared spectrum of the ISM from UKIRT (Whittet et al. 1981; using UKT6), showing a lack of strong water ice absorption toward HD 29647. The frequency range is 0.24–0.36 inverse microns. *Right*: The optical depth of the 3.0-μm ice band (y-axis) vs. visual extinction in the Taurus Molecular Cloud (Whittet et al. 1983) showing the threshold for the presence of ice on dust grains

Infrared spectroscopy is the most important astrophysical tool for the study of a vast array of interstellar phenomena. A short review cannot possibly do justice to the breadth of the science that the diverse array of spectrographs at UKIRT have contributed. Instead this paper focuses on the studies of several phenomena in which UKIRT has made particularly key contributions. These are (1) the physics of shock waves, (2) fluorescent molecular hydrogen, (3) grain mantles in dense clouds, (4) hydrocarbons in diffuse clouds, and (5) properties of dense and diffuse clouds revealed by H_3^+.

The Physics of Shock Waves in Molecular Clouds

Molecular hydrogen (H_2) is by far the most abundant molecule in interstellar space. Spectroscopically its behavior is atypical of most molecules. First, it is so lightweight that its excited vibrational and rotational levels, which are widely spaced in energy ($v = 1$ lies $\sim$6,000 K above $v = 0$ and $J = 1$ is $\sim$600 K above $J = 0$), are not populated, even in warm interstellar gas. Second, due to the structural symmetry of H_2, the lifetimes against spontaneously radiation of those first several vibrational and rotational levels are very long (months to years). It is possible to excite the ro-vibrational levels of H_2 by collisions, if the collisions are sufficiently energetic, or by downward transitions from excited electronic states if an electronic state is excited and the molecule does not dissociate as a result. However, because of both the low likelihood of H_2 becoming excited in molecular clouds and the long lifetimes of the excited states, it was not anticipated that H_2 would produce bright line emission there.

Three years before UKIRT arrived on the scene, however, bright H_2 emission lines were discovered in OMC-1, the Orion Molecular Cloud (Gautier et al. 1976). It was quickly clear that the line-emitting H_2, whose temperature is $\sim 2,000$ K, is being heated by a shock wave propagating within the cloud. Gas is "shocked" whenever it is impacted by a high-pressure region at speed exceeding the speed of sound – faster than the speed at which disturbances can otherwise propagate through it. OMC-1 was known to contain a high velocity gas flow. As OMC-1 also was known to contain a number of young stars, it was obvious that the H_2 line emission, subsequently detected in innumerable molecular clouds, is connected with events that take place during star formation.

Most observations of shock-excited H_2, including the multitude of those made at UKIRT, have been acquired with the intention of locating and investigating sites of star formation. There is, however, a second impetus for studying the H_2 line emission in molecular clouds: investigating shock physics. That is the research area that Peter Brand and his graduate students and outside collaborators worked on for many years. In the 1980s and 1990s Peter's group led the world in confronting theory with observations. Some of the issues that they uncovered still remain to be resolved.

One of the major problems in understanding shock physics in molecular clouds is if and how the molecular hydrogen survives to radiate in the cooling, post-shock gas. Collisions between two H_2 molecules at more than about 24 km s^{-1} dissociate the molecules. Thus simple shocks (called jump shocks, or J-shocks) traveling at more than that speed should destroy all H_2, leaving none to radiate the strong lines that are seen. Yet several outflows, such as the one in OMC-1, are known in which the wind or jet from the newly formed star is traveling outward at 100 km s^{-1} or higher. How does the H_2 survive?

Theorists came up with an explanation in the early 1980s – the continuous shock (C-shock) with a magnetic precursor. This type of shock occurs in a cloud that contains a small ionic fraction and a sufficiently strong magnetic field. In that case ions in the cloud are accelerated ahead of the main shock and impart some of their momentum to the neutral cloud material. As a result the acceleration and heating of the cloud are not as violent and there is much less dissociation. In fact, if the magnetic field is strong enough there is no J-shock at all. This more gentle process has a significantly different effect on the temperatures in the post-shock gas. In a J-shock gas is excited suddenly to high temperature and then cools off. In a C-shock the gas is heated more gradually and to a lower maximum temperature over a long path length and one can almost speak of an average temperature in the heated gas.

In a series of four papers published in 1988 and 1989 Peter and his students (Alan Moorhouse, Mike Burton, and Mark Toner) together with their collaborators tested the C-shock models for the prototypical cloud, Orion and a few others, in a variety of ways (Brand et al. 1988, 1989a, b; Burton et al. 1989). Their results were uniformly conclusive that C-shocks with precursors, in which the H_2 survives, cannot explain what is observed.

Figure 14.2, from Brand et al. (1988) illustrates some of their work. On the left is a composite spectrum covering H_2 lines of widely varying excitations at one location

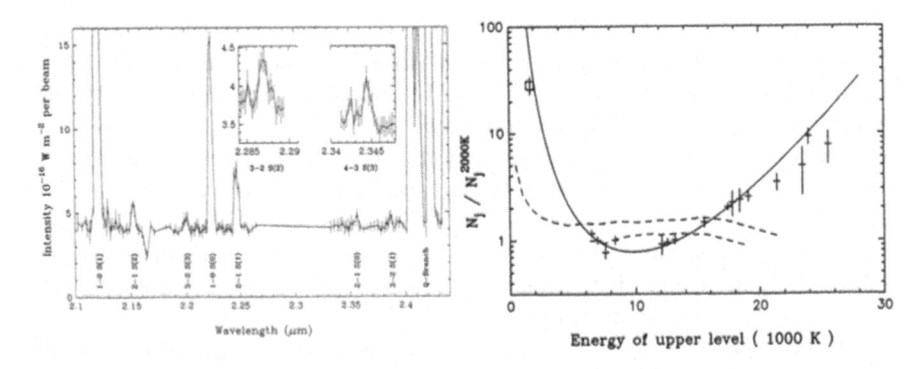

Fig. 14.2 Spectrum of H_2 lines at OMC-1 Peak 1 (*left*) and plot of level populations, compared to those for LTE at 2,000 K, vs. energy of emitting level (*right*) for a J-shock (*continuous line*) and two C-shock models (*dashed lines*), from Brand et al. (1988) (Data and uncertainties are shown as crosses)

in OMC-1. One of the detected lines originates in the $v = 4$ $J = 5$ level, $\sim$25,000 K above ground. The strengths of the highly excited lines relative to those from lower energy levels are very sensitive to the maximum post-shock temperature. The right hand plot shows attempted fits for a variety of models to the level populations derived from the observed line strengths ratioed by the populations expected at a temperature of 2,000 K. The dashed lines are predictions for a C-shock – in which a more or less constant temperature would be maintained until the C-shock passes. The continuous line is for the cooling following a J-shock, in which the gas is suddenly heated to a much higher temperature and then rapidly cools. As can be seen the latter type of cooling provides a much better fit.

Peter's group did not suggest that one abandon C-shocks with precursors. There are many measurements, obtained at UKIRT and elsewhere, which are otherwise consistent with such shocks. However, the questions raised by Peter's observations are yet to be answered, especially for the higher velocity outflows. It is not possible that J-shocks are the explanation in Orion, because the velocity limit for H_2 survival is inviolable. One possible explanation is that the spectra sample a range of unresolved C-shocks with different maximum temperatures. This might be tested by a similar experiment as the one shown in Fig. 14.2, but at much higher angular resolution, using for example a modern integral field spectrograph to "dissect" a bow shock for which different excitation conditions should exist at different locations along the shock. Another possible explanation is that H_2 is destroyed, and rapidly reforms on dust grains behind the shock front, and is released into the post-shock gas in highly excited states. A third possibility is that UV-excited fluorescence (see section "Fluorescent H_2") contributes to the emission and skews the level populations so that their ratios are more consistent with higher excitation temperatures.

As a final point, perhaps one should ask if there is any direct and unequivocal observational evidence that shock precursors exist? The answer is yes, and UKIRT is to thank for that answer. Figure 14.3 is a beautiful composite image of a portion of

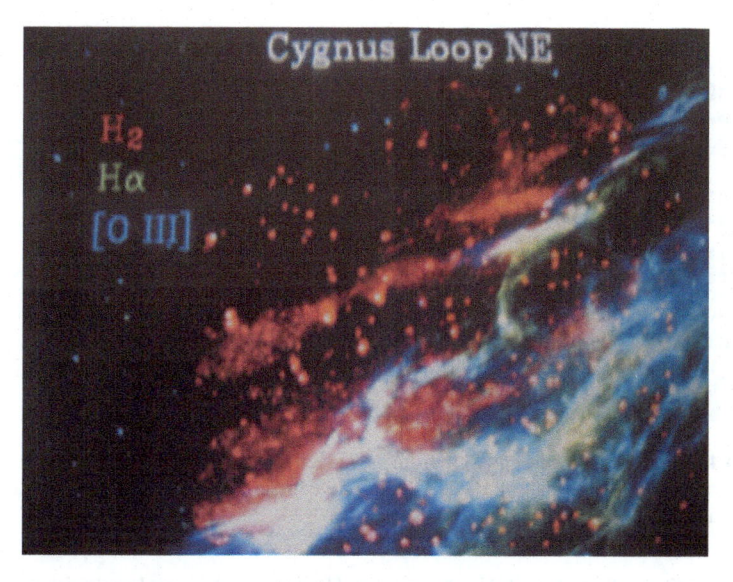

Fig. 14.3 Composite three-color image of line emission from H_2, H I, and O III in a portion of the Cygnus Loop (Graham et al. 1991). The expansion of this portion of the supernova remnant is from *bottom right* to *top left*

the Cygnus Loop supernova remnant, showing Hα and O III emission together with faint H_2 line emission obtained by James Graham, Gillian Wright, and collaborators, using IRCAM with its largest (2.4 arcsec) pixels (Graham et al. 1991). The H_2 line emission lies well in front of the main shock front, which compresses and ionizes the gas and excites the optical emission. Graham et al. showed spectroscopically that the H_2 is collisionally excited. A magnetic precursor as the exciter of the H_2 is the only viable explanation for these data. Because of the low density of the ambient gas and the high speed of the supernova ejecta, 150–200 km s^{-1}, the separation between the H_2 line emission and the Hα is very large.

Fluorescent H_2

Molecular clouds are usually located near intense sources of UV radiation, the young and hot stars that they often spawn. The remnant gas immediately surrounding these hot stars is ionized, but at sufficiently large distances the natal cloud is often largely unaffected and fully molecular. Between these two extremes lies a transition region which has come to be known as a Photo-Dissociation Region or Photon-Dominated Region (PDR). Reflection nebulae surrounding O and B stars also are examples of PDRs. In a PDR the hydrogen-ionizing UV from the hot stars has been used up, but there remains longer wavelength UV capable of exciting or dissociating molecules, and even ionizing some species.

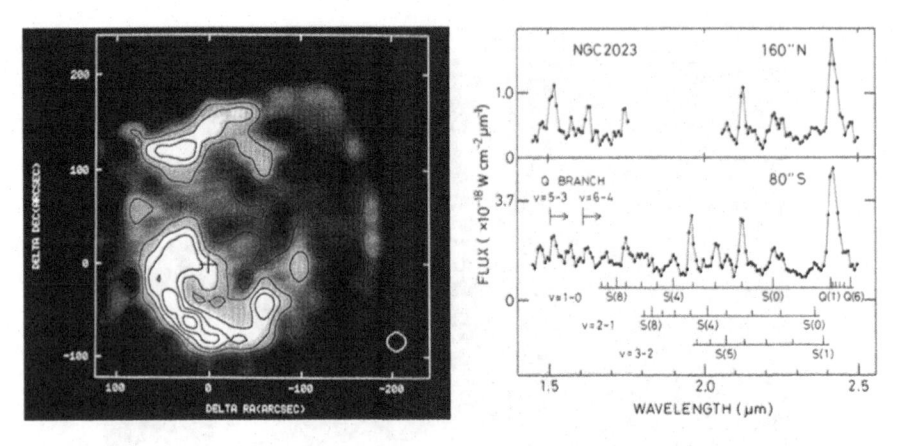

Fig. 14.4 Fluorescent H_2 line emission from the reflection nebula NGC 2023 (Gatley et al. 1987). *Left*: Image and contour map of H_2 1−0 S(1) line emission obtained with UKT9 + FP. *Right*: Spectra at two locations, obtained with UKT9(CVF)

Absorption of such a UV photon by H_2 leaves that molecule in an excited electronic state, which results either in dissociation of the molecule or radiative decay into an excited ro-vibrational level. In the latter case, if the density of the gas is not too high the vibrationally excited H_2 molecule cascades spontaneously (and very slowly) down towards the ground state. The physics of this UV-excited fluorescence is relatively simple. The branching ratios and Einstein A-coefficients in the cascade are known, as are the collisional de-excitation rates, and thus the spectrum of H_2 can be accurately predicted. It is distinctly different from that of collisionally excited H_2, with line ratios consistent with high vibrational temperatures and low rotational temperatures.

Fluorescent H_2 line emission is expected to be faint and very extended on the sky, and is not well suited to the standard techniques of infrared astronomy. From the late 1960s to the early1970s searches made for it were unsuccessful. But about 10 years later Ian Gatley and his collaborators were successful at UKIRT. In this era before detector arrays UKIRT had an advantage for detecting faint extended signals that no other similar-size telescope had: UKT6 and UKT9 with their abilities to image light from 20″ apertures onto single sensitive detectors. I believe that this product of beam size and telescope aperture on a single infrared pixel has never been exceeded. It probably never will be.

With the aid of this large beam size and a Fabry-Perot (FP) interferometer in front of UKT9, Ian and his collaborators first mapped the emission from one of the strongest H_2 lines, using the technique of "frequency-chopping" (rapidly measuring on and off the line frequency), rather than spatially chopping (the spatial distribution of the line emission was unknown and surely very extended). They identified locations where both the emission was strong and it was also possible to spatially chop off the source, thereby obtaining spectra covering enough lines to demonstrate from the line ratios that the line emission was fluorescent in a cloud of density $\sim 10^4$ cm^{-3} (Fig. 14.4; Gatley et al. 1987).

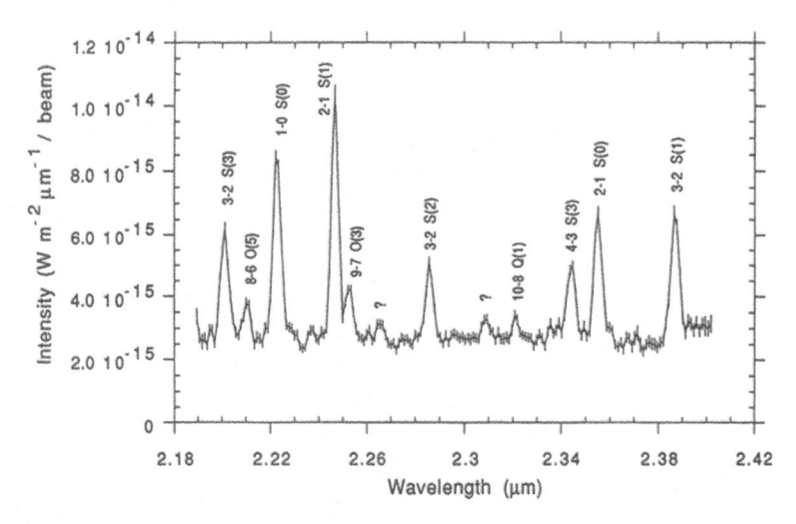

Fig. 14.5 Figure on the cover of the UKIRT Newsletter of August 1993, showing the spectrum of fluorescent H_2 in the proto-planetary nebula Hubble 12 obtained with CGS4. The associated journal paper is Ramsay et al. (1993)

Three seminal papers resulted from their observations of H_2 fluorescence reflection nebula NGC 2023 and in the Orion Molecular Cloud (Hayashi et al. 1985; Gatley et al. 1987; Hasegawa et al. 1987) and all are still being cited. Fluorescent H_2 is now know to be common in planetary nebulae and proto-planetary nebulae, as well as in reflection nebulae and the PDRs associated with H II regions. The ratio of ortho to para molecular hydrogen in fluorescing regions is almost always in the range 1–2, significantly lower than the statistical weighting of H_2 ortho/para levels, which is 3. Two viable explanations for this are (1) the UV transitions eventually that lead to the cascades down the vibrational ladder are self-shielded to different extents for ortho and para levels, and (2) H_2 is formed on dust grains at sufficiently low temperature that the $v = 0$ $J = 0$ para rotational level is favored over the $v = 1$ $J = 1$ ortho level, which is at 600 K higher energy. In most cases it seems that both effects contribute.

As instrumental sensitivities have improved better and more extensive spectra of fluorescent H_2 have been obtained. Noteworthy among these is the CGS4 spectrum of Hubble 12, a proto-planetary nebula, obtained by Suzie Ramsay and collaborators (Ramsay et al. 1993). One of their spectra is shown in Fig. 14.5; it contains several lines from very high vibrational states. The line ratios imply pure fluorescence – the density in the nebula is so low that even the intensities of the slowest transitions involving the lowest vibrational levels are unaffected by collisional relaxation.

Dust in Dense Clouds

UKIRT has long been a leading contributor to our knowledge of the properties of dust within dense clouds, and particularly so prior to the advent of ISO and Spitzer. The most significant advances achieved by astronomers using UKIRT have been in understanding the icy mantles of dust particles – their chemical make-up and the details of the distributions of the molecular species in the mantles.

The grains found in dark clouds have cores made of silicates, which have strong bands centred near 10 and 18 μm. In cold clouds gas molecules such as CO, H_2O, NH_3 freeze onto the grain surfaces, forming icy mantles. Some species, like CO, require very low temperatures (<20 K) to stick to the grains; others like H_2O are much less volatile. Chemical reactions in the mantles, caused by absorption of a visible or UV photon, or some other heating process, create additional species, including CH_3OH, OCS, and the famous "XCN," whose precise identification is still controversial. Each of these has characteristic spectral features. Thus the absorption spectra of mantled grains can be complex. The interplay between astronomical observations and laboratory simulations of dust grains is critical and is illustrated in a drawing by Mark Toner (Fig. 14.6) that served as the cover illustration for a workshop largely organized by UKIRT staff in the early days of this field.

UKIRT's important contributions began very early with Whittet's 1983 Nature paper, which demonstrated that a linear relationship between the depth of the 3-μm

Fig. 14.6 Mark Toner's cover illustration for the proceedings of the international workshop, "Laboratory and Observational Spectroscopy of Interstellar Dust"

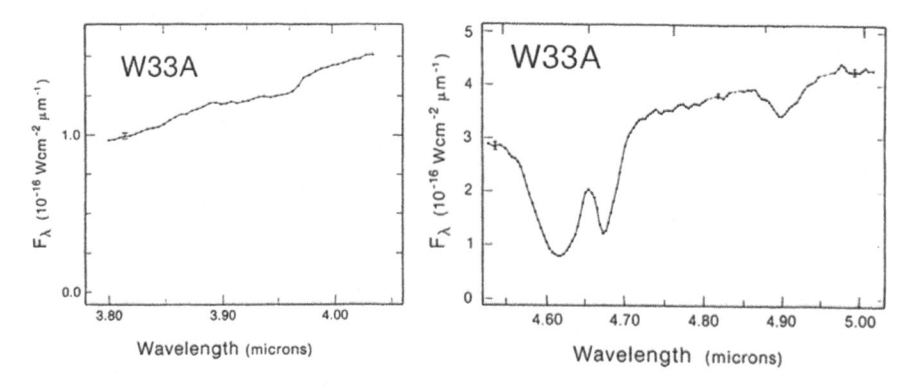

Fig. 14.7 Two new absorption bands discovered at UKIRT at 3.95 μm (frozen CH$_3$OH, originally identified as H$_2$S) and 4.92 μm (frozen OCS) in the spectrum of the young stellar object, W33A IR observed with CGS2 (Geballe et al. 1985). The other bands, at 4.61 and 4.68 μm were already known and are due to "XCN" (a solid state species containing CN) and solid CO, respectively

water ice absorption and the visual extinction shown in Fig. 14.1 exists in the Taurus Dark Cloud. Note that the linear fit does not pass through the origin. This revealed for the first time an extinction threshold for the presence of water ice on the dust grains. In other words, water ice does not survive on grains near the surface of the cloud; it only is found on the surfaces of dust particles shielded by at least three magnitudes of visual extinction and thus well within the dark cloud. Much later Doug's student Jean Chiar and collaborators established a threshold for solid CO of six magnitudes of visual extinction in the same Taurus Cloud (Chiar et al. 1995), a cloud in which only low mass stars have formed. Later others would find even higher thresholds in other clouds such as Ophiucus, which have populations of more massive and hotter stars.

A few of the known molecular species frozen on grains were discovered at UKIRT and several new absorption bands of already known mantle constituents also were identified there (Fig. 14.7). Often UKIRT's role was to obtain higher quality data than obtained at other telescopes or to go where others did not, such as 4.9 μm (Fig. 14.7) or 9.7 μm (Fig. 14.8), wavelengths where telluric interference is a serious problem, especially at low altitude sites.

The spectra of solid methanol in AFGL 2136 (Fig. 14.8) are an interesting example of this kind of work and also illustrate the importance of comparing laboratory and astronomical data Skinner et al. (1992) found that methanol must be highly concentrated in water ice. However, the ratio of the strengths of the water ice band at 3 μm and the methanol band indicates that toward AFGL 2136 water ice is ten times more abundant than methanol ice. The only way to reconcile these results is if the two species are not uniformly mixed with one another along the line of sight. There are two possibilities: (1) the regions where methanol is enhanced could be small layers of individual grain mantles, with the rest of each grain mantle being virtually methanol free; (2) grains with enhanced methanol could be found in

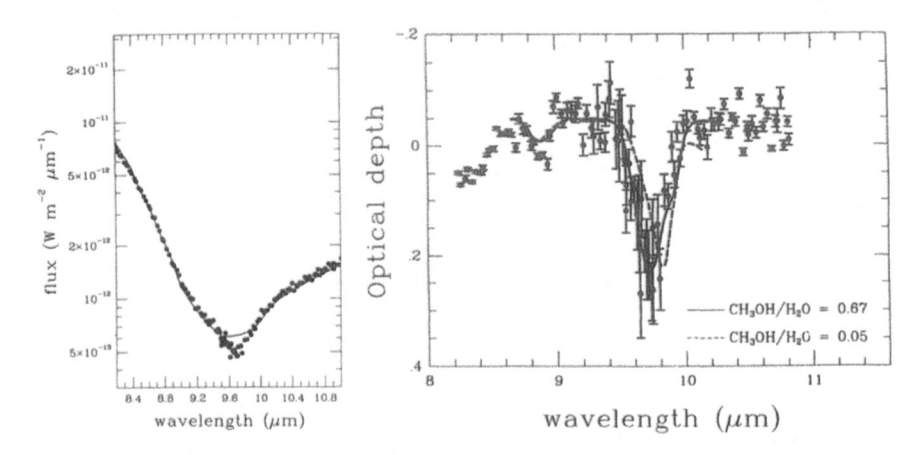

Fig. 14.8 *Left*: CGS3 spectrum of the young stellar object AFGL 2136 at the center of the silicate band (Skinner et al. 1992). *Solid line* is best model fit to the silicate feature. *Right*: Optical depth of the residual, identified as solid CH_3OH. *Solid* and *dashed lines* are laboratory spectra of methanol–water ice mixtures

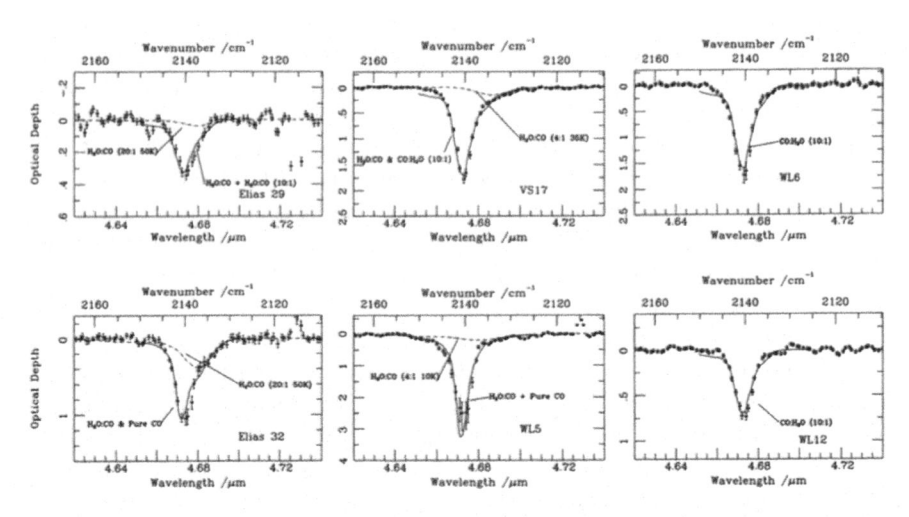

Fig. 14.9 Spectra of solid CO in six objects in the ρ Ophiucus dark cloud compared with synthetic spectra based on laboratory data (Kerr et al. 1993), showing varying physical and chemical conditions in the cloud

small portions of the line of sight though the cloud. Either of these appears plausible, as methanol freezes out on grains at much lower temperatures than H_2O. However, they imply different grain mantle development histories.

The most elegant work in this field has been on the 4.7 μm solid CO absorption. The precise wavelength peak of this absorption and the shape of the band, depend on the immediate neighbors of the CO molecules. Figure 14.9, from Kerr et al. (1993)

reveals the diversity of the CO absorption profiles and shows how laboratory spectra allow one to effectively "dissect" the frozen ices and discover the manner in which they are mixed with one another and in some cases isolated from one another. The spectra reveal the inherent segregation of the mantling process, which is caused by different species freezing out at different temperatures.

"Dust" Features in Diffuse Clouds

The infrared spectra of diffuse clouds are much less rich than those of dense clouds, because dust grains in diffuse clouds do not have icy mantles. There are only four known "dust" features, those at 3.3 and 3.4 μm, which are presumably due to hydrocarbons, and the two silicate features at 9.7 and 18 μm). The first high quality spectrum of the 3.4 μm absorption band (Fig. 14.10) was obtained at UKIRT (Butchart et al. 1986). This feature is a universal signature of the diffuse ISM. It was discovered in extragalactic sources at UKIRT (Bridger et al. 1994; Wright et al. 1996) and in them is often diagnostic for the dominance of an AGN over nuclear star formation. The 3.3 μm absorption feature was found at UKIRT by careful removal of the wing of the ice band absorption toward sources in GC (Chiar et al. 2002). The existence of this feature elsewhere in the diffuse ISM besides the GC, let alone its universality, is yet to be demonstrated.

One of the key contributions of UKIRT in the study of the diffuse ISM has been to effectively kill a long held assumption about the 3.4 μm feature. The identity of this feature with aliphatic hydrocarbons is well accepted, but the nature and history of the absorber has been a long-standing question, and it had generally been assumed that it is a refractory material on dust grains created by photochemical processing of mantles originally produced in molecular clouds. However, this assumption has been shown via spectroscopy and spectro-polarimetry

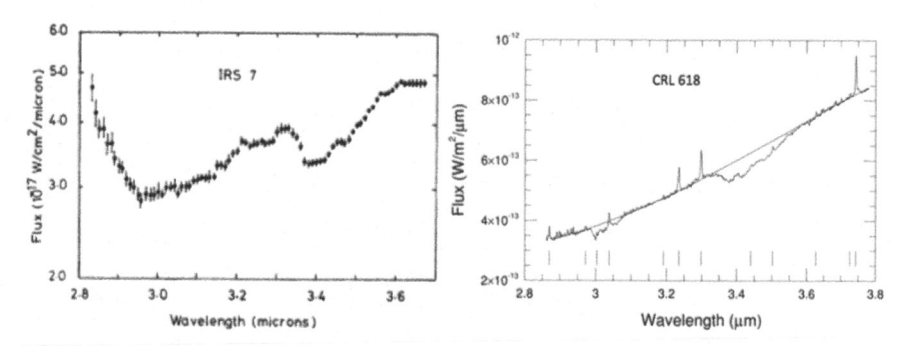

Fig. 14.10 Spectra of the Galactic center source IRS 7 (Butchart et al. 1986), showing the 3.4 μm feature in foreground diffuse gas, and the carbon-rich proto-planetary nebula CRL 618 (Chiar et al. 1998), showing the same feature

to be incorrect (Shenoy et al. 2003; Chiar et al. 2006). The source of absorbing material may be the ejecta of carbon stars and UKIRT has provided most of the evidence in support of this (e.g., Lequeux and Jourdain de Muizon 1990; Chiar et al. 1998).

H_3^+ in Dense and Diffuse Clouds

The discovery of H_3^+ in the ISM, made at UKIRT (Geballe and Oka 1996) confirmed a long accepted theory that this molecular ion is the starting point for almost all gas phase chemistry. In one sense this discovery was a breakthrough. However, there was plenty of other evidence for H_3^+ – driven ion-molecule chemistry in the ISM. Thus, H_3^+, which is produced in molecular clouds following cosmic-ray ionization of H_2, had to be there, and the detection itself (Fig. 14.11) should not have been surprising, except perhaps as a technical achievement.

What is now most significant about H_3^+ is its value in understanding interstellar environments. This has been established largely from observations made at UKIRT. Perhaps the best example stems from the surprising discovery (at UKIRT) that the H_3^+ in diffuse clouds is more than an order of magnitude more abundant than was predicted. Obtaining the explanation for this mystery (McCall et al. 2003) involved (1) observing very weak H_3^+ lines in a diffuse cloud (Fig. 14.11) sufficiently optically thin that UV spectra from space were available to provide information on the abundances of H_2 and free electrons, and (2) laboratory measurements at cloud temperatures of the dissociation rate of H_3^+ on electrons. The conclusion is that the cosmic-ray ionization rate, generally thought to be more or less constant across the Galaxy, is on average an order of magnitude or more higher in diffuse clouds than in dark clouds. The most likely explanation is that there is a large population of low energy cosmic rays that penetrates diffuse clouds and ionizes their H_2, but only affects the surfaces of dense clouds.

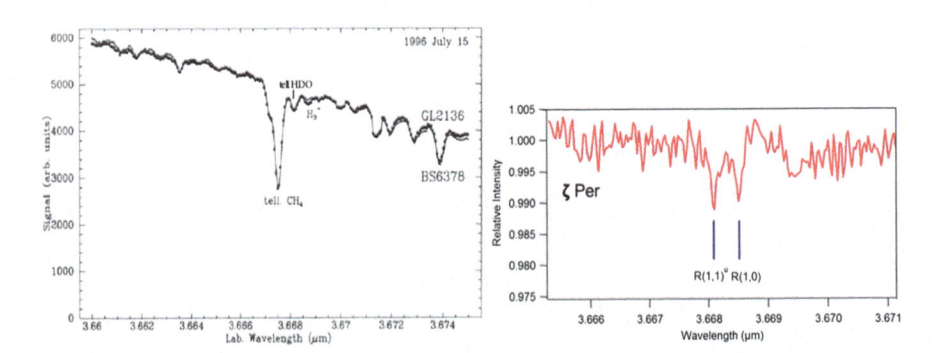

Fig. 14.11 *Left*, discovery of H_3^+ in dense clouds; the raw spectra of GL 2136 and the telluric standard, BS 6378 at the wavelength of the H_3^+ ortho/para doublet are shown with the locations of the doublet interfering *telluric lines* indicated. *Right*, ratioed spectrum of the ζ Per showing the H_3^+ doublet absorption in a diffuse cloud with $A_V = 2.8$ mag

Some Concluding Thoughts

I arrived at UKIRT late in 1981, as one of two scientists hired by The Netherlands after it secured a 15 % share of UKIRT time and was fortunate to be retained by UKIRT when The Netherlands ended its involvement 6 years later. The environment at UKIRT was ideal for me – someone who loved observing, but didn't have a formal astronomy education and was eager to continue learning astronomy both by doing it and by interacting with real astronomers. Most importantly, I had access to continuous stream of state-of-the-art facility spectroscopic capabilities. As the majority of my research up to 1981 had involved spectroscopy of the interstellar medium, UKIRT provided me with a natural way to continue and expand my research. UKIRT's first Hilo office was a rented building on Leilani Street, where on the second floor most of us were crowded together into one big room – staff scientists, engineers of all types, visiting astronomers, and visiting grad students. The collaborative atmosphere was great. In fact, I worried about the effect of moving to the new building on A'ohoku Place, where scientists would have private offices with doors. Happily the sense of shared adventure continued in the much more spacious working environment at A'ohoku Place.

In its discoveries and advances in the field of spectroscopy of the interstellar medium, as in so many other areas, UKIRT has surely met or exceeded the expectations of those farsighted individuals who conceived this telescope. I would only add a couple of points. First, UKIRT has not only provided new discoveries, but as befits its status as the largest dedicated infrared telescope in the world, it often has provided the first high quality infrared spectra of phenomena, including those discovered at other telescopes located on inferior sites. Second, removing my 'UKIRT hat' and viewing UKIRT as a slight outsider, I still find it amazing as well as totally admirable (and I know other "foreigners" do too) that in the highly competitive field of astronomy UKIRT from the start was unconditionally open to scientists from other countries. There can be no doubt that this policy led to UKIRT being at the forefront of much more science than it would have been otherwise. It created cross-fertilizations that strengthened astronomy in other countries, as well as in the United Kingdom. Speaking as a member of the rest of the world, I thank British astronomy for making UKIRT readily available to scientists everywhere.

References

Brand, P.W.J.L., Moorhouse, A., Burton, M.G., Geballe, T.R., Wade, R.: Ratios of molecular hydrogen line intensities in shocked gas – evidence for cooling zones. Astrophys. J. **334**, L103 (1988)

Brand, P.W.J.L., Toner, M.P., Geballe, T.R., Webster, A.S., Williams, P.M., Burton, M.G.: The constancy of the ratio of the molecular hydrogen lines at 3.8 microns in Orion. Mon. Not. R. Astron. Soc. **236**, 929 (1989a)

Brand, P.W.J.L., Toner, M.P., Geballe, T.R., Webster, A.S.: The velocity profile of the 1-O S(1) line of molecular hydrogen at Peak 1 in Orion. Mon. Not. R. Astron. Soc **237**, 1009 (1989b)

Bridger, A., Wright, G.S., Geballe, T.R.: Dust absorption in NGC1068. In: Mclean, I.S. (ed.) Infrared astronomy with arrays: the next generation, p. 537. Kluwer, Dordrecht (1994)

Burton, M.G., Brand, P.W.J.L., Geballe, T.R., Webster, A.S.: Molecular hydrogen line ratios in four regions of shock-excited gas. Mon. Not. R. Astron. Soc. **236**, 409 (1989)

Butchart, I., McFadzean, A.D., Whittet, D.C.B.W., Geballe, T.R., Geenberg, J.M.: Three micron spectroscopy of the galactic centre source IRS 7. Astron. Astrophys. **154**, L5 (1986)

Chiar, J.E., Adamson, A.J., Kerr, T.H., Whittet, D.C.B.: High-resolution studies of solid CO in the taurus dark cloud: characterizing the ices in quiescent clouds. Astrophys. J. **455**, 234 (1995)

Chiar, J.E., Pendleton, Y.J., Geballe, T.R., Tielens, A.G.G.M.: Near-infrared spectroscopy of the proto-planetary nebula CRL 618 and the origin of the hydrocarbon dust component in the interstellar medium. Astrophys. J. **507**, 281 (1998)

Chiar, J.E., Adamson, A.J., Pendleton, Y.J., Whittet, D.C.B., Caldwell, D.A., Gibb, E.L.: Hydrocarbons, ices, and 'XCN' in the line of sight toward the galactic center. Astrophys. J. **570**, 198 (2002)

Chiar, J.E., Adamson, A.J., Whittet, D.C.B., Chrysostomou, A., Hough, J.H., Kerr, T.H., Mason, R.E., Roche, P.F., Wright, G.: Spectropolarimetry of the 3.4 μm feature in the diffuse ISM toward the galactic center quintuplet cluster. Astrophys. J. **651**(268) (2006)

Gautier III, T.N., Fink, U., Treffers, R.P., Larson, H.P.: Detection of molecular hydrogen quadrupole emission in the Orion Nebula. Astrophys. J. **207**, L129 (1976)

Gatley, I., Hasegawa, T., Suzuki, H., Garden, R., Brand, P., Lightfoot, J., Glencross, W., Okuda, H., Nagata, T.: Fluorescent molecular hydrogen emission from the reflection nebula NGC 2023. Astrophys. J. **318**, L73 (1987)

Geballe, T.R., Baas, F., Greenberg, J.M., Schutte, W.: New infrared absorption features due to solid phase molecules containing sulfur in W 33 A. Astron. Astrophys. **146**, L6 (1985)

Geballe, T.R., Oka, T.: Detection of H_3^+ in interstellar space. Nature **384**, 334 (1996)

Graham, J.R., Wright, G.S., Hester, J.J., Longmore, A.J.: H_2 excitation by magnetic shock precursors in the cygnus loop supernova remnant. Astron. J. **101**, 175 (1991)

Hasegawa, T., Gatley, I., Garden, R.P., Brand, P.W.J.L., Ohishi, M., Hayashi, M., Kaifu, N.: Level population and para/ortho ratio of fluorescent H_2 in NGC 2023. Astrophys. J. **318**, L77 (1987)

Hayashi, M., Hasegawa, T., Gatley, I., Garden, R., Kaifu, N.: The molecular hydrogen emission associated with the orion bright star. Mon. Not. R. Astron. Soc. **215**, 31P (1985)

Kerr, T.H., Adamson, A.J., Whittet, D.C.B.: Infrared spectroscopy of solid CO – the Rho Ophiuchi molecular cloud. Mon. Not. R. Astron. Soc. **262**, 1047 (1993)

Lequeux, J., Jourdain de Muizon, M.: The 3.4 and 12 micrometer absorption bands in the protoplanetary nebula CRL 618. Astron. Astrophys. **240**, L19 (1990)

McCall, B.J., et al.: An enhanced cosmic-ray flux towards ζ Persei inferred from a laboratory study of the H_3^+ – e$^-$ recombination rate. Nature **422**, 500 (2003)

Ramsay, S.K., Chrysostomou, A., Geballe, T.R., Brand, P.W.J.L., Mountain, M.: Pure fluorescent H_2 emission from hubble 12. Mon. Not. R. Astron. Soc. **263**, 695 (1993)

Shenoy, S.S., Whittet, D.C.B.W., Chiar, J.E., Adamson, A.J., Roberge, W.G., Hassel, G.E.: A test case for the organic refractory model of interstellar dust. Astrophys. J. **591**, 962 (2003)

Skinner, C.J., Tielens, A.G.G.M., Barlow, M.J., Justtanont, K.: Methanol ice in the protostar GL 2136. Astrophys. J. **399**, L79 (1992)

Whittet, D.C.B., Bode, M.F., Longmore, A.J., Baines, D.W.T., Evans, A.: Interstellar ice grains in the Taurus molecular clouds. Nature **303**, 218 (1983)

Whittet, D.C.B., Evans, A., Bode, M.F., Butchart, I.: Ice mantles and the anomalous ultraviolet extinction of HD 29647. Mon. Not. R. Astron. Soc. **196**, 81P (1981)

Wright, G.S., Bridger, A., Geballe, T.R., Pendleton, Y.: Studies of NIR dust absorption features in the nuclei of Active and IRAS galaxies. In: Block, D.L., Greenberg, J.M. (eds.) Cold dust and galaxy morphology. Astrophysics and space sciences library, vol. 209, p. 207. Kluwer, Dordrecht (1996)

Chapter 15
UKIRT and the Brown Dwarfs: From Speculation to Classification

Sandy K. Leggett

Abstract This paper reviews the contributions of UKIRT to understanding the initial mass function of brown dwarfs.

Introduction

Brown dwarfs were postulated in 1963 by Kumar (1963) and Hayashi and Nakano (1963) in two independent theoretical papers. These authors showed that during the formation of a low mass star, a mass would be reached below which electron degeneracy would prevent the core from collapsing further and heating to the temperatures required for hydrogen nuclear fusion. Thus these low mass objects never enter a period of long-lived stable temperature and luminosity and instead they fade and cool over time. Now known as brown dwarfs, such objects have masses below 80 Jupiter masses.

A convenient feature of these objects is that all brown dwarfs older than about 200 Myr have a radius R within 20 % of Jupiter's (e.g. Burrows et al. 2001). The small dependency of radius on mass and age means that there is a tight relationship between luminosity L and effective temperature T_{eff}, as $L = \sigma \pi R^2 T_{eff}^4$. Through the rest of this review temperature and luminosity can be thought of as effectively interchangeable.

Figure 15.1 shows cooling tracks as a function of mass for low mass stars and brown dwarfs. Because brown dwarfs cool with age, T_{eff} alone cannot constrain mass, so age (or surface gravity) also needs to be known. This review deals with low luminosity, low temperature, field brown dwarfs which are typically older than 0.5 Gyr – objects in the lower right region of Fig. 15.1. UKIRT has also made large

---

S.K. Leggett (✉)
Gemini Observatory, Northern Operations Center, 670 N A'ohoku Pl.
University Park, Hilo, HI 96720, USA
e-mail: sleggett@gemini.edu

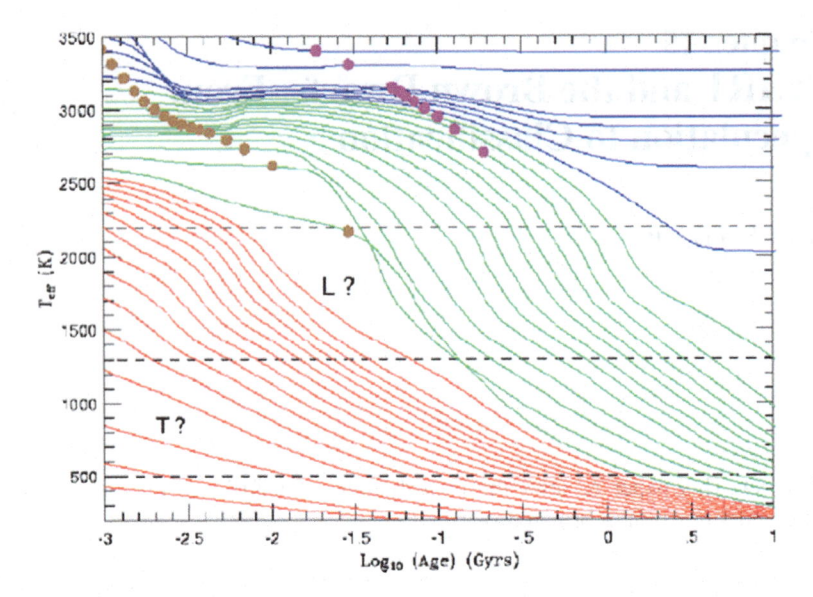

Fig. 15.1 Cooling tracks for low mass stars (*blue lines*), brown dwarfs (*green lines*) and planet-like mass brown dwarfs (*red lines*), from Burrows et al. (2001). The brief periods of deuterium and lithium burning for the stars and higher mass brown dwarfs are indicated. The T_{eff} range for L and T spectral type dwarfs suggested here have now been confirmed, except for the lower temperature limit of the T spectral type

contributions to studies of warmer low mass objects that would lie in the central region of Fig. 15.1 – young brown dwarfs in open clusters (e.g. Hodgkin et al. 1999; Jameson et al. 2002; Lodieu et al. 2006) and in star forming regions (e.g. Lucas and Roche 2001).

Discovering and Classifying the Brown Dwarf Population

It took several decades for brown dwarfs to be found, following their definition in 1963. During this time the lowest region of the main sequence was defined by the M dwarfs with $T_{eff} \sim 3{,}000$ K and $L \sim 10^{-3.5}$ $L_\odot$. In 1988 a 10^{-4} $L_\odot$ 2,000 K object was found, and in 1995 a 10^{-5} $L_\odot$ 1,000 K object. Starting in 1997 dwarfs were found with $L \sim 10^{-5.7} - 10^{-3.6}$ $L_\odot$ and the local population of L and T dwarfs with $2{,}500 > T_{eff}$ K > 800 began to be defined.

UKIDSS has recently found the faintest dwarfs known, with $L \sim 10^{-6}$ $L_\odot$ and Teff ~ 600 K. Thus UKIRT has now been involved in the discovery or confirmation of half the objects known with $L < 10^{-5}$ $L_\odot$. This fraction will increase rapidly with more, imminent, UKIDSS discoveries (Warren/Burningham these proceedings.)

As cooler objects were found, spectral classifications schemes had to be defined. The L dwarfs with $1{,}350 < T_{eff}$ K $< 2{,}200$ have strong neutral alkali and metal

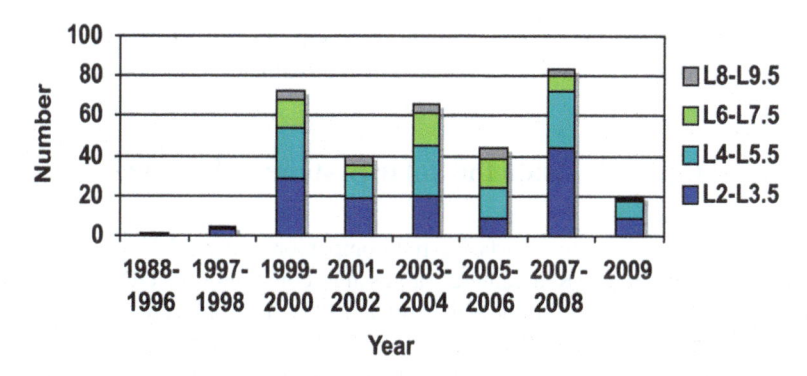

15 UKIRT and the Brown Dwarfs: From Speculation to Classification

Fig. 15.2 Number of L2 to L9.9 dwarfs discovered as a function of date

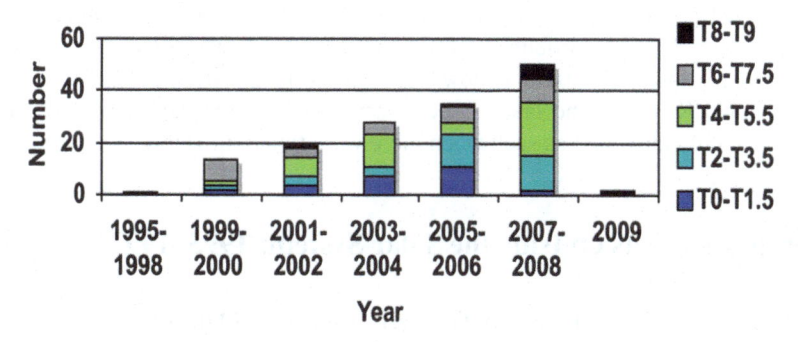

Fig. 15.3 Number of T0 to T9 dwarfs discovered as a function of date

hydride features in the red, and strong H$_2$O and CO absorption bands in the near-infrared. Both regions have been used to classify these dwarfs (Kirkpatrick et al. 1999; Geballe et al. 2002). The T dwarfs are cool enough with T$_{eff}$ < 1,350 K that CH$_4$ bands are seen at 1.6 and 2.2 µm, and these, together with near-infrared H$_2$O features, are used to define the T spectral types (Burgasser et al. 2006). T dwarfs are extremely red in the far-red, with very little flux emitted shortwards of 1 µm.

Figures 15.2 and 15.3 illustrate the rate of discovery of L and T dwarfs, and the coverage with type (based on information provided at *DwarfArchives*.org as of August 25 2009). The L and T dwarfs have predominantly been found as a result of three sky surveys. The near-infrared Two Micron All Sky Survey (2MASS) ran from 1997 to 2003, and used 2-m telescopes in Arizona and Chile to survey the entire sky at *JHK* (Skrutskie et al. 2006). The red/optical Sloan Digital Sky Survey (SDSS) ran from 1999 to 2006, and used a 2-m telescope in New Mexico to map 8,000 deg.^{2} of the Northern sky with *ugriz* filters (York et al. 2000). The near-infrared UKIRT Infrared Deep Sky Survey (UKIDSS) started in May 2005, and will cover several thousands of degrees of sky in *YJHK*, going 3–4 magnitudes fainter than 2MASS (Lawrence et al. 2007).

The following sections describe UKIRT's contributions to the discovery, classification and analysis of this recently discovered population below the main sequence, a population that is a factor of 10^3 fainter in luminosity and a factor of 5 cooler in T_{eff} than the M dwarfs. Topics are covered in approximately chronological order.

Faintest Objects Outside the Solar System: 1988–1995

During a survey of white dwarfs for red companions, using UKT9 and IRCAM as well as other Mauna Kea cameras, a possible companion to GD 165 was seen in UH 88″ images (Becklin and Zuckerman 1988). The object was confirmed as a proper motion companion by Becklin and Zuckerman (1992). CGS4 data were used by Tinney et al. (1993) to determine a bolometric luminosity and confirm the object as the lowest luminosity dwarf then known. CGS4 was also used by Jones et al. (1994) in their investigation of the correlation between near-infrared spectral features with $T_{eff,}$ (in particular H_2O), which showed GD 165B to be the coolest dwarf then known. GD 165B is now classified as an L3, with $L = 10^{-4.1}$ $L_\odot$ and $T_{eff} = 1,850$ K. It was the first L-type dwarf, and the only one known until 1997 (although only classified as such when more were found, in 1999).

Faintest Objects Outside the Solar System: 1995–1997

Nakajima et al. (1995) discovered GL 229B by imaging M dwarfs for companions. An early spectrum indicated it was extremely cool, however the CGS4 spectrum by Geballe et al. (1996) became the definitive measurement, and this showed strong H_2O and CH_4 absorption bands, and similarity to spectra of Saturn and Titan (Fig. 15.4). Later Noll et al. (1997) used CGS4 at 5 μm to demonstrate that CO is enhanced over chemical equilibrium models by a factor $>10^3$. They suggested that mixing occurs in the atmosphere, and that, as in Jupiter, CO is dredged up from deeper layers.

GL 229B is now classified as a T7p, with $L = 10^{-5.2}$ $L_\odot$ and $T_{eff} = 950$ K. It was the first T-type dwarf, and the only one known until 1999 (although only classified as such in 2000).

UKIRT and the L Dwarfs: 1997–2000

More GD 165B-like objects were finally discovered in 1997; five examples were published that year. IRCAM and CGS4 data were used by Ruiz et al. (1997) to classify the high proper motion object Kelu-1 as one of these new beyond-M class of objects. The other four objects were found as red objects in early 2MASS and the Deep Near-Infrared Survey (DENIS; Epchtein 1994) data. In 1999 2MASS found many more L dwarfs; Kirkpatrick et al. (1999) defined an optical classification scheme for L dwarfs, and used CGS4 to demonstrate that no CH_4 is seen in their

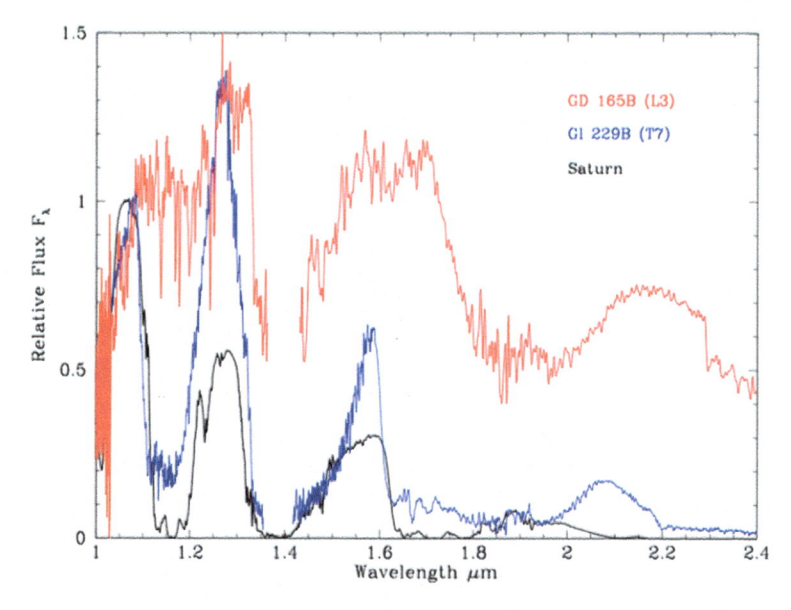

15 UKIRT and the Brown Dwarfs: From Speculation to Classification 177

Fig. 15.4 CGS4 spectra of the prototype L and T dwarfs GD165B and Gl 229B, and the planet Saturn (Geballe et al. 1996 and unpublished, Leggett et al. 2001)

near-infrared spectra. Noll et al. (2000) used CGS4 to show that the 3 μm CH_4 feature is seen however in L5 and later type dwarfs. Leggett et al. (1998) used IRCAM data to demonstrate that grains must redden the atmospheres of the L dwarfs, and that these dwarfs have $T_{eff} \sim 2{,}000$ K (Fig. 15.5).

The formation of cloud decks of condensates in brown dwarf atmospheres is a crucial part of their spectral evolution. At temperatures around 1,500 K iron and silicate grains condense, forming liquid or solid iron and silicates like olivine ($(Mg,Fe)_2SiO_4$) or enstatite ($MgSiO_3$). Ackerman and Marley (2001) model horizontal decks of condensates, with the vertical extent determined by the balance between upward turbulent mixing and downward sedimentation. A parameter f_{sed} is used, which is the ratio of sedimentation to convection velocities. Smaller values of f_{sed} imply more extensive cloud decks. For L dwarfs $1 < f_{sed} < 3$, for T dwarfs $f_{sed} = 4$ to infinity (cloud free). At T dwarf temperatures the grains condense below the photosphere and do not impact the spectral distribution, apart from removing the condensate species as opacity sources.

UKIRT and the Second 1,000 K Methane Dwarf: 1999

In April 1999 the Princeton group of the Sloan Survey contacted T. Geballe and S. Leggett to obtain infrared follow up observations of a uniquely (optically) red object they had found in SDSS commissioning data. IRCAM and CGS4 data rapidly

Fig. 15.5 The extreme redness of the near-infrared colours of the beyond-M class of dwarfs demonstrates that grains redden their atmosphere (Leggett et al. 1998)

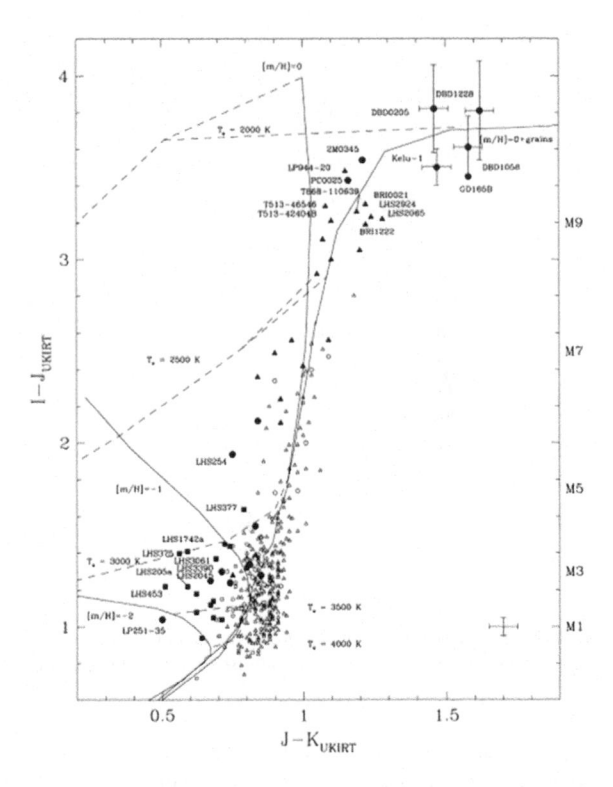

confirmed this to be the second example of a GL 229B-like object (Strauss et al. 1999). The spectrum was almost identical to that of GL 299B. Finally, after more than 30 years, we were on our way to discovering and studying the local population of cold brown dwarfs, objects that spectrally appeared much more similar to planets than stars (Fig. 15.4).

UKIRT Identifies the T Spectral Sequence: 2000

In the year after the 1999 SDSS + UKIRT discovery of the second methane or T dwarf, several more were found by 2MASS and SDSS. All looked like GL 229B, with strong CH_4 features in the near-infrared. In an SDSS brown dwarf candidate follow up observing run at UKIRT in March 2000, in one night with CGS4, examples were finally found of early-type T dwarfs. Three objects were found with weak *H*-band CH_4 features, and these provided the spectral link between the late-type L dwarfs and the mid-type Ts (Leggett et al. 2000). These objects were easier to find in SDSS than 2MASS as their optical colours are distinctive while their *JHK* colours are similar to early type stars. SDSS + UKIRT had the monopoly on early T dwarfs for some time.

Fig. 15.6 Example T dwarf CGS4 spectra from Geballe et al. (2002)

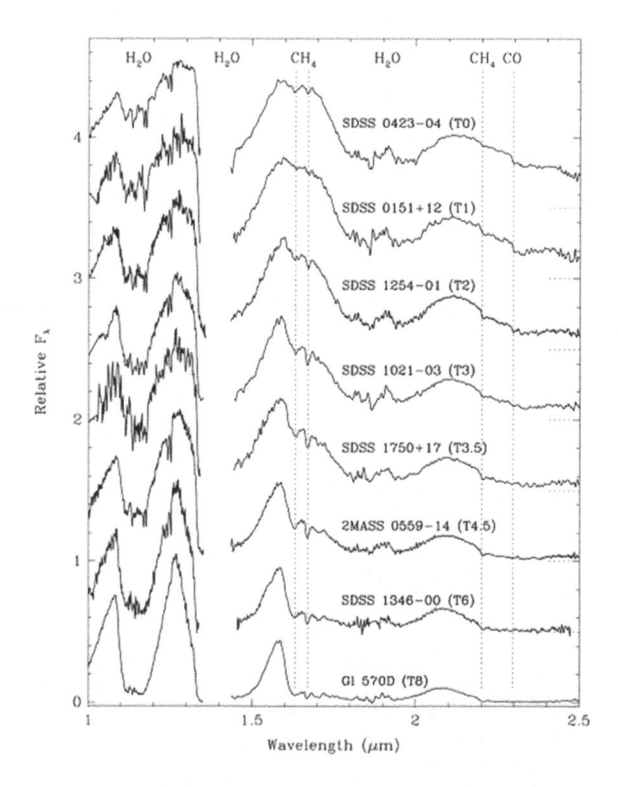

Classification of M, L and T Dwarfs: 2001–2006

High quality CGS4 spectra have consistently defined the infrared spectral classification of the lower main sequence and beyond. Reid et al. (2001) presented a scheme to classify M and L dwarfs, which was superceded by the Geballe et al. (2002) scheme for L and T types. The universally adopted scheme for T dwarfs (Burgasser et al. 2006) incorporates and builds on the Geballe et al. results (e.g. Fig. 15.6).

UKIRT and the Fast L to T Transition: 1999–2006

IRCAM, UFTI, CGS4 and UIST data for around 120 SDSS L and T dwarfs produced a well quantified sample in several publications between 1999 and 2006. One still puzzling aspect of this population is the rapid clearing of condensates from the photospheres of late-L dwarfs. Knapp et al. (2004) showed that this could be explained by an increase in sedimentation at $T_{eff} \sim 1,300$ K, dubbed the "Hilo rain" scenario (Fig. 15.7). Another possible explanation is holes between the clouds, as seen in Jupiter's atmosphere.

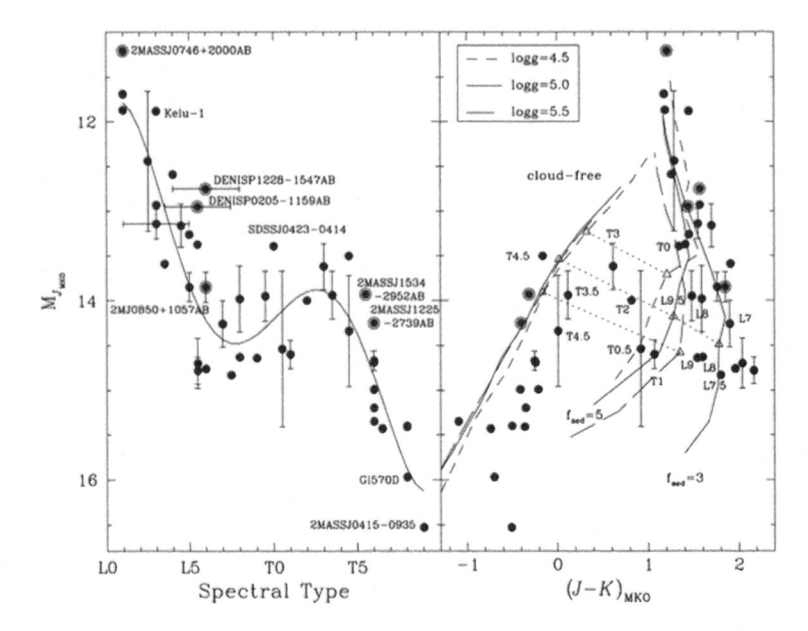

Fig. 15.7 M_J as a function of type and colour from Knapp et al. (2004). The *left panel* shows a fit to the observed trend with type. The *right panel* shows cloudy and cloud-free models and illustrates how the change in colour from late-L to early-T can be reproduced by connecting increasingly cloud-free models at a constant temperature

UKIRT and Non-equilibrium Chemistry: 1997–2004

Noll et al. (1997) used CGS4 at 5 μm to show that the photosphere of GL 229B had an unexpectedly large amount of CO. Leggett et al. (2002) and Golimowski et al. (2004) used IRCAM and UIST M-band photometry to show that this is a universal feature of T dwarfs. Enhanced CO and N_2, and diminished CH_4 and NH_3, can be explained by vertical mixing in the atmosphere. The very stable CO and N_2 can be dredged up, leading to changes in abundance of one to several orders of magnitude.

The mixing can be modelled as turbulence with timescale H^2/K where H is the pressure scale height and K is the coefficient of diffusion. The data shows that $K \sim 10^2 - 10^6$ cm^2/s, implying fast mixing timescales of 10 year to 1 h respectively. The factor ∼2 loss in flux at 5 μm is critical for mid-infrared space missions like Spitzer and WISE.

UKIRT, Benchmark Dwarfs and the T_{eff} Scale: 2001–2008

Following the pioneering work on GL 229B, UKIRT continued to make large contributions to the study of brown dwarf companions to main-sequence stars. Such objects are very important, as the main-sequence star constrains metallicity

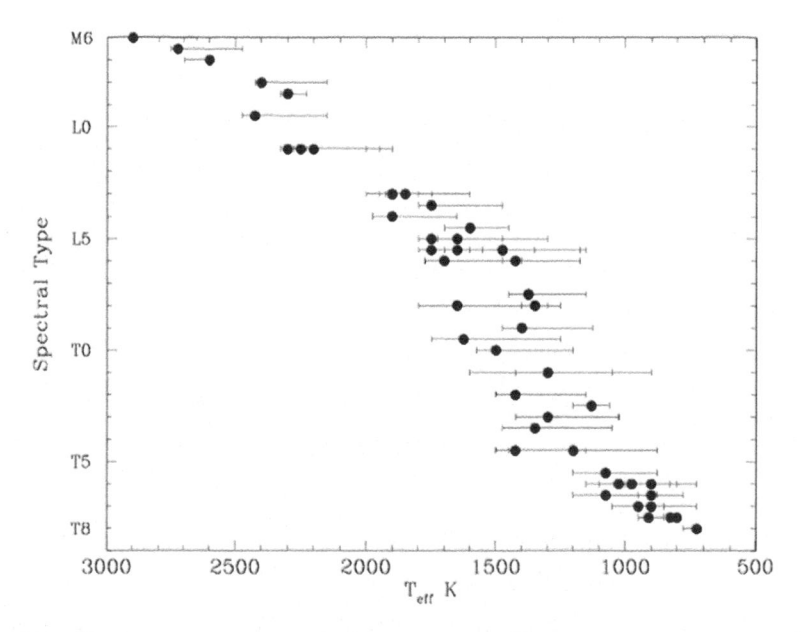

Fig. 15.8 Effective temperature scale for L and T dwarfs (Golimowski et al. 2004). Error bars reflect the uncertainty in age for each dwarf

and age, and hence mass, for the brown dwarf, which otherwise is difficult or impossible to do.

For example, Geballe et al. (2001) used UFTI, IRCAM and CGS4 to study GL 570D, a T7.5 dwarf in a K and M dwarf system. They found $T_{eff} = 800 \pm 20$ K, and mass 40–70 Jupiter masses, showing this to be the coolest object known in these early days of T dwarf discoveries. Mugrauer et al. (2006) used UFTI to confirm a proper motion companion to the K0V star HD 3651, the first brown dwarf companion to a planet-hosting star. The 40–70 Jupiter mass T7.5 dwarf is 500 AU from the star, with $T_{eff} = 810 \pm 30$ K. Burningham et al. (2009) found the coolest companion to a main sequence star to date, in UKIDSS data: the T8.5 Wolf 940B is 400 AU from the M4V star and has $T_{eff} = 575 \pm 25$ K, and mass 20–30 Jupiter masses (Warren/Burningham these proceedings).

Note the high degree of accuracy in the values of T_{eff} above. This is made possible by the tight relationship between luminosity and T_{eff} (S. 1). An accurate measurement of luminosity provides an accurate measurement of T_{eff}, thanks to the small dependency of radius on mass and age, and the well-understood structural models of brown dwarfs. Figure 15.8 shows the resulting relationship between type and T_{eff}. UKIRT data contributes significantly to this by providing 3–5 μm photometry, data difficult to get at any other ground-based telescope, and where there is significant flux (Golimowski et al. 2004).

Summary

Ten years ago, evidence of a local population of cold brown dwarfs was finally found, 36 years after their existence was hypothesized. The population has been classified into L dwarfs with $T_{eff} \sim 1{,}350$–$2{,}200$ K ($L \sim 10^{-5}$–10^{-4} $L_\odot$) and T dwarfs with $T_{eff} \sim 500$–$1{,}350$ K ($L \sim 10^{-6}$–10^{-5} $L_\odot$). UKIRT provided fast response, high quality, easily reduced, well understood and quickly published data that provided, and continues to provide, the core dataset for this rapidly advancing field. Highlights include:

- Spectroscopy which defined the classification schemes for L and T dwarfs.
- 3–5 μm data (for which Mauna Kea is ideal) which demonstrated turbulent atmospheres and unexpected chemical abundances.
- Near-infrared imaging data enabling many discoveries, including the first examples of isolated field L and T dwarfs: Kelu-1 in 1997 and SDSS 1624 + 00 in 1999. UKIDSS has recently found 500–600 K objects, which suggests that we are at the brink of discovering the next spectral class; soon this UKIRT survey will have found more T dwarfs than 2MASS and SDSS combined.

Without UKIRT, the study of the initial population of brown dwarfs would have been set back several years; it is not clear that the mid-infrared analysis would have been achieved at all prior to the launch of the *Spitzer* telescope. Without the UKIRT Infrared Deep Sky Survey we would still be waiting to find objects with luminosity a millionth that of the Sun, and with very planet-like temperatures possibly as cold as 500 K. The astronomical community, and this author, is hugely indebted to UKIRT and its staff for making these achievements possible.

Acknowledgments I am very grateful to the UKIRT staff, the Sloan Digital Sky Survey and UKIRT Infrared Deep Sky Survey teams, and to my collaborators, in particular Tom Geballe of Gemini Observatory, Dave Golimowski and Keith Noll of STScI, Jill Knapp of Princeton University, Mark Marley of NASA Ames, David Pinfield, Ben Burningham and Phil Lucas of the University of Hertfordshire, Didier Saumon of the Los Alamos National Laboratory and Steve Warren of Imperial College London. SKL is supported by the Gemini Observatory, which is operated by the Association of Universities for Research in Astronomy, Inc., on behalf of the international Gemini partnership of Argentina, Australia, Brazil, Canada, Chile, the United Kingdom, and the United States of America.

References

Ackerman, A.S., Marley, M.S.: Astrophys. J. **556**, 827 (2001)
Becklin, E.E., Zuckerman, B.: Nature **336**, 656 (1988)
Becklin, E.E., Zuckerman, B.: Astrophys. J. **386**, 260 (1992)
Burgasser, A.J., Geballe, T.R., Leggett, S.K., Kirkpatrick, J.D., Golimowski, D.A.: Astrophys. J. **637**, 1067 (2006)
Burningham, B., et al.: Mon. Not. R. Astron. Soc. **395**, 1237 (2009)
Burrows, A., Hubbard, W.B., Lunine, J.I., Liebert, J.: Rev. Mod. Phys. **73**, 719 (2001)

Epchtein, N.: Exp. Astron. **3**, 73 (1994)
Geballe, T.R., Kulkarni, S.R., Woodward, C.E., Sloan, G.C.: Astrophys. J. **467**, L101 (1996)
Geballe, T.R., Saumon, D., Leggett, S.K., Knapp, G.R., Marley, M.S., Lodders, K.: Astrophys. J. **556**, 373 (2001)
Geballe, T.R., et al.: Astrophys. J. **564**, 466 (2002)
Golimowski, D.A., et al.: Astron. J. **127**, 1733 (2004)
Hayashi, C., Nakano, T.: Prog. Theor. Phys. **30**, 460 (1963)
Hodgkin, S.T., Pinfield, D.J., Jameson, R.F., Steele, I.A., Cossburn, M.R., Hambly, N.C.: Mon. Not. R. Astron. Soc. **310**, 87 (1999)
Jameson, R.F., Dobbie, P.D., Hodgkin, S.T., Pinfield, D.J.: Mon. Not. R. Astron. Soc. **335**, 853 (2002)
Jones, H.R.A., Longmore, A.J., Jameson, R.F., Mountain, C.M.: Mon. Not. R. Astron. Soc. **267**, 413 (1994)
Kirkpatrick, J.D., et al.: Astrophys. J. **519**, 802 (1999)
Knapp, G.K., et al.: Astron. J. **127**, 3553 (2004)
Kumar, S.S.: Astrophys. J. **137**, 1126 (1963)
Lawrence, A., et al.: Mon. Not. R. Astron. Soc. **379**, 1599 (2007)
Leggett, S.K., Allard, F., Hauschildt, P.H.: Astrophys. J. **509**, 836 (1998)
Leggett, S.K., et al.: Astrophys. J. **536**, L35 (2000)
Leggett, S.K., Allard, F., Geballe, T.R., Hauschildt, P.H., Schweitzer, A.: Astrophys. J. **548**, 908 (2001)
Leggett, S.K., et al.: Astrophys. J. **564**, 452 (2002)
Lodieu, N., Hambly, N.C., Jameson, R.F.: Mon. Not. R. Astron. Soc. **373**, 95 (2006)
Lucas, P.W., Roche, P.F.: Mon. Not. R. Astron. Soc. **314**, 858 (2001)
Mugrauer, M., Seifahrt, A., Neuhauser, R., Mazeh, T.: Mon. Not. R. Astron. Soc. **373**, L31 (2006)
Nakajima, T., Oppenheimer, B.R., Kulkarni, S.R., Golimowski, D.A., Matthews, K., Durrance, S.T.: Nature **378**, 463 (1995)
Noll, K.S., Geballe, T.R., Marley, M.S.: Astrophys. J. **489**, L87 (1997)
Noll, K.S., Geballe, T.R., Leggett, S.K., Marley, M.S.: Astrophys. J. **541**, L75 (2000)
Reid, I.N., Burgasser, A.J., Cruz, K.L., Kirkpatrick, J.D., Gizis, J.E.: Astron. J. **121**, 1710 (2001)
Ruiz, M.R., Leggett, S.K., Allard, F.: Astrophys. J. **491**, L107 (1997)
Skrutskie, M.F., et al.: Astron. J. **131**, 1163 (2006)
Strauss, M.A., et al.: Astrophys. J. **522**, L61 (1999)
Tinney, C.G., Mould, J.R., Reid, I.N.: Astron. J. **105**, 1045 (1993)
York, D.G., et al.: Astron. J. **120**, 1579 (2000)

Chapter 16
White Dwarfs in UKIDSS

P.R. Steele, M.R. Burleigh, J. Farihi, B. Gänsicke, R.F. Jameson, P.D. Dobbie, and M.A. Barstow

Abstract We present a near-infrared photometric search for low mass stellar and substellar companions, and for debris disks around white dwarfs in the UKIDSS Large Area Survey DR5. A cross correlation of the SDSS DR4 and McCook and Sion catalogues of white dwarfs gave near-infrared photometry for 1,161 stars. Atmospheric models are fitted to the optical and near-infrared photometry of the matched stars to identify those with a photometric excess consistent with an unresolved companion or debris disk. In total 7 new potential white dwarf + brown dwarf binary candidates have been identified. We present follow-up spectroscopy of a previously identified candidate; PHL5038. We confirm that this system is a resolved white dwarf + L8 brown dwarf binary.

Introduction

The detection and observation of ultracool companions to white dwarfs (WDs) allows us to investigate an extreme of binary formation and evolution. WD + brown dwarf (BD) binaries also allow us to investigate the fraction of main-sequence stars with substellar companions. This will give us further insight into the possible discrepancy between the BD companion fraction at small separations

---

P.R. Steele (✉) • M.R. Burleigh • J. Farihi • R.F. Jameson • M.A. Barstow
Department of Physics and Astronomy, University of Leicester, Leicester Road, Leicester LE1 7RH, UK
e-mail: prsteele@outlook.com

B. Gänsicke
Department of Physics, University of Warwick, Coventry CV4 7AL, UK

P.D. Dobbie
Anglo-Australian Observatory, P.O. Box 296, Epping, Sydney, NSW 1710, Australia

(1–10 % at <10 AU, Marcy and Butler 2000; McCarthy and Zuckerman 2004) and large radii (10–30 % at $a > 1,000$ AU, Gizis et al. 2001). In addition there are very few observational constraints on BD evolutionary models at large ages, such as those expected for most white dwarfs (>1 Gyr). Thus wide, detached WD + BD binaries can provide benchmarks for testing these models (Pinfield et al. 2006).

Near-infrared (NIR) photometry and spectroscopy allows for the detection and study of such systems. The spectral energy distribution of a WD in this region is markedly different to that of a BD, with the latter peaking at NIR wavelengths. As the WD falls off in flux, the two components of a binary system are easily separated in broadband photometry facilitating detection and enabling straightforward spectroscopic follow-up.

However, BD companions to WDs are rare (Farihi et al. 2005), with only three systems confirmed prior to this work.; GD165 (DA + L4, Becklin and Zuckerman 1988), GD1400 (DA + L6-7, Farihi and Christopher 2004; Dobbie et al. 2005) and WD0137-349 (DA + L8, Burleigh et al. 2006; Maxted et al. 2006). Large, deep NIR surveys such as UKIDSS are therefore an excellent place to search for these objects.

Identifying White Dwarfs in UKIDSS

The UKIDSS Large Area Survey (LAS) was cross correlated with two catalogues of WDs: the on-line 2006 version of the McCook and Sion catalogue of spectroscopically identified WDs (McCook and Sion 1999), which contains 5,557 stars, and the Eisenstein et al. (2006) catalogue of 9,316 spectroscopically confirmed WDs from SDSS DR4. The cross correlation produced 1,161 matches with SDSS DR4 and 489 with the McCook and Sion WD catalogues. Only 106 of the McCook and Sion WDs were not found in the SDSS.

Identifying White Dwarfs with Near-Infrared Excesses

In order to identify WDs with NIR excesses we decided to model each star independently and then inspect each for evidence of NIR excess. Only those hydrogen (DA) and helium (DB) atmosphere WDs with both H and K-band photometry were modelled, as a substellar companion will give only a small or non-detectable excess at Y and J, depending on the temperature of the components. Also, H and K-band photometry together allow substellar companions and blackbody-like dust disk emission to be distinguished from each other.

We use Pierre Bergeron's cooling models for DA and DB WDs (Bergeron et al. 2001) to predict the WD absolute magnitudes for the SDSS i' and the

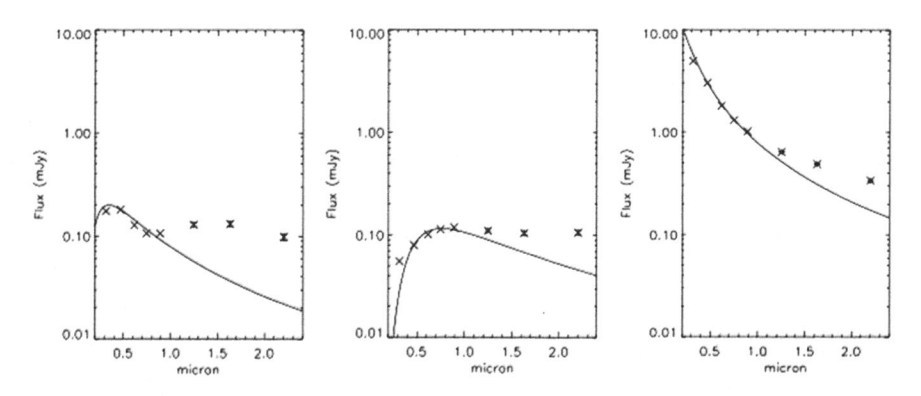

Fig. 16.1 Example output from an automated attempt to identify stars with excess emission. The *solid line* is the WD cooling model, and the crosses are the SDSS *ugriz'* and UKIDSS *JHK* photometry

2MASS JHK_S filters, as well as the estimated distance to the WD and cooling age of the system. The 2MASS magnitudes were converted to the corresponding UKIDSS filters using the colour transformations of Carpenter (2001). Out of the 1,037 WDs classified by Eisenstein et al. and McCook and Sion as DA WDs, 639 had both H and K-band photometry and fell within the DA model grid. Only 10 DB WDs had both H and K-band photometry and fell within the DB model grid.

Candidate substellar companions and dust disks were identified as stars showing >3σ excess emissions in both the UKIDSS H and K-bands, or the K-band only, when compared to the model predicted values. 235 such excesses were identified, with 206 of these accounted for as previously identified DA + dM and DB + dM type binaries from optical excess in SDSS (or other optical data) (Fig. 16.1).

Spectral Typing Candidate Substellar Companions

In order to estimate a spectral type for the putative companions, empirical models for ultracool objects were added to synthetic WD spectra and compared to the UKIDSS photometry. For the WDs pure-H or pure-He synthetic spectra were constructed using the TLUSTY and SYNSPEC atmospheric and spectral synthesis codes. The empirical models for the ultracool companions were constructed using the NIR spectra of late M-dwarfs from the IRTF spectral library (Cushing et al. 2005, Rayner et al. 2009), and L and T-dwarfs presented by McLean (2003). Examples are shown in Figs. 16.2, 16.3, and 16.4.

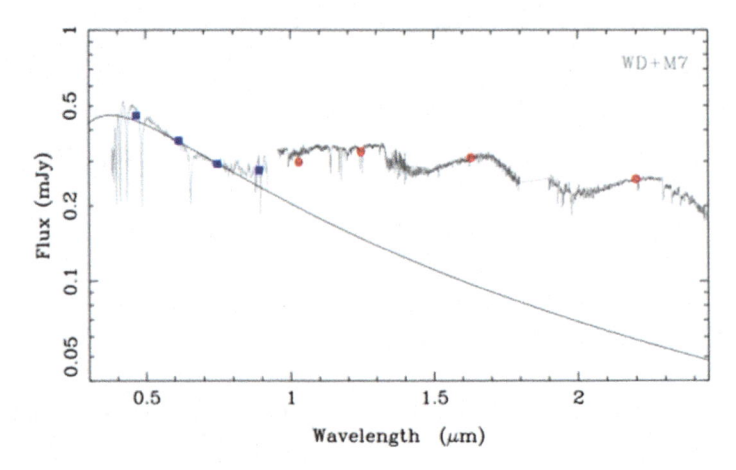

Fig. 16.2 An example of a candidate WD + late M-dwarf binary system (Graph shows the SDSS griz' (*blue*) and UKIDSS YJHK (*red/purple*) photometry, a WD synthetic spectrum (*black*) and the SDSS optical spectrum (*grey*))

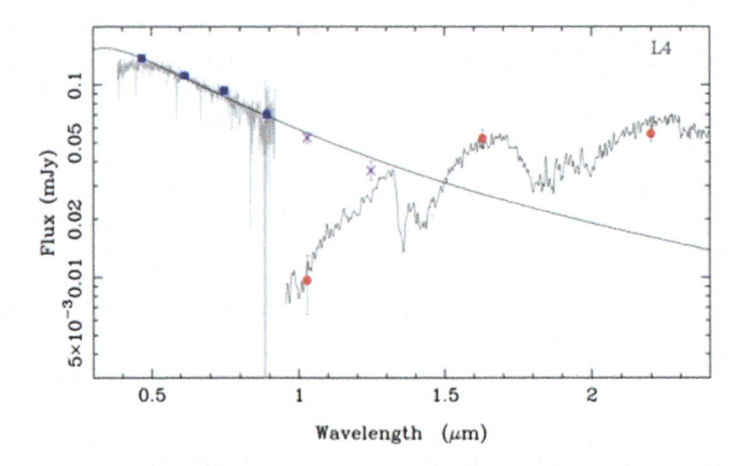

Fig. 16.3 An example of a candidate WD + L-dwarf binary system (Graph shows the SDSS griz' (*blue*) and UKIDSS YJHK (*red/purple*) photometry, a WD synthetic spectrum (*black*) and the composite WD + Companion model (*grey*))

Summary of Results from UKIDSS DR5

Out of the 29 candidate WDs with an apparent NIR excess, 24 had multiple excesses indicative of an ultracool companion. Seven of these having a predicted mass in the range associated with BDs, and a further 13 likely very low mass stellar companions.

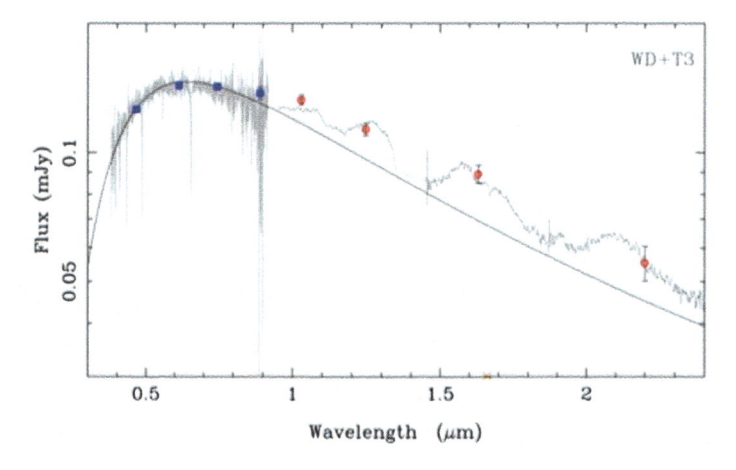

Fig. 16.4 An example of a candidate WD + T-dwarf binary system

Two WDs were identified with putative companions that are likely spectral class K. Four of the sample showed evidence of background or foreground contamination. The remaining three WDs each had a K-band excess indicative of a debris disk.

PHL5038: A Resolved White Dwarf + Brown Dwarf Binary

PHL5038 was first reported by Haro and Luyten (1962) in a photographic search for faint blue stars near the South Galactic Pole. The star was recovered in the SDSS as SDSSJ222030.68-004107.3, a DA WD with $T_{\text{eff}} \approx 8{,}037$ K and $\log g = 8.28$. The UKIDSS DR5 photometry for PHL5038 gave an excess of 7-10σ in the H and K-bands when compared to the predicted values for the WD (Fig. 16.5). Thus, this object was included in a Gemini programme to clarify the nature of the NIR excess.

NIR grism spectroscopy was obtained of PHL5038 at Gemini North using NIRI (Hodapp et al. 2003). During the first acquisition images, the science target was spatially resolved into two components separated by 0.94″ (Fig. 16.6), and the slit was placed over both sources for the subsequent observations. For a detailed description of all observations and data reduction see Steele et al. (2009).

Figure 16.7 shows the fully processed spectrum of PHL5038B. To determine the spectral type, the standard spectral indices for late L and T-dwarfs (Burgasser et al. 2006) were measured (Table 2 of Steele et al. 2009). These values indicated the spectral type is most likely L8-L9.

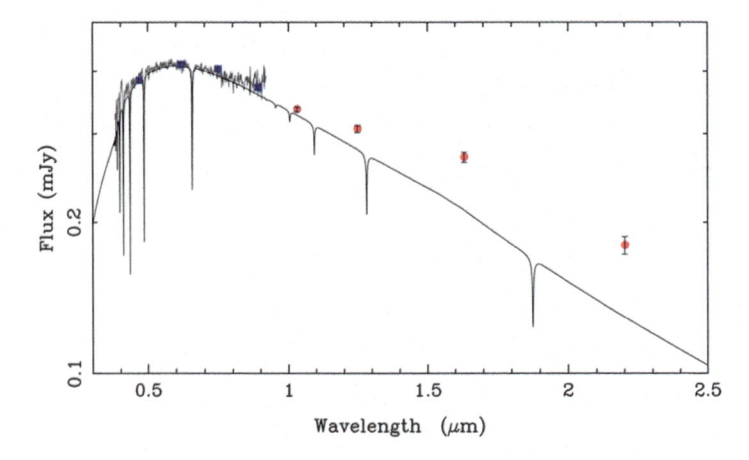

Fig. 16.5 The SDSS and UKIDSS photometry which initially identified PHL5038 as a candidate binary system. Symbols and colours are as in Figs. 16.2 and 16.3

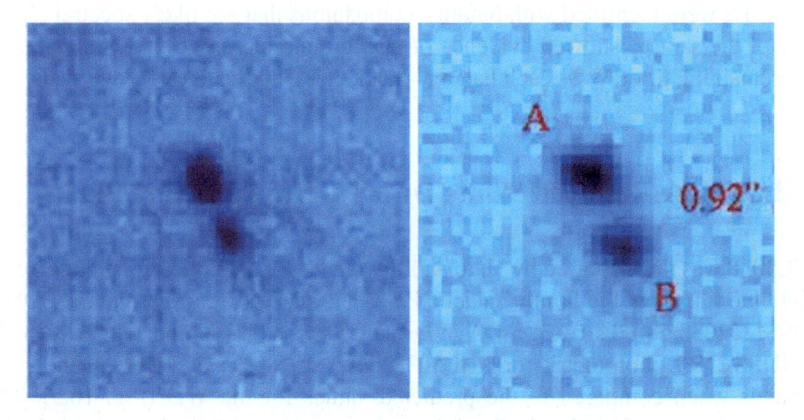

Fig. 16.6 NIR *H* (*left*) and *K*-band (*right*) acquisition images of PHL5038 taken with Gemini + NIRI immediately prior to grism spectroscopy. Frames are 6″ across with North up and East left. The BD companion, PHL5038B is located 0.94″ from A at position angle 293.2″

The 0.94 arcsec angular separation of the binary, at the estimated distance to the WD of 64pc, equates to a projected orbital separation of 55 AU. This makes PHl5038 only the second resolved WD + BD binary to be discovered, the first being GD165 over 20 years ago. At an age of 2.3 ± 0.4 Gyr, PHL5038B is a potential benchmark BD for evolutionary and atmospheric models at older ages (Steele et al. 2009). We have recently obtained high S/N optical spectra of the WD to more precisely constrain this age estimate.

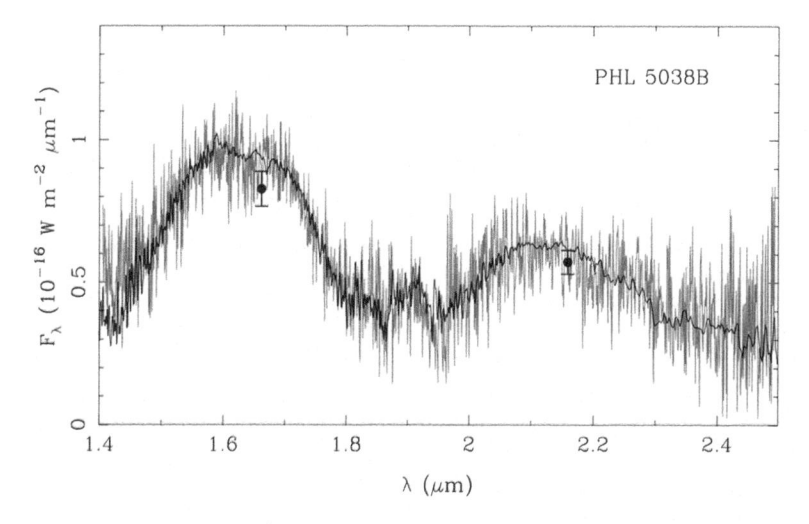

Fig. 16.7 Observed *HK* spectrum (*grey*) of PHL5038B plotted with photometry measured from the acquisition images. An L8 spectrum (2MASS1632 + 1904) has been scaled to match the flux of the secondary and over-plotted (*black*) for comparison

References

Baraffe, I., Chabrier, G., Allard, F., Hauschildt, P.H.: Evolutionary models for low-mass stars and brown dwarfs: uncertainties and limits at very young ages. Astron. Astrophys. **382**, 563 (2002)
Becklin, E.E., Zuckerman, B.: A low-temperature companion to a white dwarf star. Nature **336**, 656 (1988)
Bergeron, P., Leggett, S.K., Ruiz, M.T.: Photometric and spectroscopic analysis of cool white dwarfs with trigonometric parallax measurements. Astrophys. J. Suppl. **133**, 413 (2001)
Burgasser, A.J., Geballe, T.R., Leggett, S.K., Kirkpatrick, J.D., Golimowski, D.A.: A unified near-infrared spectral classification scheme for T dwarfs. Astrophys. J. **637**, 1067 (2006)
Burleigh, M.R., Hogan, E., Dobbie, P.D., Napiwotski, R., Maxted, P.F.L.: A near-infrared spectroscopic detection of the brown dwarf in the post common envelope binary WD0137-349. Mon. Not. R. Astron. Soc. **373**, L55 (2006)
Carpenter, J.M.: Color transformations for the 2MASS second incremental data release. Astron. J. **121**, 2851 (2001)
Chabrier, G., Baraffe, I., Allard, F., Hauschildt, B.: Evolutionary models for very low-mass stars and brown dwarfs with dusty atmospheres. Astrophys. J. **542**, 464 (2000)
Cushing, M.C., Rayner, J.T., Vacca, W.D.: An infrared spectroscopic sequence of M, L, and T dwarfs. Astrophys. J. **623**, 1115 (2005)
Dobbie, P.D., et al.: A near-infrared spectroscopic search for very-low-mass cool companions to notable DA white dwarfs. Mon. Not. R. Astron. Soc. **357**, 1049 (2005)
Eisenstein, D.J., et al.: A catalog of spectroscopically confirmed white dwarfs from the sloan digital sky survey data release 4. Astrophys. J. Suppl. **167**, 40 (2006)
Farihi, J., Christopher, M.: A possible brown dwarf companion to the white dwarf GD 1400. Astron. J. **128**, 1868 (2004)
Farihi, J., Becklin, E.E., Zuckerman, B.: Low-luminosity companions to white dwarfs. Astrophys. J. Suppl. **161**, 394 (2005)

Gizis, J.E., Kirkpatrick, J.D., Burgasser, A., Reid, I.N., Monet, D.G., Liebert, J., Wilson, J.C.: Substellar companions to main-sequence stars: no brown dwarf desert at wide separations. Astrophys. J. **551**, 163 (2001)

Haro, G., Luyten, W.J.: Faint blue stars in the region near the South Galactic Pole. Boletin del Instituto de Tonantzintla **3**, 37 (1962)

Hodapp, K.W., et al.: The Gemini near-infrared imager (NIRI). Publ. Astron. Soc. Pac. **155**, 1388 (2003)

Marcy, G.W., Butler, R.P.: Planets orbiting other suns. Publ. Astron. Soc. Pac. **112**, 137 (2000)

Maxted, P.F.L., Napiwotzki, R., Dobbie, P.D., Burleigh, M.R.: Survival of a brown dwarf after engulfment by a red giant star. Nature **442**, 543 (2006)

McCarthy, C., Zuckerman, B.: The brown dwarf desert at 75-1200 AU. Astron. J. **127**, 2871 (2004)

McCook, G.P., Sion, E.M.: A catalog of spectroscopically identified white dwarfs. Astrophys. J. Suppl. **121**, 1 (1999)

McLean, I.S.: Near-infrared spectroscopic survey of brown dwarfs using NIRSPEC on the Keck II Telescope. Proc. SPIE **4834**, 111 (2003)

Pinfield, D.J., Jones, H.R.A., Lucas, P.W., Kendall, T.R., Folkes, S.L., Day-Jones, A.C., Chapelle, R.J., Steele, I.A.: Finding benchmark brown dwarfs to probe the substellar initial mass function as a function of time. Mon. Not. R. Astron. Soc. **368**, 1281 (2006)

Rayner, J.T., Cushing, M.C., Vacca, W.D.: The Infrared Telescope Facility (IRTF) spectral library: cool stars. Astrophys. J. Suppl. **185**, 289 (2009)

Steele, P.R., Burleigh, M.R., Farihi, J., Gansicke, B.T., Jameson, R.F., Dobbie, P.D., Barstow, M.A.: PHL 5038: a spatially resolved white dwarf + brown dwarf binary. Astron. Astrophys. **500**, 1207 (2009)

Chapter 17
Discovery of Variables in WFCAM and VISTA Data

Nicholas Cross, Nigel Hambly, Ross Collins, Eckhard Sutorius, Mike Read, and Rob Blake

Abstract The advent of WFCAM on UKIRT has allowed high-resolution (sub-arcsecond), multi-epoch surveys of medium sized (tens of square degrees) areas of sky in the near infrared for the first time. At the Wide Field Astronomy Unit in Edinburgh we have built a science archive to allow users easy access to all WFCAM data: UKIDSS surveys, large campaigns and PI data. As part of our work we have designed and built tables and software that allow users to work with multi-epoch data: a table of motions and statistics that allows users to select objects that are variable in different ways and a table that allows easy generation of light-curves. We will process VISTA data in much the same way, but VISTA will extend these surveys to cover hundreds of square degrees.

The WFCAM Science Archive

The WFCAM Science Archive[1] (Hambly et al. 2008) is the main access for users of WFCAM data. WFCAM data forms the 5 UKIDSS surveys (Lawrence et al. 2007), various medium sized campaigns and multiple small PI programmes. Data from WFCAM (and VISTA) are processed through the VISTA Data Flow System (VDFS), which comprises pipeline processing and archiving. The first step is done by the Cambridge Astronomy Survey Unit (CASU).[2] CASU do the

---

[1] http://surveys.roe.ac.uk/wsa
[2] http://www.ast.cam.ac.uk/vdfs/

N. Cross (✉) • N. Hambly • R. Collins • E. Sutorius • M. Read • R. Blake
Scottish Universities' Physics Alliance; Institute for Astronomy, University of Edinburgh, Royal Observatory, Edinburgh, EH9 3HJ, UK

Institute for Astronomy, University of Edinburgh, Royal Observatory, Edinburgh EH9 3HJ, UK
e-mail: njc@roe.ac.uk

data reduction, sky subtraction, catalogue extraction on an observing block basis and do the main calibration of WFCAM data via comparison to the 2MASS survey (Hodgkin et al. 2009). The data are then transferred to the Wide Field Astronomy Unit in Edinburgh (WFAU) for ingestion into the archive. We then put together the data for the same programme taken in different observing blocks, often on different nights and produce deep stacks, merged filter source tables, neighbour tables (Hambly et al. 2008) and synoptic tables that bring together all the observations of an object and the statistical properties of those observations (Cross et al. 2009).

All programmes with multiple observations of the same pointing in the same filter are now processed as synoptic data-sets (Cross et al. 2009): the main surveys are processed manually, with much input from the survey teams, whereas campaigns and PI programmes are completely automated (Collins et al. 2009).

Science from Multi-epoch Data

The scientific uses of multi-epoch data are vast. Multi-epoch WFCAM data has been taken for calibration purposes (Hodgkin et al. 2009), to find planets around M-dwarf stars, to understand young stellar objects in star-forming regions (Alves de Oliviera and Casali 2008) and will be used to separate brown dwarfs from high-redshift quasars since the former have measurable proper motion. VISTA has more ambitious surveys, which expect to detect supernovae (VIDEO) in distant galaxies, and to determine the distance to structures of the Magellanic Clouds (VMC) and the galactic plane and bulge (VVV) using periodic variables such as RR-Lyrae and Cepheid stars. Users will no doubt use WFCAM and VISTA data to do science not envisaged in the original proposals, so the science archives need to be able to support the scientific requirements of the surveys as well as be flexible enough for users to make discoveries of new unknown objects. Unfortunately, transient objects that need rapid follow-up, such as supernovae and microlenses, will need to be identified at the telescope, because there is a delay of at least 6 weeks between observations and access to data through the science archives. The delay is due to the careful calibration and quality control that takes place between observations and access to images.

The synoptic archive is designed to be flexible and play to the strengths of the data-flow system, which include good quality control and calibration. The design made sure that all observations were tracked, because missing observations can indicate important changes (motion, fading, or deblending with another source). We match sources based on astrometry only: no magnitude information, because objects may vary in brightness in a variety of ways by many magnitudes. However, the motion is simpler, since stars and galaxies are either static or have approximately linear motion for objects beyond the Solar System. Objects within the Solar System will follow well prescribed arcs. Sufficiently nearby objects will also show signs of parallax, but this is well defined, and the only unknown is the distance to the object. Finally, since our data-flow system will pick out transients too late for rapid follow up, but does have good calibration, we concentrate our efforts on statistics to pick out low-amplitude variables and separating these from noise.

Synoptic Tables

The synoptic archive has been available since UKIDSS Data Release 5. The Deep Extragalactic Survey and Ultra Deep Survey each came with three new synoptic tables to allow users to more easily select objects with interesting time variability and to more easily view the light curves. These three tables are:

- *SourceXDetectionBestMatch* – this table links the Source table, a list of unique sources produced from deep stacks to each individual observation, recorded in the Detection table. If an observation does not include a detection at the correct position, this table includes a default row, so that the user knows that there should have been an observation.
- *Variability* – this table gives statistical information on the observations in the BestMatch table. It includes both astrometric and photometric quantities, data on the number of good, flagged and missing detections, data on the typical interval between observations and classification of the source.
- *VarFrameSetInfo* – this table gives information on the noise characteristics of the data in a particular pointing.

A more detailed description and implementation of these tables can be found in Cross et al. (2009). We use the statistical attributes and noise model from Sesar et al. (2007) and the aperture selection and relative photometry used in Irwin et al. (2007). We calculate basic statistical attributes of the good photometric data for each source such as the mean, median, standard deviation, median-absolute-deviation, skewness, minimum and maximum, but we also calculate the expected-RMS from a noise model fitted to the median-RMS of data binned by magnitude. From this we calculate the intrinsic-RMS of the source by subtracting the expected-RMS from the RMS in quadrature. We also calculate the χ^2 for the hypothesis that the data are not-variable and the probability of variability. We classify objects in each band based on the ratio of intrinsic-RMS to expected-RMS and probability of variability, and classify across bands using the ratio of intrinsic-RMS to expected-RMS and the number of good observations.

We have some programmes where all the filters are observed close together in time, so that colour variations can be monitored. The WFCAM standard star programme is an example of this, where the ZYJHK and occasionally narrow-band data are observed within 10–15 min of each other. Typically the same field would then be re-observed a day later. In these programmes, we merge the different filter observations, using the same code as we use to produce the Source table, except that frames are merged with a maximum duration between the observations. We produce two new tables:

- *SynopticMergeLog* – this is the table of matching frames for each epoch
- *SynopticSource* – the table of merged filter detections at each epoch.

In this case we replace the SourceXDetectionBestMatch with SourceXSynopticSourceBestMatch. Users can produce colour light curves, which can help distinguish between different types of variables such as microlensing which has no colour variation, pulsating variables, which have changes in opacity and colour associated with different parts of the cycle.

During processing, we recalibrate all individual epoch measurements compared to the deep stack to improve the relative photometry.

Using the Science Archives

The magnitude-RMS plot in Fig. 17.1 shows the K-band data in the DXS. This plot demonstrates how we fit a noise model and classify objects as variable or not. Plots like this and the light curves from the WFCAM standard star data (Fig. 17.2) can easily be produced using SQL queries of the database.[3]

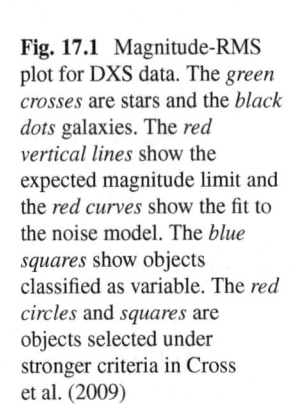

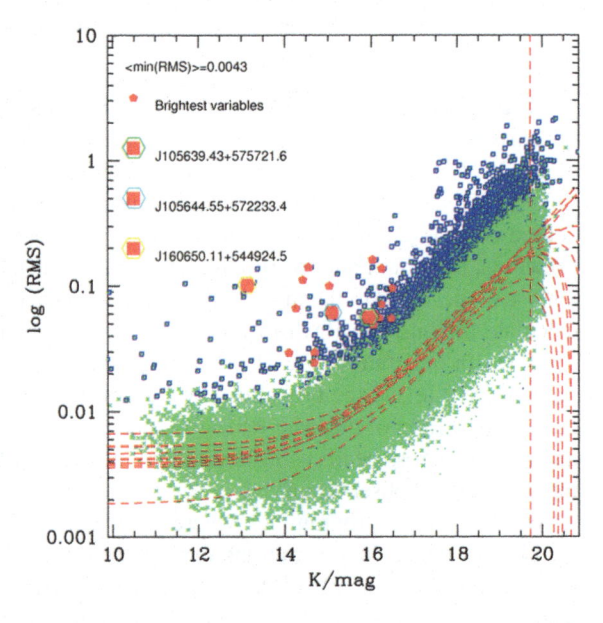

Fig. 17.1 Magnitude-RMS plot for DXS data. The *green crosses* are stars and the *black dots* galaxies. The *red vertical lines* show the expected magnitude limit and the *red curves* show the fit to the noise model. The *blue squares* show objects classified as variable. The *red circles* and *squares* are objects selected under stronger criteria in Cross et al. (2009)

---

[3]http://surveys.roe.ac.uk/wsa/sqlcookbook.html#Multi-epoch

Fig. 17.2 *Light curves* of a variable from the WFCAM standard star data. The variations are correlated as expected for a real variable

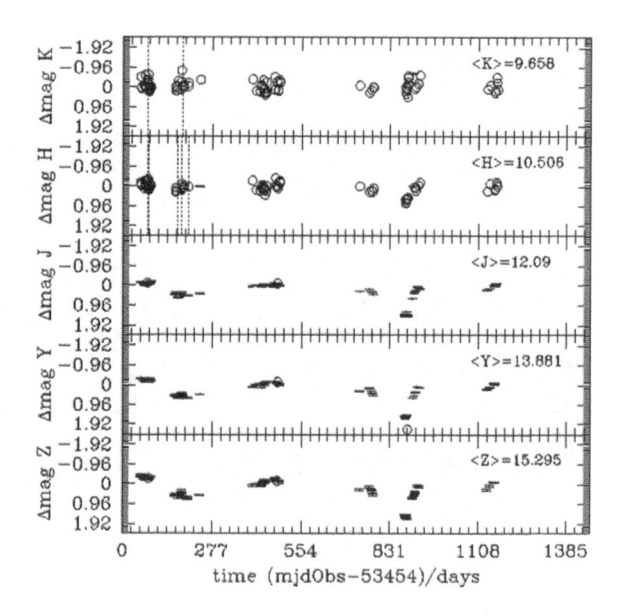

To select objects based on their variability properties, use the Variability table: e.g.

SELECT v.sourceID FROM dxsVariability as v,dxsVarFrameSetInfo as i WHERE i.frameSetID = v.frameSetID and knGoodObs > =20 and variableClass = 1 and kMeanMag < (i.kexpML-3.)

This selects sources that are classed as variable, have at least 20 K-band observations and have a mean K-band magnitude that is at least 3-magnitudes brighter than the expected magnitude limit. The light curves can easily be produced using the BestMatch table along with Multiframe and the Detection table:

SELECT m.mjdObs,d.aperMag3,d.aperMag3Err,d.ppErrBits,b.seqNum,b.flag,b.modelDistSecs FROM udsSourceXDetectionBestMatch as b, udsDetection as d, Multiframe as m WHERE m.multiframeID = d.multiframeID and b.sourceID = 450971566150 and b.multiframeID = d.multiframeID and b.extNum = d.extNum and b.seqNum = d.seqNum order by m.mjdObs

The ppErrBits, flag, modelDistSecs and seqNum attributes give additional information, such as whether the object has data quality issues, or is missing.

VISTA Surveys

The VISTA telescope will survey at more than three times the speed of WFCAM. The VISTA Public Surveys (Fig. 17.3) will commence in January 2010. The VISTA Variables in Via Lactea (VVV), VISTA Magellanic Clouds (VMC) and VISTA

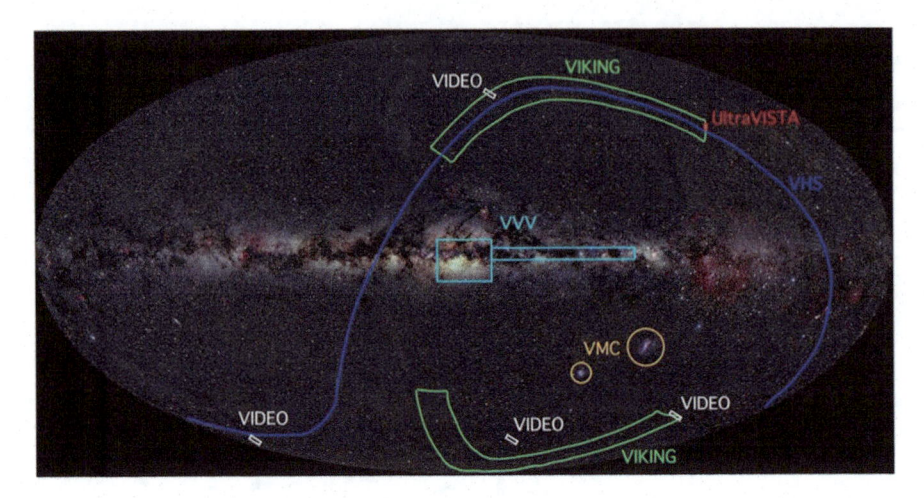

Fig. 17.3 Plot of VISTA Public Surveys

Deep Extragalactic Observations (VIDEO) are all synoptic surveys which will be archived by WFAU. VVV in particular will cover 500$\Box^\circ$ and will include 109 sources observed in 100 epochs, so the BestMatch table may include up to 1,011 rows. These surveys will test the synoptic pipeline, but we are confident that our methods will be fast enough and reliable enough to cope. For surveys of these sizes, it is vital to be able to select out different types of variable reliably.

Summary and Future Work

Archiving of multi-epoch data has been accomplished in the WFCAM Science Archive since Data Release 5 of UKIDSS (April 2009). This uses a new scheme and pipeline (Cross et al. 2009), which will be used on deep surveys and PI programmes alike and will be available for VISTA data from the outset.

Future improvements to this work will include the full astrometric model that hasn't been applied yet. We will also improve the classification so that types of variable can be differentiated from each other. This will use some of the different statistical measures already in place and may eventually use Fourier analysis. We will also look at using difference imaging in crowded regions.

References

Collins, R.S., Cross, N.J.G., Hambly, N.C., Sutorius, E.T.W., Blake, R.P., Read, M.A.: Automated data releases from the WFCAM science archive. In: Bohlender, D.A., Durand, D., Dowler, P. (eds.) Astronomical Data Analysis Software and Systems XVIII ASP Conference Series, vol. 411, proceedings of the conference held 2–5 November 2008 at Hotel Loews Le Concorde, Québec City, QC, Canada, p. 226. Astronomical Society of the Pacific, San Francisco (2009)

Cross, N.J.G., et al.: Archiving multi-epoch data and the discovery of variables in the near-infrared. Mon. Not. R. Astron. Soc. **399**, 1730 (2009)

Hambly, N.C., et al.: The WFCAM science archive. Mon. Not. R. Astron. Soc. **384**, 637 (2008)

Hodgkin, S.T., Irwin, M.J., Hewett, P.C., Warren, S.J.: The UKIRT wide field camera ZYJHK photometric system: calibration from 2MASS. Mon. Not. R. Astron. Soc. **394**, 675 (2009)

Irwin, J., Irwin, M., Aigrain, S., Hodgkin, S., Hebb, L., Moraux, E.: The monitor project: data processing and light curve production. Mon. Not. R. Astron. Soc. **375**, 1449 (2007)

Lawrence, A., et al.: The UKIRT infrared deep sky survey (UKIDSS). Mon. Not. R. Astron. Soc. **379**, 1599 (2007)

Oliviera, C., Casali, M.: Deep near-IR variability survey of pre-main sequence stars in rho-Ophiucus. Astron. Astrophys. **485**, 155 (2008)

Sesar, B., Ivezic, Z., Lupton, R., Juric, M., et al.: Exploring the variable sky with the Sloan Digital Sky Survey. Astron. J. **134**, 2236 (2007)

Chapter 18
Near Infrared Extinction at the Galactic Centre

Andrew J. Gosling, Reba M. Bandyopadhyay, and Katherine M. Blundell

Abstract We present new results from UKIDSS on the Nuclear Bulge of our Galaxy that shows the previous determinations of the extinction parameter need revising.

Introduction

Study of the Nuclear Bulge (NB) is extremely difficult as it is one of the most highly obscured regions of the Galaxy. The extinction towards the NB is so great that it is almost impossible to observe at visual wavelengths. It is necessary to observe in the near-infrared (NIR) to obtain imaging similar to that normally obtained in the visual. Previous measures of the NIR extinction towards the NB (Catchpole et al. 1990; Schultheis et al. 1999; Dutra et al. 2003) found the extinction to be highly spatially variable. We have mapped the extinction across a region of approximately $2° \times 2°$ centred on Sgr A* using NIR data from the United Kingdom Infrared Deep Sky Survey (UKIDSS) Galactic Plane Survey (GPS) (Lucas et al. 2008). Using a 2:1 over-sampling of the data, we were able to obtain a resolution of $5''$ over the entire region of the map. The previous measurement of the extinction across the NB was based on the 2MASS catalogue as detailed in Dutra et al. (2003). As can be seen in Fig. 18.1, the resolution of this extinction map is insufficient for accurate correction of the photometry of the sources in the UKIDSS catalogue.

A.J. Gosling (✉) • K.M. Blundell
Department of Astrophysics, University of Oxford, Keble Road, Oxford OX1 3RH, UK
e-mail: andrew.gosling@oulu.fi

R.M. Bandyopadhyay
Department of Astronomy, University of Florida, Gainesville, FL 32611, USA

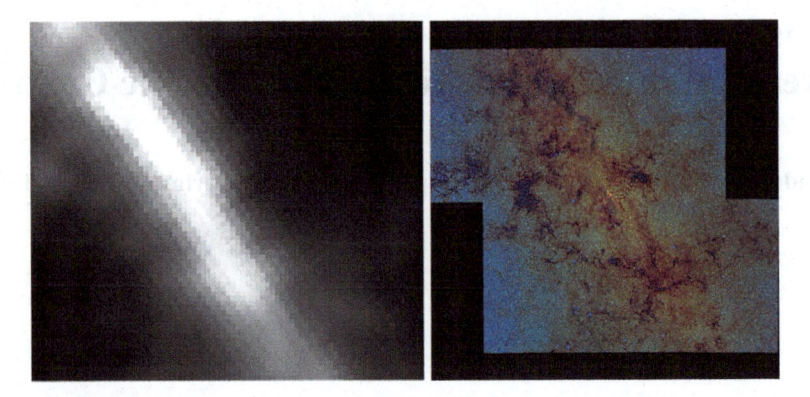

Fig. 18.1 (*Left*) Extinction map based on 2MASS data of an area approximately 2° × 2° centred on Sgr A*. *Lighter areas* correspond to higher levels of extinction. Each resolution element in the extinction map is 400″. (*Right*) An image of the NB created using UKIDSS-GPS data. The resolution of the image is 0.2″ and in the inner region, the average stellar separation is <2″. A higher resolution version of the image can be found at http://andrewgosling.pbworks.com/Research

Extinction Mapping

Extinction in the NIR can generally be described as having a power-law relation to wavelength, parameterized in the form: $A_\lambda \propto \lambda^{-\alpha}$. In previous studies, the extinction law parameter has been assumed/set to be a constant with $\alpha \sim 2$ (Martin and Whittet 1990; Rieke and Lebofsky 1985; Nishiyama et al. 2006) across the region of the NB in which we are working. When trying to produce a new, higher resolution extinction map based on UKIDSS-GPS data, we discovered that we were unable to make extinction corrections resulting in consistent photometry for the NB stars using a fixed value of α.

Our technique for the calculation of extinction is one that takes into account variation in the extinction law as well as the amount of absolute extinction. Using data from three bands, *J*, *H*, and *K*, we are able to calculate three values of colour *J-H*, *J-K* and *H-K*. Subtracting values for the intrinsic colours of giant type stars, using values from the UKIRT standards catalogue, from these measured colours we obtain a measurement of the value of the extinction law parameter as well as the absolute extinction at each position as functions of the colour excess. Full details of this method are in Gosling et al. (2009).

Using the above equation, and knowing that the observed magnitude of a star is a combination of its intrinsic magnitude and the extinction ($m_{obs} = m_{int} + A_\lambda$), we can express the extinction law parameter α as a function of two-colour excess and wavelength:

$$\frac{\langle E(\lambda_1 - \lambda_2)\rangle}{\langle E(\lambda_2 - \lambda_3)\rangle} = \frac{(\lambda_2/\lambda_1)^\alpha - 1}{1 - (\lambda_2/\lambda_3)^\alpha}$$

The extinction in any of the bands can then be calculated using this value of α with:

$$A_{\lambda_1} = \frac{\langle E(\lambda_1 - \lambda_2)\rangle}{1 - (\lambda_1/\lambda_2)^\alpha} \qquad A_{\lambda_2} = \frac{\langle E(\lambda_1 - \lambda_2)\rangle}{(\lambda_2/\lambda_1)^\alpha - 1}$$

We mapped the extinction across a region of approximately $2° \times 2°$ centred on Sgr A*. Using a 2:1 over-sampling, we were able to obtain a resolution of 5″ over the entire region, an order of magnitude better than previous maps. At each position on the map we use the median colour-indices of the stars within a 10″ box to calculate the local extinction law parameter α, and using the value of α measured, we then calculate the values of absolute extinction in all three bands J, H and K.

Measured Extinction

Figure 18.2 shows the histogram of the values of extinction law parameter α as measured from the measured colour ratios. The black line denotes a best fit Gaussian used to characterise the data. We find that the extinction law has a significant amount

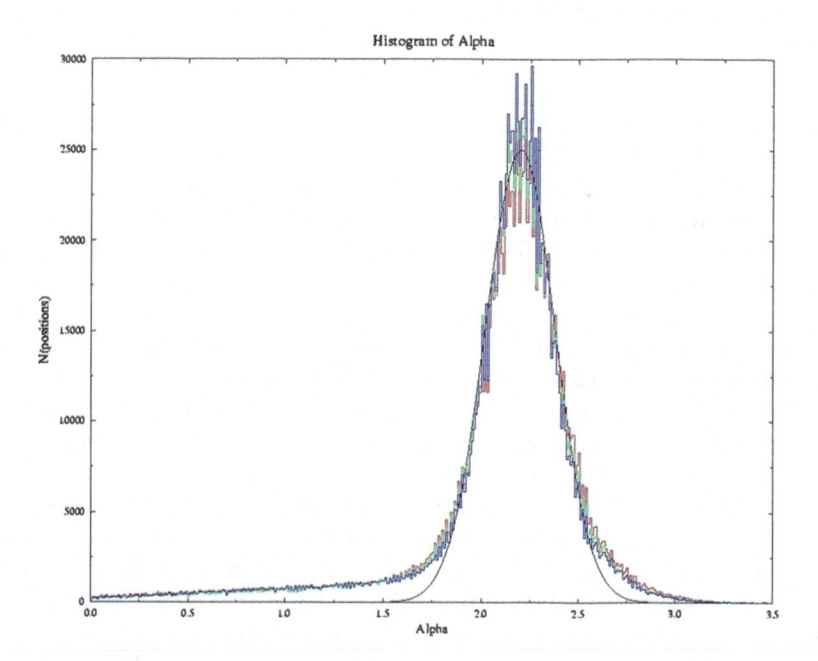

Fig. 18.2 Histogram of the measured values of α from UKIDSS-GPS data. The three colours represent the use of three differently sized sample sizes to check for selection effects. We find in all three cases the data can be fit by a Gaussian with central value of $\alpha = 2.20$ and standard deviation of 0.18

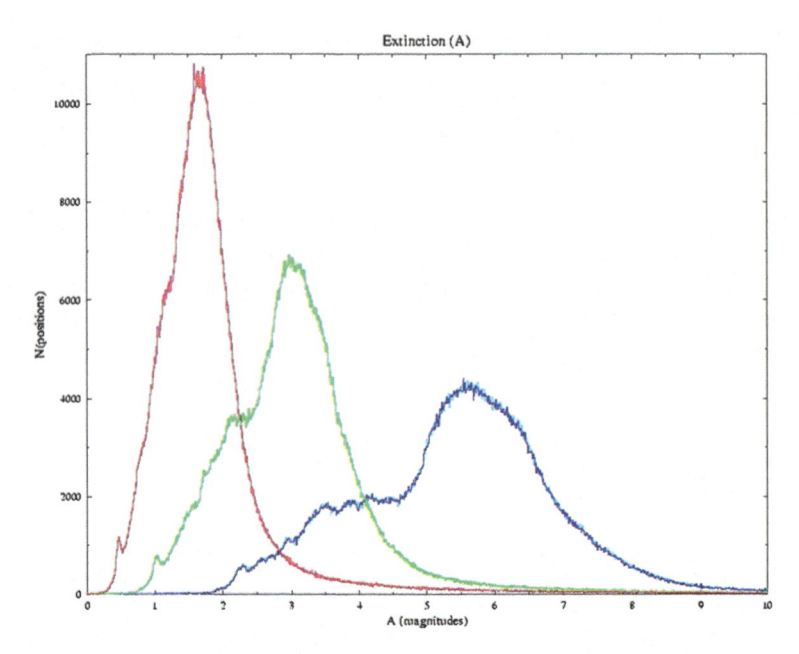

Fig. 18.3 Histograms of the value of absolute extinction measured in the three UKIDSS band J, H and K. The *red curve* represents values measured for A_K, the *green curve* are those for A_H and the *blue curve* for A_J

of variation with values centred on $\alpha = 2.20$ with a 3σ spread of 0.54. In addition to this there is a significant tail to low values of α.

Figure 18.3 shows histograms of the measured values of extinction for J, H and K. We measured extinction in magnitudes to be $1.8 < A_J < 10.0$, $0.7 < A_H < 7.0$ and $0.3 < A_K < 6.0$. The distributions of extinction in all three bands have common features. They have a small peak in the distribution at a relatively low value of extinction (compared to the main peak in the distribution). They have a main peak, the majority value of measured extinction. Finally, all three distributions have high extinction tails extending to much higher values than the ranges given above.

We repeated the extinction calculation described using a single, fixed extinction law to compare with previous studies. A χ_ν^2-test comparison with our variable extinction law method showed that we produce statistically consistent extinction in all three bands (differences within 3σ for the calculated A_λ from different colour indices), whereas using a single extinction value produced statistically inconsistent extinction between bands (differences in measured extinction: $10-50\sigma$). Figure 18.4 shows the resultant extinction map in false colour produced by our work. When compared to the previous extinction map shown in Fig. 18.1, it is evident the improvement in the resolution of the new extinction map. In addition, we probe to higher levels of extinction with the darkest areas on the map corresponding to extinction values $A_K > 3.5$. In the areas of highest extinction we even see the effect

Fig. 18.4 Extinction map based on the UKIDSS NIR data. Each resolution element of the map corresponds to 5″ on sky. The false colour scale corresponds to 0.0 < AK < 3.5 (*white to black*). The *square shaped blocks* of *black* are areas currently without data in the extinction map

of complete obscuration of the background sources and so we have a "turn-over" in the measured extinction (light areas within dark areas in Fig. 18.4) where we are only measuring extinction towards foreground sources.

The variable extinction law, α, follows a similar distribution to the absolute extinction (when mapped it appears similar to Fig. 18.4). From this, we speculate that the dominant cause of the variation in the extinction law is the density of the extinguishing material and possible variations in composition resulting from this.

As we showed in Gosling et al. (2009), measured values of extinction are strongly dependent on the photometric system in which the data is obtained and so caution should be exercised when applying these and other results to datasets. Where possible, the extinction should be calculated intrinsically to a dataset to avoid these issues, or if not possible, then a method that allows for variable extinction law such as our method should be used.

Conclusions

We have presented a new, highly detailed map of the NIR extinction towards the NB using data from the UKIDSS-GPS. We have found both variations in the degree of absolute extinction, and in the extinction law parameter α (Gosling et al. 2009). We show that the extinction law is not "universal" as had previously been thought (Martin and Whittet 1990; Rieke and Lebofsky 1985), but is highly variable, and that only when this variation is taken into account can absolute extinction be calculated consistently at different wavelengths. We recommend that this variable extinction law be taken into account in all future extinction corrections, especially in regions of particularly high levels of extinction. The ability to produce such a high-resolution extinction map is only possible due to the high resolution, deep NIR UKIDSS survey as carried out on UKIRT. In future, we hope to extend our extinction mapping further out into the Galactic Plane, and explore options for increasing the resolution of the map in the most crowded and complex regions of extinction. We also intend to compare our extinction maps with other wavelength datasets to explore the variation in the extinction law.

References

Catchpole, R.M., Whitelock, P.A., Glass, I.S.: The distribution of stars within two degrees of the Galactic centre. Mon. Not. R. Astron. Soc. **247**, 479 (1990)

Dutra, C.M., Santiago, B.X., Bica, E.L.D., Barbuy, B.: Extinction within 10° of the Galactic centre using 2MASS. Mon. Not. R. Astron. Soc. **338**, 253 (2003)

Gosling, A.J., Bandyopadhyay, R.M., Blundell, K.M.: The complex, variable near-infrared extinction towards the Nuclear Bulge. Mon. Not. R. Astron. Soc. **394**, 2247 (2009)

Lucas, P.W., et al.: The UKIDSS Galactic plane survey. Mon. Not. R. Astron. Soc. **391**, 136 (2008)

Martin, P.G., Whittet, D.C.B.: Interstellar extinction and polarization in the infrared. Astrophys. J. **357**, 113 (1990)

Nishiyama, S., et al.: Interstellar extinction law in the J, H, and K_s Bands toward the Galactic Center. Astrophys. J. **638**, 839 (2006)

Rieke, G.H., Lebofsky, M.J.: The interstellar extinction law from 1 to 13 microns. Astrophys. J. **288**, 618 (1985)

Schultheis, M., et al.: Interstellar extinction towards the inner Galactic Bulge. Astron. Astrophys. **349**, L69 (1999)

Chapter 19
Observations of PAHs and Nanodiamonds with UKIRT

Peter J. Sarre

Abstract I describe various contributions made by UKIRT to studies of the structure and chemistry of polycyclic hydrocarbon molecules in the interstellar medium.

Introduction

PAHs – Polycyclic Aromatic Hydrocarbon molecules – are ubiquitous in the interstellar medium and play a major role in its astrochemistry and astrophysics. They are considered to be responsible for the so-called 'Unidentified' Infrared (UIR) emission bands which arise from vibrational transitions following absorption of UV/visible radiation. Major challenges remain in understanding their origin despite a huge research effort. It has so far proved impossible to identify a single specific PAH or even to define the PAH mass distribution. This is very unfortunate as the spectra are known to vary both between and within astronomical objects and so hold considerable potential as a diagnostic tool for interstellar conditions and processes. The lack of detailed assignment contrasts strongly with most spectroscopic observational work in the radio and optical regions where individual molecules are generally easily distinguished. An exception is the hundreds of optical absorption lines seen along lines of sight towards reddened stars and known as the diffuse interstellar bands (DIBs). These too are thought to arise from PAHs but there is no correspondence between astrophysical and laboratory data to date. Infrared (UIR) studies of PAHs are therefore important not only in their own right but in a wider context of molecular astrophysics.

P.J. Sarre (✉)
School of Chemistry, The University of Nottingham, University Park, Nottingham NG7 2RD, UK
e-mail: peter.sarre@nottingham.ac.uk

Fig. 19.1 The Red Rectangle (Credit: NASA; ESA; Hans Van Winckel (Catholic University of Leuven, Belgium); and Martin Cohen (University of California, Berkeley))

Fig. 19.2 Spectrum of the 3.3-μm feature on-star (*solid line*) with the fit to a Lorentzian profile (*dashed line*). The residual Pfε line at 3.039 μm is marked. The Pfδ line at 3.297 μm has been removed and the plot extrapolated over that region. For comparison the inset shows the relatively poor fit achieved with a Gaussian function (*dot–dashed line*)

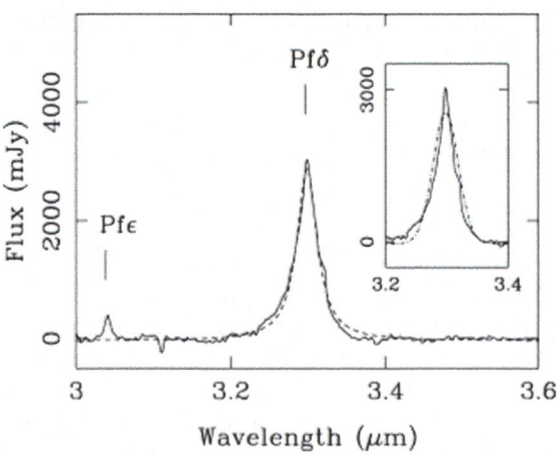

The UIR and DIB problems are in all probability brought together in one object, the Red Rectangle (see Fig. 19.1), and possibly also in the R Coronae Borealis star V854 Cen observed at minimum light. The UIR bands are exceptionally strong in the Red Rectangle nebula and are accompanied by prominent unidentified optical emission bands, some of which are close in wavelength to a sub-set of the diffuse interstellar absorption features. It is very likely but not yet proven that the same carriers give rise to both diffuse interstellar band absorption and optical emission spectral features. The definitive answer to this tantalising question will probably have to wait until laboratory spectra are obtained.

One of the main PAH emission features appears at 3.3 μm and is attributed to the C–H stretching motion in the plane of the aromatic structure. Spectra recorded at UKIRT with CGS4 are shown in Figs. 19.2 and 19.3. For these observations the slit was positioned on the central star HD 44179 (Fig. 19.2), and along the NW whisker with the slit offset 2 arcsec from the star (Fig. 19.3). Notable results are the near-

Fig. 19.3 Spectrum showing the 3.3 and 3.4 μm features at 2-arcsec offset from HD 44179 along the north-western interface. Development of a shoulder on the short wavelength side of the 3.3 μm feature is evident

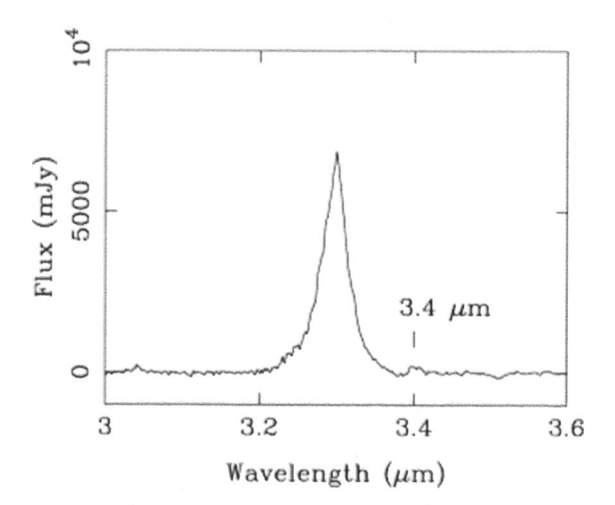

perfect Lorentzian shape of the on-star feature (maximum at 3.30 μm) with the development of a short-wavelength shoulder with increasing offset (Fig. 19.3), and the appearance of a (probable) aliphatic emission feature at ∼3.4 μm (Song et al. 2003).

The discovery of a gradual change of 3.3 μm shape with offset within a single object was a first for UKIRT and is illustrated Fig. 19.4 together with Lorentzian fitting of the profiles. It is found that the behaviour can be modelled in terms of two components centred near 3.30 and 3.28 μm.

The origin of the two components is not known with certainty. Possible interpretations are that these arise from different structural forms, sizes or hydrogenation/ionization states. Laboratory gas-phase *absorption* spectra taken at elevated temperatures show a shift of the 3.3-μm peak maximum to shorter wavelength with increase in PAH size. This amounts to about 15 cm^{-1} (0.015 μm) from pyrene ($C_{16}H_{10}$) to ovalene ($C_{32}H_{14}$) (Joblin et al. 1995), and is similar to the ∼0.02 μm difference between the 3.30 and 3.28 μm features. van Diedenhoven et al. (2004) comment that the observed 3.3 μm feature is attributable to the smallest PAHs, and reported that laboratory absorption spectra in an inert gas matrix taken at the NASA Ames laboratory yield a C–H stretch frequency distribution which is *bimodal* with the bands of the smallest PAHs centred near 3,060 cm^{-1} (3.27 μm) and transitions of larger PAHs occurring near ∼3,090 cm^{-1} (3.24 μm). Applying a shift to longer wavelength of ∼0.03 μm to these laboratory data, as is commonly invoked on account of the higher carrier temperature in the ISM, gives bands at 3.30 μm ($N_C \leq 40$) and 3.27 μm ($N_C \geq 40$), where N_C is the number of carbon atoms, in good agreement with the two-component picture.

It is notable that although laboratory data indicate that the wavelength of the 3.3-μm feature varies with PAH size and temperature, calculated DFT B3LYP/4-31G C–H stretching wavelengths are virtually independent of size for compact symmetrical PAHs and range between 3.262 μm for $C_{24}H_{12}$ (coronene) and 3.266 μm for $C_{130}H_{28}$ (Bauschlicher et al. 2008). However, recent calculations by

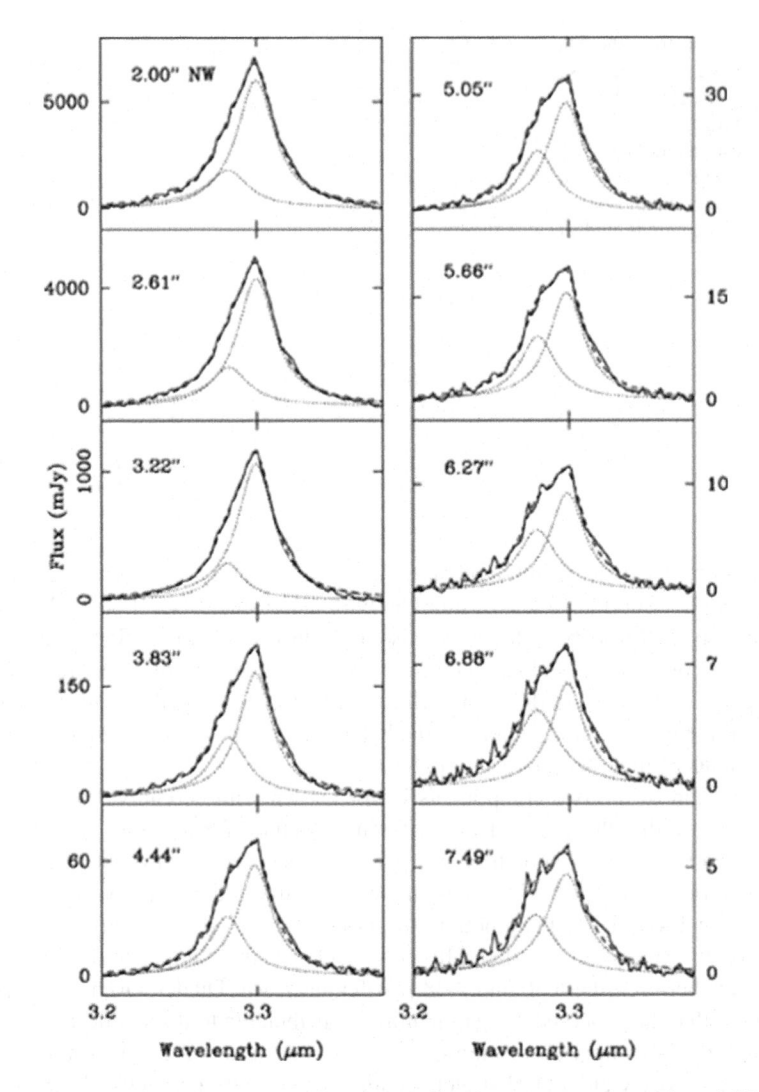

Fig. 19.4 Fitted Lorentzian profiles as a function of offset from the central star. The *solid lines* are the observed spectra, the *dotted lines* are individual Lorentzian profiles and the *dashed lines* are the sum of the two profiles. The 3.28/3.30 μm intensity ratio increases with offset

Bauschlicher et al. (2009) introduce an additional aspect. It is found that there is a wavelength difference of ∼0.03 μm between 'bay' and 'non-bay' hydrogen sites on large PAH structures such as $C_{120}H_{36}$. The origin of the difference in wavelength is a steric restriction for bay-type hydrogen stretching motion which leads to a higher vibrational frequency. Could the 3.30 μm component be due to PAHs with non-bay hydrogens and the 3.28 μm component to PAHs with some 'bay' hydrogens? There is an urgent need for PAH-specific laboratory spectra to resolve this issue.

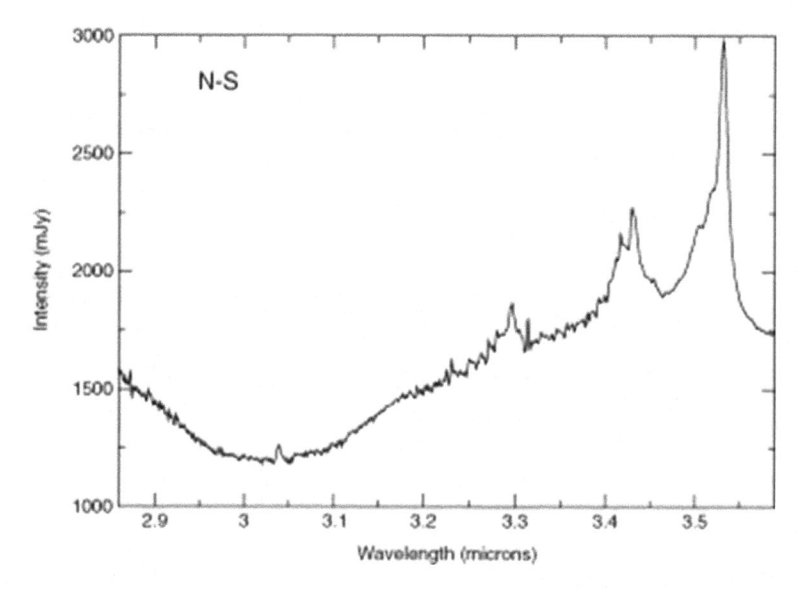

Fig. 19.5 Spectra of Elias 1 in the 2.9–3.6 μm region recorded with slit alignment N–S. The data were summed over the inner ±0.3 arcsec. The features at 3.43 and 3.53 μm arise from nanodiamonds. The 3.3 μm feature is due to PAHs

Although this brief overview concentrates on 3.3 μm emission from the ubiquitous PAHs, a few astrophysical objects display infrared emission from another form of carbon – nanodiamonds – recorded at UKIRT using UIST. One such spectrum for the pre-main-sequence star Elias 1 is shown in Fig. 19.5 where the slit was aligned in the N–S direction (Topalovic et al. 2006). The spectral signatures arise from various C–H groups on the diamond surface. It is notable that the 3.3 μm PAH feature is quite weak compared with the longer wavelength 'diamond' features between 3.4 and 3.6 μm. Long-slit spectroscopy established that the PAH and nanodiamond spatial distributions are significantly different in this Herbig Ae/Be star, the PAH distribution being more extended.

Discussion of detection of astronomical diamonds with UKIRT in this volume is apposite; it is 30 years since the inauguration of UKIRT and a (modern) 30th anniversary gift is a diamond. UKIRT surely is the jewel in the crown of British (and international) infrared astronomy!

Acknowledgements The work described here involved a number of people. First I want to pay particular credit to Tom Kerr who was a postdoctoral research fellow at Nottingham and joined UKIRT well over 10 years ago. Our collaboration has continued ever since. Nottingham's loss was most certainly UKIRT's gain! June McCombie also contributed greatly to the science and was co-supervisor to In-Ok Song who worked particularly on the 3.3 μm PAH feature. Radmila Topalovic was the second PhD student involved and her thesis contains UKIRT studies of nanodiamond spectra. Financial support from STFC, PPARC, the RAS and The University of Nottingham is gratefully acknowledged. Some of the figures included here were first published in Monthly Notices of the Royal Astronomical Society.

References

Bauschlicher, C.W., Peeters, E., Allamandola, L.J.: The infrared spectra of very large, compact, highly symmetric, polycyclic aromatic hydrocarbons (PAHs). Astrophys. J. **678**, 316 (2008)

Bauschlicher, C.W., Peeters, E., Allamandola, L.J.: The infrared spectra of very large irregular polycyclic aromatic hydrocarbons (PAHs): observational probes of astronomical PAH geometry, size, and charge. Astrophys. J. **697**, 311 (2009)

Joblin, C., Boissel, P., Léger, A., d'Hendecourt, L., Defourneau, D.: Infrared spectroscopy of gas-phase PAH molecules. II. Role of the temperature. Astron. Astrophys. **299**, 835 (1995)

Song, I.-O., McCombie, J., Kerr, T.H., Sarre, P.J.: Evolution of the 3.3-μm emission feature in the Red Rectangle. Mon. Not. R. Astron. Soc **380**, 979 (2003)

Topalovic, R., Russell, J., McCombie, J., Kerr, T.H., Sarre, P.J.: Diamonds and polycyclic aromatic hydrocarbons in the circumstellar environment of the Herbig Ae/Be star Elias 1. Mon. Not. R. Astron. Soc. **372**, 1299 (2006)

van Diedenhoven, B., Peeters, E., Van Kerckhoven, C., Hony, S., Hudgins, D.M., Allamandola, L.J., Tielens, A.G.G.M.: The profiles of the 3–12 micron polycyclic aromatic hydrocarbon features. Astrophys. J. **611**, 928 (2004)

Chapter 20
Nearby Galaxies with UKIRT: Uncovering Star Formation, Structure and Stellar Masses

Phil James

Abstract I review the central role played by UKIRT observations in furthering our understanding of the formation, evolution, structure and stellar contents of nearby galaxies. Highlights include uncovering the true nature of the infrared-luminous galaxy population, and exploration of the starburst phenomenon; pioneering the use of near-infrared imaging to uncover the true structures of disks, bars, arms, bulges, nuclei and spheroids, free from the obscuring effects of dust; and moving infrared astronomy into the modern survey era and thus providing high-quality imaging for tens of thousands of nearby galaxies.

Introduction

UKIRT has played a central role in the establishment of infrared (IR) observations at the forefront of galaxy studies. In the 1970s, the IR was considered a specialist area, the province of 'infrared astronomers'; now, any study of galaxy morphology, evolution or star formation would be considered sadly deficient without IR observations as a central component. This revolution was due in no small part to UKIRT, which has been maintained at the forefront of galaxy studies through the continued development of world-leading instrumentation. In this review I will try to indicate how the successive generations of instruments have been used to build our understanding of galaxy structure and evolution. This is very much a personal view, with no attempt at completeness, and I give apologies to those whose work is omitted.

P. James (✉)
Astrophysics Research Institute, Liverpool John Moores University, Twelve Quays House, Egerton Wharf, Birkenhead, Wirral CH41 1LD, UK
e-mail: paj@astro.livjm.ac.uk

Infrared-Luminous Galaxies ('IRAS Galaxies', ULIRGs etc.)

The early years of UKIRT fortuitously coincided with the IRAS discovery of a population of galaxies with prodigious far-IR luminosities (Lonsdale et al. 1984), and UKIRT played a central role in determining the true nature of these objects. It was quickly realized that the IRAS sources were almost always associated with peculiar galaxies, and although these often had substantial optical luminosities, their bolometric luminosities were completely dominated by their mid- and far-IR emission, as illustrated in Fig. 20.1 (Joseph and Wright 1985). This emission peaked within the IRAS bands, typically at 50–100 μm, indicating thermal dust emission at a few 10s of Kelvin, and a vigorous debate took place over the underlying heating source of this dust. Was it stellar processes or Active Galactic Nuclei, 'Starbursts' or 'Monsters'? Ultimately, of course, there was no clear winner, with AGN now thought to play a key role in the formation of all large galaxies, and merger-driven starbursts being a central part of current models of hierarchical galaxy formation. Given that none of this was known in 1980, it is interesting to see how much of the 'starburst' picture had become apparent by 1985, when Bob Joseph and Gillian

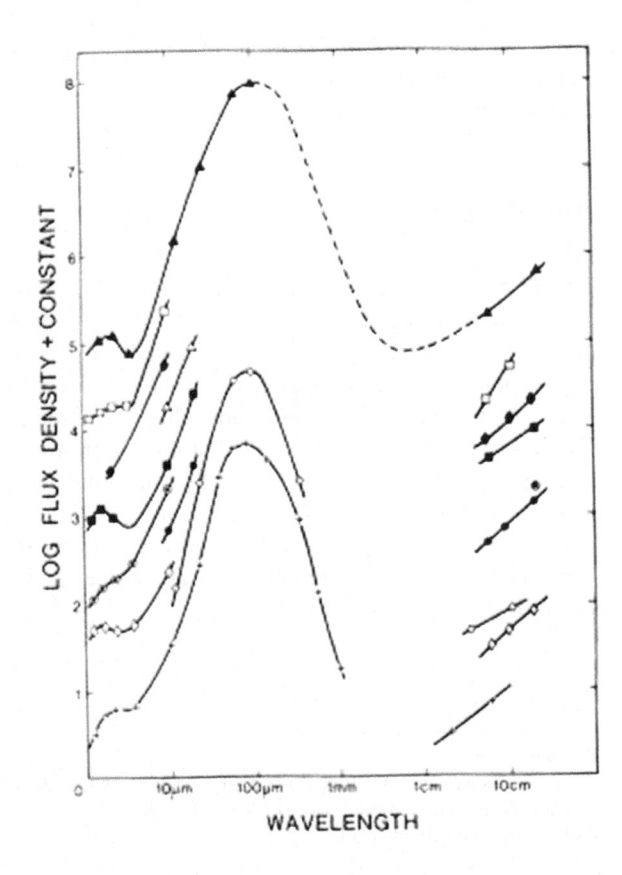

Fig. 20.1 Spectral energy distributions of merging galaxies, showing the dominant far-IR peaks (Joseph and Wright 1985)

Wright wrote *"mergers... are easily understood in terms of a burst of star formation of extraordinary intensity and spatial extent: they are 'super starbursts'. We argue that such super starbursts occur in the evolution of most mergers, and discuss the implications of super starbursts for the suggestion that mergers evolve into elliptical galaxies. Finally, we note that merger-induced shocks are likely to leave the gas from both galaxies in dense molecular form which will rapidly cool, collapse and fragment."*

Much of this understanding, and its further elucidation, came from UKIRT observations. Joseph et al. (1984a) demonstrated the high efficiency of mergers in triggering starburst activity, from the high fraction of disturbed galaxies found to show a K-L excess. The case for a spatially extended starburst as the dominant heating source came from challenging observations of the extent of the 10 μm emission (e.g. Wright et al. 1988) with a variety of early mid-IR instruments of varying sensitivity and stability. (Those of the AGN persuasion who might still doubt this conclusion are referred to recent work by Symeonidis et al. 2010, who found all sources within a sample of Spitzer 70-μm selected galaxies to be predominantly powered by star formation. But my personal biases are showing...). Joseph et al. (1984b) detected shocked molecular hydrogen in two much-studied mergers, Arp 220 and NGC 6240, with the estimated gas masses being consistent with the quantities required to fuel the central super-starbursts.

A very exciting development came with the ability to image these galaxy mergers in the near-IR, following the advent of IRCAM1 in 1986. It was known that almost all of the mergers were highly disturbed and asymmetric in the optical, but given the strong and chaotic dust distributions, and the amount of young highly luminous stars, there was great interest in revealing the underlying stellar structure. Figure 20.2 shows some examples of these early IRCAM images, reduced by the author (who takes full responsibility for the deficiencies in the data reduction!) and Gillian Wright.

In some cases, the near-IR morphologies were found to be highly disturbed. In others, however, the near-IR structure was found to be dominated by smooth, symmetric, centrally-concentrated light distributions, with light profiles resembling those of elliptical galaxies (Wright et al. 1990; Fig. 20.3).

Spectroscopic observations of the 2.29 μm CO absorption feature demonstrated that some also showed kinematic properties placing them on the 'Fundamental Plane', characteristic of elliptical galaxies (see Fig. 20.4, taken from Doyon et al. 1994; also James et al. 1999).

The burst histories and stellar energetics of merger-induced starbursts were also probed spectroscopically, with CGS2 (Prestwich et al. 1994) and CGS4 (Doyon et al. 1992; Coziol et al. 2001), providing indications of top-heavy initial mass functions in these systems. UKIRT was also used in combination with the JCMT to determine the IR-submillimetre spectral energy distributions of starbursting galaxies, e.g. the interacting pair VV114 by Frayer et al. (1999).

Pioneering work on spectral diagnostics of AGN and star formation at 8–13 μm, on the properties of dust grains and on the determination of mid-IR extinction resulted in an extensive series of papers by Dave Aitken, Pat Roche and

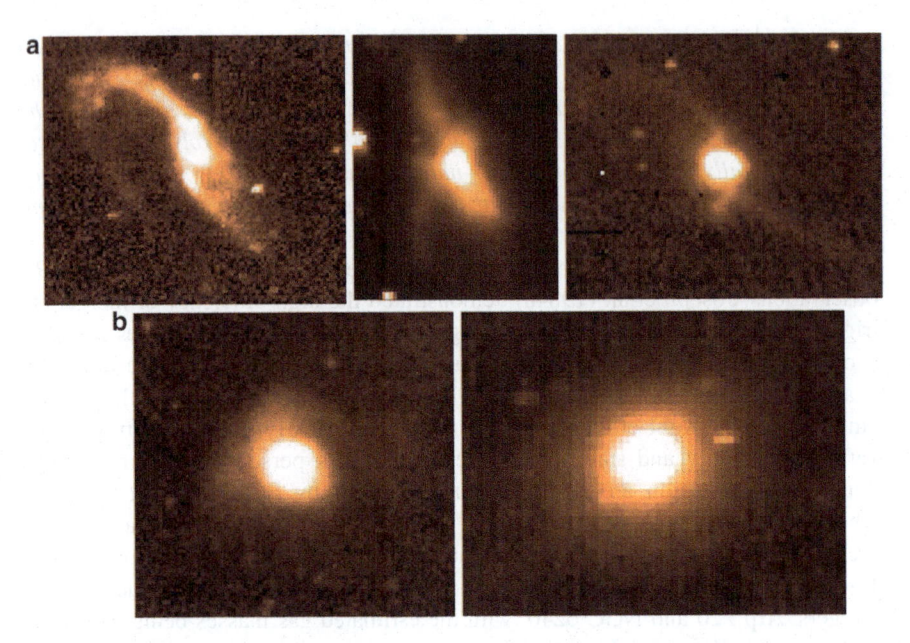

Fig. 20.2 (**a**) IRCAM1 K-band images of (*left* to *right*) NGC 3509, NGC 6240 and NGC 2623. (**b**) Arp 220 (*left*) and NGC 7252

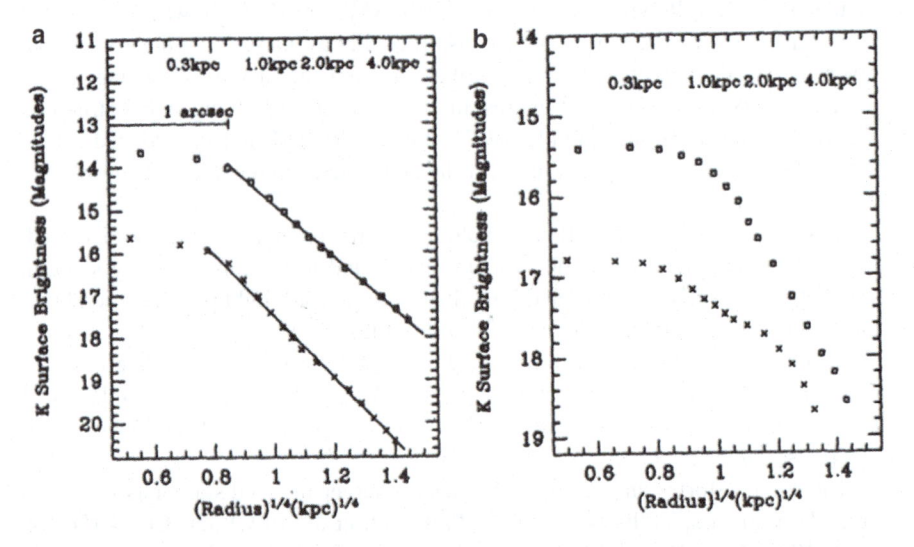

Fig. 20.3 *K*-band light profiles of mergers. (**a**) Arp 220 (above) and NGC 2623 are fairly well fitted by a de Vaucouleurs profile; (**b**) IC 883 (above) and NGL 6052 do not follow this profile

collaborators, e.g. Aitken et al. (1981), Phillips et al. (1984), Roche et al. (1984), Aitken and Roche (1985), Roche and Aitken (1985) and Smith et al. (1989). The resulting spectral energy distributions for 60 galaxies in the 8–13 μm window, and in some cases the more challenging 17–23 μm window, were assembled as a spectral

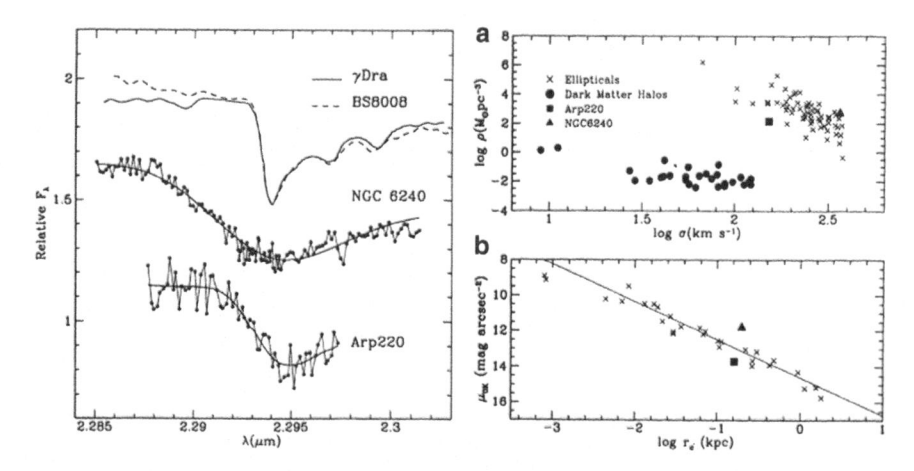

Fig. 20.4 Derivation of merger velocity dispersions from CGS4 observations of the CO bandheads (*left*) and the location of two mergers relative to the parameter correlations found for elliptical galaxies (*right-hand* frames; (**a**) Mass density vs. velocity dispersion, and (**b**) K-band surface brightness vs. core radius; all diagrams taken from Doyon et al. 1994)

atlas (Roche et al. 1991) which demonstrated the diversity of mid-IR properties of AGN. This contrasted strongly with the surprising uniformity of mid-IR properties found by these authors for HII-region dominated nuclei.

Other work on AGN made use of near-IR spectroscopy, which can be used to search for broad emission lines, such as the broad Pa-β lines found using CGS2 spectroscopy of two narrow-line X-ray sources by Blanco et al. (1990). The near-IR also contains the important molecular hydrogen features used by Fernandez et al. (1999) in a study of the central gas disks of Seyfert galaxies NGC 3227 and NGC 4151.

Structural Properties of Normal Galaxies

Bars

The ability of near-IR imaging to uncover structures that are not apparent at other wavelengths is particularly clearly demonstrated by studies of barred galaxies. An early example of this was the work of Shaw et al. (1993) who used IRCAM1 imaging of the central regions of strongly barred galaxies to uncover isophote twists (Fig. 20.5 below). Modeling of these light distributions revealed the associated potentials to have the form needed to centrally concentrate disk gas. This finding was of relevance for later work on secular theories of bulge formation in isolated galaxies, and may have relevance for gas supply to AGN (Shlosman et al. 1989).

Knapen et al. (1995) used IRCAM3 K-band imaging of M100 to discover a central bar, again visible only in the near-IR. This bar was found to be associated

Fig. 20.5 K-band imaging of the bar in NGC 5728 (Shaw et al. 1993)

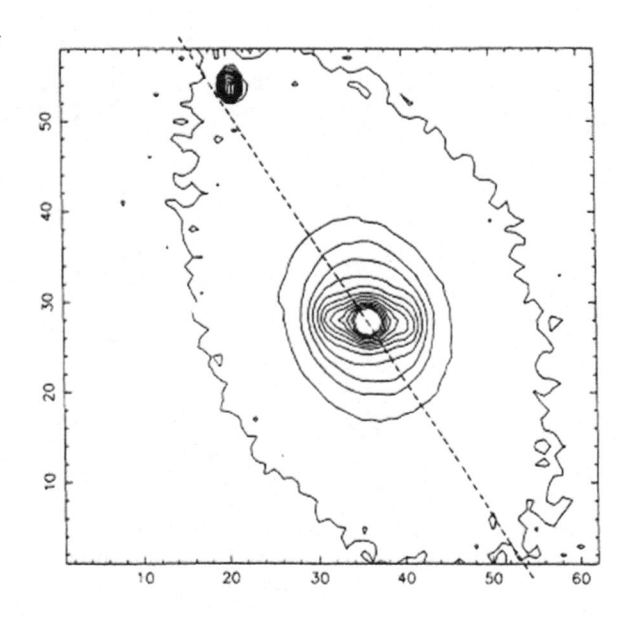

with a pair of inner Lindblad resonances, and again was found to be plausibly linked to central gas supply related to AGN and nuclear star formation activity, with the latter being investigated further by Ryder and Knapen (1999). Seigar and James (1998a) looked at the frequency of bars and more moderate 'oval distortions' in IRCAM3 imaging of spiral galaxies, again finding such features to be frequently revealed in the near-IR.

Spiral Arms

Seigar and James (1998b) studied the K-band properties of the spiral arms in 45 Sa – Sd galaxies, calculating arm strengths as a function of radius, and arm cross-sections, to quantify the strength of the perturbation in the disk stellar mass distribution. Follow-up of the same sample by Seigar and James (2002) used the concentration of Hα emission in K-band-defined arm regions (see Fig. 20.6 below) to show that SF must be triggered within arms; they cannot simply be regions where the local density in all stellar populations is increased by the same factor.

Disks and Bulges

A highly important dataset of near-IR and optical imaging of nearby disk galaxies was assembled by Roelof de Jong in the 1990s. The imaging of 86 face-on galaxies

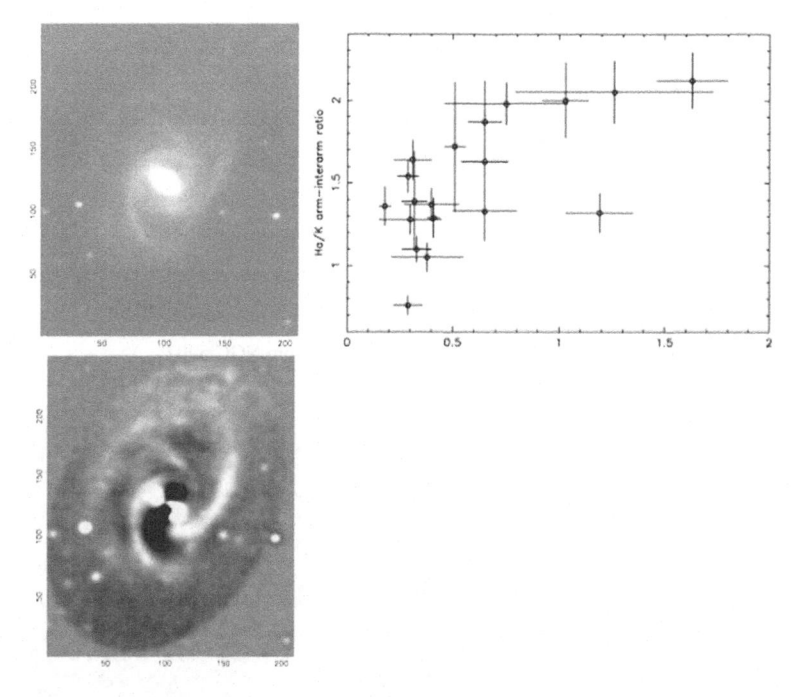

Fig. 20.6 (*Left* and *centre*) IRCAM3 K-band imaging of a spiral galaxy from Seigar and James (2002); (*Right*) Hα/K-band concentrations within arm regions, with values above one indicating triggering of star formation within K-band defined arms

was obtained in 1991–1992 with IRCAM2 in 1.2 arcsec/pixel mode, giving a field of view just over 1 arcmin on a side, complemented by optical CCD imaging from the 1-m Jacobus Kapteyn Telescope (de Jong and van der Kruit 1994). The images were analysed in a series of papers (de Jong 1996a, b, c) looking at the characteristic surface brightnesses and profiles of disks and bulges, stellar contents and the impact of dust on observed properties. The same data were also used as input for the population synthesis modeling of galaxies performed by Bell and de Jong (2000), which was the basis for some of the most widely used methods for estimating stellar masses from the optical and near-IR colours and luminosities of unresolved stellar populations.

Haloes

Mark Casali and I made a brave but ultimately unsuccessful attempt to detect diffuse near-IR emission from putative 'dim matter' haloes of low-mass stars around edge-on nearby galaxies. The small area coverage of the IRCAM3 detector led Mark

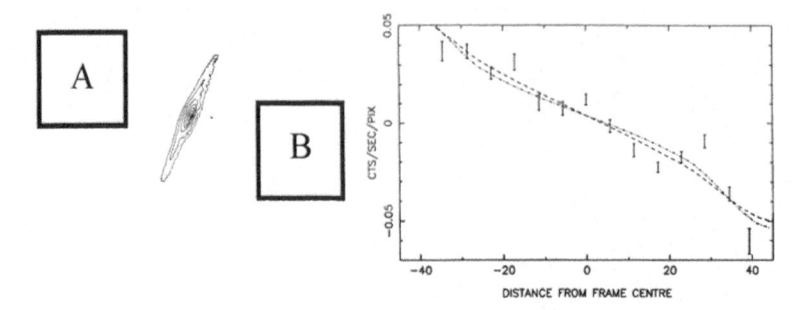

Fig. 20.7 Apparent detection of diffuse near-IR emission from NGC 5907

Fig. 20.8 NGC 5907 imaged with the 0.5 m telescope at the BlackBird Remote Observatory (Martinez-Delgado et al. 2008) showing spectacular disrupted satellite trails

to suggest a differential detection method, where we looked for a gradient in a difference image of two fields on opposite sides of the galaxy (Fig. 20.7, left-hand diagram).

Emission was detected in both the J and K bands (James and Casali 1996, 1998; Fig. 20.7 right-hand plot) which attracted significant interest at the time. However, our initial paper in the Spectrum newsletter acknowledged that we could not rule out an alternative explanation of emission from tidally-disturbed satellite galaxies; this has been spectacularly reinforced by subsequent optical imaging (e.g. Martinez-Delgado et al. 2008; Fig. 20.8).

Dwarf Galaxies

In the early 1990s, I undertook what was, I think, the first near-IR imaging study of dwarf galaxies. In James (1991) I presented IRCAM1 *JHK* imaging of 13 Virgo cluster dwarfs, and concluded that dwarf irregulars (dIs) do not resemble dwarf elliptical galaxies (dEs) in their near-IR properties, which argues against the simplest evolutionary links between these two types. The sample size was later doubled and extended to starbursting blue compact dwarfs (James 1994), with these BCDs being found to have different near-IR light distributions to both dIs and dEs, and very disturbed morphologies even in the older populations traced by near-IR emission. This is somewhat surprising given the short dynamical timescales of these small galaxies.

Hunt et al. (2003) used IRCAM3 *JHK* imaging of the extremely low-metallicity blue compact dwarf galaxy IZw18, combined with HST WFPC2 and NICMOS archival images, to determine the spatially-resolved star formation history and put an upper limit of 22 % on the mass fraction in an old stellar population (>500 Myr). This confirmed results obtained by Vanzi et al. (2000), who studied another BCD, SBS 0335-052, using IRCAM3 *JHK* imaging plus NTT near-IR imaging and spectroscopy, and finding that no more than 15 % of the near-IR luminosity can come from an old population. More recently, Thuan et al. (2008) used UKIRT IRCAM3 *JHK* imaging of the BCD Mrk 996, in addition to Spitzer mid-IR imaging and spectroscopy and MMT optical spectroscopy, to determine the properties of the extremely dense nuclear SF region in this galaxy.

Statistical Properties of the Local Galaxy Population

Since the near-IR more accurately traces stellar mass than any other photometric passband, it was quickly realized that an accurate K-band galaxy luminosity function is of fundamental importance for determining the current stellar content of the Universe and hence constraining cosmological models. The first serious attempt at this was made by Mobasher et al. (1993), who overcame the limitations of the single-element aperture photometry provided by UKT9 to produce K-band photometry for a sample of 181 field galaxies selected from the Anglo-Australian Redshift Survey (Fig. 20.9 left). The modern equivalent, a K-band LF for 40,000 galaxies from the UKIDSS Large Area Survey (Smith et al. 2009) using WFCAM data is shown in the right-hand frame of Fig. 20.9. The different Hubble constants used and the reversed x-axes make direct comparison difficult, but Bahram Mobasher has assured me, with some satisfaction, that they overlay exactly.

UKIRT has also contributed to our understanding of early-type, non-star-forming galaxies, including a classic early study of what has now come to be known as the 'Red Sequence' of galaxies. Bower et al. (1992a, b) used UKT9 aperture photometry plus CCD imaging from the Isaac Newton Telescope to explore the optical:near-IR

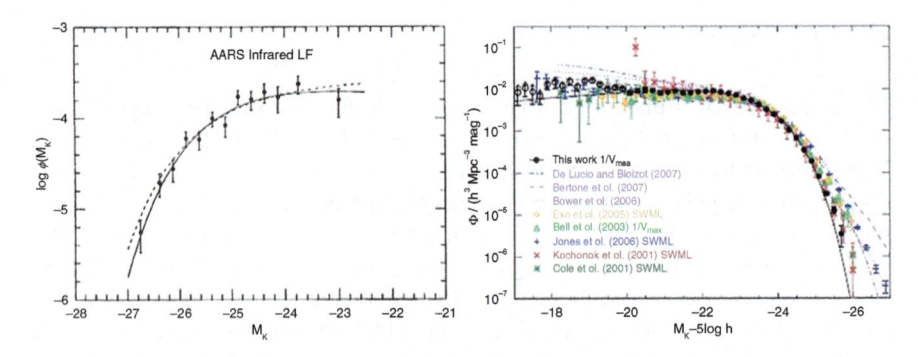

Fig. 20.9 Luminosity functions old and new – for 181 galaxies with UKT9 on the *left* (Mobasher et al. 1993), and for 40,000 galaxies with WFCAM on the *right* (Smith et al. 2009)

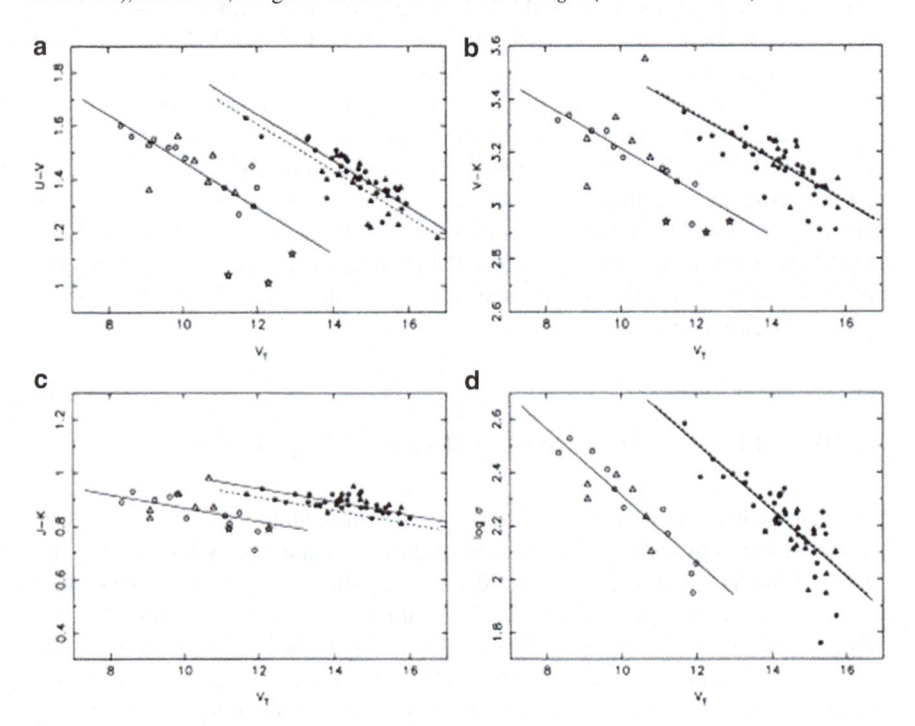

Fig. 20.10 Early-type galaxy colour-magnitude relations in (**a**) U-V vs. V (**b**) V-K vs. V and (**c**) J-K vs. V (**d**) shows the Faber-Jackson relation. Open symbols denote galaxies in the Virgo Cluster, filled symbols the Coma Cluster (Bower et al. 1992b)

colour-magnitude relation of early-type galaxies. The similarity of the relations found for galaxies from the Virgo and Coma clusters (Fig. 20.10) hinted at the universality of these correlations, providing one of the most powerful methods for identifying cluster populations of red galaxies at high redshift (note that the two Bower et al. papers have over 850 citations between them at the time of writing).

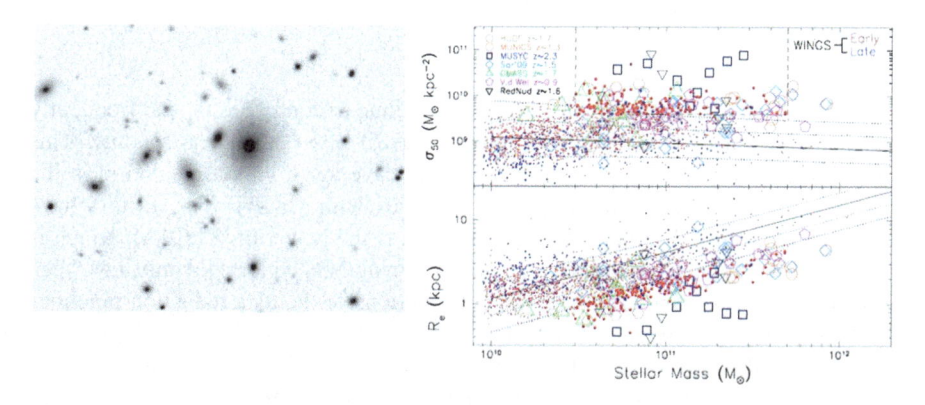

Fig. 20.11 (*Left*) K-band WFCAM imaging of the nearby cluster A2124 from the WINGS survey (Valentinuzzi et al. 2009). (*Right*) Parameter correlations for WINGS cluster galaxies, highlighting the compact galaxy population (Valentinuzzi et al. 2010)

To show this continuing interest and to bring the UKIRT red sequence story fully up-to-date, Stott et al. (2009) use WFCAM imaging in a comparison of the red sequence of early-type galaxies in local galaxy clusters with that seen at intermediate redshifts; local clusters show an excess of galaxies at the faint end of the red sequence, demonstrating a surprising amount of evolution in this population at a relatively recent epoch.

Other studies have made use of UKIRT K-band photometry to provide stellar mass information in galaxy parameter correlation studies. Khosroshahi et al. (2000) looked at the K-band 'photometric plane' (Sersic index vs. effective radius vs. surface brightness) of 42 Coma elliptical galaxies and the bulges of 26 field spirals. The near-IR Fundamental Plane, which additionally incorporates velocity dispersion information, was analysed by Mobasher et al. (1999) for 48 Coma galaxies; and with much better statistics (UKIDSS photometry for ∼40,000 early-type galaxies) by La Barbera et al. (2008, 2010a, b).

Eminian et al. (2008) have used UKIDSS galaxy photometry and stellar population synthesis techniques to study correlations between nuclear near-IR colours and specific SF rate, stellar age, metallicity and dust attenuation. La Barbera and de Carvalho (2009) investigated metallicity and colour gradients in early-type galaxies, finding evidence for age gradients that are consistent with size evolution and may explain the population of compact galaxies found by some groups at intermediate redshifts. However, in another approach to the compact galaxies problem, Valentinuzzi et al. (2010) have searched for directly for such galaxies in the 28 nearby clusters of the WINGS survey, all with deep WFCAM imaging (Valentinuzzi et al. 2009). Figure 20.11 (left) shows an example of the WFCAM K-band imaging used in this study, while the right-hand frame shows the evidence that there are compact galaxies in the local cluster galaxy population.

Conclusions

UKIRT has shown a remarkable ability to continue to contribute to, and frequently lead, the advances being made in many areas of observational astronomy. This process is still occurring, and the last 5 years have again seen a revolution in the type of science being done on nearby galaxies with UKIRT. Two factors have contributed to this. The first and most obvious is the power of WFCAM, which is providing high quality photometry for galaxies in numbers that could only have been dreamt of 20 years ago (and here the contributions of the data reduction pipelines and archiving systems provided by the Cambridge Astronomical Survey Unit and the Edinburgh Wide Field Astronomy Unit respectively are absolutely vital). The second is the recent improvement in population synthesis modeling, particularly as applied to near-IR stellar populations, which means that it is no longer necessary, or indeed justified, to assume that near-IR luminosity directly traces stellar mass. Thus, much or all of the early work discussed here can be substantially improved.

The central role of near-IR in population studies has been further reinforced by the discovery that optical-NIR colours help to break the age-metallicity degeneracy of unresolved stellar populations, noting that it is essential to use models that accurately incorporate all relevant evolutionary phases.

Acknowledgements I had many enjoyable and inspiring discussions on galaxies, UKIRT and infrared astronomy with Tim Hawarden, the last being immediately after the talk which formed the basis of this paper. He is sadly missed.

References

Aitken, D.K., Roche, P.F.: 8-13 microns spectrophotometry of galaxies. IV – six more Seyferts and 3C345. V – the nuclei of five spiral galaxies. Mon. Not. R. Astron. Soc. **213**, 777–797 (1985)

Aitken, D.K., Roche, P.F., Phillips, M.M.: The question of extinction in active galactic nuclei – infrared spectral observations of NGC 1614, NGC 7469 and NGC 1275. Mon. Not. R. Astron. Soc. **196**, 101P–107P (1981)

Bell, E.F., de Jong, R.S.: The stellar populations of spiral galaxies. Mon. Not. R. Astron. Soc. **312**, 497–520 (2000)

Bell, E.F., McIntosh, D.H., Katz, N., Weinberg, M.D.: The optical and near-infrared properties of galaxies. I. Luminosity and stellar mass functions. Astrophys. J. Suppl. Ser. **149**, 289–312 (2003)

Bertone, S., De Lucia, G., Thomas, P.A.: The recycling of gas and metals in galaxy formation: predictions of a dynamical feedback model. Mon. Not. R. Astron. Soc. **379**, 1143–1154 (2007)

Blanco, P.R., Ward, M.J., Wright, G.S.: Broad infrared line emission from the nuclei of Seyfert 2 galaxies. Mon. Not. R. Astron. Soc. **242**, 4P–8P (1990)

Bower, R.G., Benson, A.J., Malbon, R., Helly, J.C., Frenk, C.S., Baugh, C.M., Cole, S., Lacey, C.G.: Breaking the hierarchy of galaxy formation. Mon. Not. R. Astron. Soc. **370**, 645–655 (2006)

Bower, R.G., Lucey, J.R., Ellis, R.S.: Precision photometry of early-type galaxies in the Coma and Virgo clusters: a test of the universality of the colour-magnitude relation. I – the data. Mon. Not. R. Astron. Soc. **254**, 589–600 (1992a)

Bower, R.G., Lucey, J.R., Ellis, R.S.: Precision photometry of early type galaxies in the Coma and Virgo clusters – a test of the universality of the colour- magnitude relation. II – analysis. Mon. Not. R. Astron. Soc. **254**, 601–613 (1992b)

Cole, S., and 26 colleagues.: The 2dF galaxy redshift survey: near-infrared galaxy luminosity functions. Mon. Not. R. Astron. Soc. **326**, 255–273 (2001)

Coziol, R., Doyon, R., Demers, S.: Near-infrared spectroscopy of starburst galaxies. Mon. Not. R. Astron. Soc. **325**, 1081–1096 (2001)

de Jong, R.S.: Near-infrared and optical broadband surface photometry of 86 face-on disk dominated galaxies. III. The statistics of the disk and bulge parameters. Astron. Astrophys. **313**, 45–64 (1996a)

de Jong, R.S.: Near-infrared and optical broadband surface photometry of 86 face-on disk dominated galaxies. II. A two-dimensional method to determine bulge and disk parameters. Astron. Astrophys. Suppl. Ser. **118**, 557–573 (1996b)

de Jong, R.S.: Near-infrared and optical broadband surface photometry of 86 face-on disk dominated galaxies. IV. Using color profiles to study stellar and dust content of galaxies. Astron. Astrophys. **313**, 377–395 (1996c)

de Jong, R.S., van der Kruit, P.C.: Near-infrared and optical broadband surface photometry of 86 face-on disk dominated galaxies. I. Selection, observations and data reduction. Astron. Astrophys. Suppl. Ser. **106**, 451–504 (1994)

De Lucia, G., Blaizot, J.: The hierarchical formation of the brightest cluster galaxies. Mon. Not. R. Astron. Soc. **375**, 2–14 (2007)

Doyon, R., Puxley, P.J., Joseph, R.D.: The He I 2.06 microns/Br-gamma ratio in starburst galaxies – an objective constraint on the upper mass limit to the initial mass function. Astrophys. J. **397**, 117–125 (1992)

Doyon, R., Wells, M., Wright, G.S., Joseph, R.D., Nadeau, D., James, P.A.: Stellar velocity dispersion in ARP 220 and NGC 6240: elliptical galaxies in formation. Astrophys. J. **437**, L23–L26 (1994)

Eke, V.R., Baugh, C.M., Cole, S., Frenk, C.S., King, H.M., Peacock, J.A.: Where are the stars? Mon. Not. R. Astron. Soc. **362**, 1233–1246 (2005)

Eminian, C., Kauffmann, G., Charlot, S., Wild, V., Bruzual, G., Rettura, A., Loveday, J.: Physical interpretation of the near-infrared colours of low-redshift galaxies. Mon. Not. R. Astron. Soc. **384**, 930–942 (2008)

Fernandez, B.R., Holloway, A.J., Meaburn, J., Pedlar, A., Mundell, C.G.: Excited molecular hydrogen around the Seyfert nuclei of NGC 3227 and 4151. Mon. Not. R. Astron. Soc. **305**, 319–324 (1999)

Frayer, D.T., Ivison, R.J., Smail, I., Yun, M.S., Armus, L.: Submillimeter imaging of the luminous infrared galaxy pair VV 114. Astron. J. **118**, 139–144 (1999)

Hunt, L.K., Thuan, T.X., Izotov, Y.I.: New light on the stellar populations in I Zw 18: deep near-infrared imaging. Astrophys. J. **588**, 281–298 (2003)

James, P., Casali, M.: Near-infrared emission from a spiral galaxy halo. IEEE Spectr. **9**, 14–16 (1996)

James, P.A.: Near infrared imaging of dwarf ellipticals irregulars and blue compact galaxies in the Virgo cluster. Mon. Not. R. Astron. Soc. **269**, 176 (1994)

James, P.A., Casali, M.M.: Confirmation of a faint red halo around NGC 5907. Mon. Not. R. Astron. Soc. **301**, 280–284 (1998)

James, P., Bate, C., Wells, M., Wright, G., Doyon, R.: Do galaxy mergers form elliptical galaxies? A comparison of kinematic and photometric properties. Mon. Not. R. Astron. Soc. **309**, 585–592 (1999)

James, P.: An infrared study of dwarf galaxies in the Virgo cluster. Mon. Not. R. Astron. Soc. **250**, 544–554 (1991)

Jones, D.H., Peterson, B.A., Colless, M., Saunders, W.: Near-infrared and optical luminosity functions from the 6dF Galaxy Survey. Mon. Not. R. Astron. Soc. **369**, 25–42 (2006)

Joseph, R.D., Meikle, W.P.S., Robertson, N.A., Wright, G.S.: Recent star formation in interacting galaxies. I – evidence from JHKL photometry. Mon. Not. R. Astron. Soc. **209**, 111–122 (1984a)

Joseph, R.D., Wade, R., Wright, G.S.: Detection of molecular hydrogen in two merging galaxies. Nature **311**, 132 (1984b)

Joseph, R.D., Wright, G.S.: Recent star formation in interacting galaxies. II – super starburst in merging galaxies. Mon. Not. R. Astron. Soc. **214**, 87–95 (1985)

Khosroshahi, H.G., Wadadekar, Y., Kembhavi, A., Mobasher, B.: A near-infrared photometric plane for elliptical galaxies and bulges of spiral galaxies. Astrophys. J. **531**, L103–L106 (2000)

Knapen, J.H., Beckman, J.E., Shlosman, I., Peletier, R.F., Heller, C.H., de Jong, R.S.: The striking near-infrared morphology of the inner region in M100. Astrophys. J. **443**, L73–L76 (1995)

Kochanek, C.S., Pahre, M.A., Falco, E.E., Huchra, J.P., Mader, J., Jarrett, T.H., Chester, T., Cutri, R., Schneider, S.E.: The K-band galaxy luminosity function. Astrophys. J. **560**, 566–579 (2001)

La Barbera, F., Busarello, G., Merluzzi, P., de la Rosa, I.G., Coppola, G., Haines, C.P.: The SDSS-UKIDSS Fundamental Plane of early-type galaxies. Astrophys. J. **689**, 913–918 (2008)

La Barbera, F., de Carvalho, R.R.: The origin of color gradients in early-type systems and their compactness at high-z. Astrophys. J. **699**, L76–L79 (2009)

La Barbera, F., de Carvalho, R.R., de La Rosa, I.G., Lopes, P.A.A.: SPIDER – II. The Fundamental Plane of early-type galaxies in grizYJHK. Mon. Not. R. Astron. Soc. **408**, 1335–1360 (2010a)

La Barbera, F., de Carvalho, R.R., de La Rosa, I.G., Lopes, P.A.A., Kohl-Moreira, J.L., Capelato, H.V.: SPIDER – I. Sample and galaxy parameters in the grizYJHK wavebands. Mon. Not. R. Astron. Soc. **408**, 1313–1334 (2010b)

Lonsdale, C.J., Persson, S.E., Matthews, K.: Infrared observations of interacting/merging galaxies. Astrophys. J. **287**, 95–107 (1984)

Martinez-Delgado, D., Penarrubia, J., Gabany, R.J., Trujillo, I., Majewski, S.R., Pohlen, M.: The ghost of a dwarf galaxy: fossils of the hierarchical formation of the nearby spiral galaxy NGC 5907. Astrophys. J. **689**, 184–193 (2008)

Mobasher, B., Guzman, R., Aragon-Salamanca, A., Zepf, S.: The near-infrared Fundamental Plane of elliptical galaxies. Mon. Not. R. Astron. Soc. **304**, 225–234 (1999)

Mobasher, B., Sharples, R.M., Ellis, R.S.: A complete galaxy redshift survey – Part Five – infrared luminosity functions for field galaxies. Mon. Not. R. Astron. Soc. **263**, 560 (1993)

Phillips, M.M., Aitken, D.K., Roche, P.F.: 8-13 micron spectrophotometry of galaxies. I – galaxies with giant H II region nuclei. Mon. Not. R. Astron. Soc. **207**, 25–33 (1984)

Prestwich, A.H., Joseph, R.D., Wright, G.S.: Starburst models of merging galaxies. Astrophys. J. **422**, 73–80 (1994)

Roche, P.F., Aitken, D.K.: 8-13-micron spectrophotometry of galaxies – Part Five – the nuclei of five spiral galaxies. Mon. Not. R. Astron. Soc. **213**, 789 (1985)

Roche, P.F., Whitmore, B., Aitken, D.K., Phillips, M.M.: 8-13 micron spectrophotometry of galaxies. II – 10 Seyferts and 3C 273. Mon. Not. R. Astron. Soc. **207**, 35–45 (1984)

Roche, P.F., Aitken, D.K., Smith, C.H., Ward, M.J.: An atlas of mid-infrared spectra of galaxy nuclei. Mon. Not. R. Astron. Soc. **248**, 606–629 (1991)

Ryder, S.D., Knapen, J.H.: High-resolution UKIRT observations of circumnuclear star formation in M100. Mon. Not. R. Astron. Soc. **302**, L7–L12 (1999)

Seigar, M.S., James, P.A.: The structure of spiral galaxies – I. Near-infrared properties of bulges, discs and bars. Mon. Not. R. Astron. Soc. **299**, 672–684 (1998a)

Seigar, M.S., James, P.A.: The structure of spiral galaxies – II. Near-infrared properties of spiral arms. Mon. Not. R. Astron. Soc. **299**, 685–698 (1998b)

Seigar, M.S., James, P.A.: A test of arm-induced star formation in spiral galaxies from near-infrared and H alpha imaging. Mon. Not. R. Astron. Soc. **337**, 1113–1117 (2002)

Shaw, M.A., Combes, F., Axon, D.J., Wright, G.S.: Isophote twists in the nuclear regions of barred spiral galaxies. Astron. Astrophys. **273**, 31 (1993)

Shlosman, I., Frank, J., Begelman, M.C.: Bars within bars – a mechanism for fuelling active galactic nuclei. Nature **338**, 45–47 (1989)

Smith, A.J., Loveday, J., Cross, N.J.G.: Luminosity and surface brightness distribution of K-band galaxies from the UKIDSS Large Area Survey. Mon. Not. R. Astron. Soc. **397**, 868–882 (2009)

Smith, C.H., Aitken, D.K., Roche, P.F.: The nature of the infrared luminous galaxies ARP 220 and NGC 6240. Mon. Not. R. Astron. Soc. **241**, 425–431 (1989)

Stott, J.P., Pimbblet, K.A., Edge, A.C., Smith, G.P., Wardlow, J.L.: The evolution of the red sequence slope in massive galaxy clusters. Mon. Not. R. Astron. Soc. **394**, 2098–2108 (2009)

Symeonidis, M., Rosario, D., Georgakakis, A., Harker, J., Laird, E.S., Page, M.J., Willmer, C.N.A.: The central energy source of 70 micron-selected galaxies: starburst or AGN? Mon. Not. R. Astron. Soc. **403**, 1474–1490 (2010)

Thuan, T.X., Hunt, L.K., Izotov, Y.I.: The Spitzer view of low-metallicity star formation. II. Mrk 996, a blue compact dwarf galaxy with an extremely dense nucleus. Astrophys. J. **689**, 897–912 (2008)

Valentinuzzi, T., and 14 colleagues.: Superdense massive galaxies in wings local clusters. Astrophys. J. **712**, 226–237 (2010)

Valentinuzzi, T., and 14 colleagues.: WINGS: a WIde-field nearby Galaxy-cluster survey. III. Deep near-infrared photometry of 28 nearby clusters. Astron. Astrophys. **501**, 851–864 (2009)

Vanzi, L., Hunt, L.K., Thuan, T.X., Izotov, Y.I.: The near-infrared view of SBS 0335-052. Astron. Astrophys. **363**, 493–506 (2000)

Wright, G.S., Joseph, R.D., Robertson, N.A., James, P.A., Meikle, W.P.S.: Recent star formation in interacting galaxies. III – evidence from mid-infrared photometry. Mon. Not. R. Astron. Soc. **233**, 1–23 (1988)

Wright, G.S., James, P.A., Joseph, R.D., McLean, I.S.: Elliptical-like light profiles in infrared images of merging spiral galaxies. Nature **344**, 417–419 (1990)

Chapter 21
WFCAM Surveys of Local Group Galaxies

Mike J. Irwin

Abstract From mid 2005 to 2008 we have been using WFCAM on UKIRT to probe the near-infrared properties of Local Group galaxies lying at distances of 0.5–1 Mpc. In addition to the dwarf galaxies accessible from UKIRT, which are mainly M31 satellites, we have also completed large area surveys of the Local Group spirals M31 and M33. The combination of near-infrared data from WFCAM with our existing optical data across the entirety of M31, M33 and the Local Group dwarf satellites, is enabling an unprecedented census and study of the structure of their stellar populations. I show some examples of the near-infrared results obtained which demonstrate the power of WFCAM as a wide field survey imaging system.

Introduction

The Local Group provides a unique opportunity for detailed study of the structure and evolution of a wide variety of "typical" galaxies. Recent significant advances in instrumentation have opened up analysis of the global properties of these galaxies facilitating a better understanding of their construction and evolution from detailed studies of their stellar content. Such studies are relevant to a wide variety of astronomical problems, including stellar evolution, stellar population synthesis, galactic structure and, increasingly, the predictions of cosmological models. Hierarchical formation scenarios (e.g. White and Rees 1978; Searle and Zinn 1978), and in particular Λ-CDM cosmologies, are now sufficiently well developed theoretically that the Local Group provides a means of directly testing and constraining these

M.J. Irwin (✉)
Cambridge Astronomy Survey Unit, Institute of Astronomy, Madingley Road,
Cambridge CB3 0HA, UK
e-mail: mike@ast.cam.ac.uk

theories, particularly with regard to the amount, age and metallicity distribution of the substructure predicted to be found in the outer parts of galaxy disks and in galaxy halos.

Photometric knowledge of a variety of galaxies in the Local Group that differ in luminosity, mass and metallicity is of prime importance if we wish to understand the structure and evolution of galaxies. WFCAM has provided a unique opportunity to use JHK photometry to study the AGB populations for a large number of Local Group galaxies. These stars are excellent tracers of the structural, chemical and kinematic evolution of the intermediate-age stellar populations in these galaxies. Furthermore, by surveying to a depth of $K = 19$ we can isolate and trace the AGB component of each galaxy, in particular the cool carbon star and M-giant populations, and also probe the upper parts of the RGB. In combination with optical data this also enables mapping the spatial variation of extinction within each galaxy.

Why Near-Infrared Imaging?

Near-infrared imaging of nearby galaxies comes with several benefits and a few handicaps. The near-infrared regime is of course much better able to penetrate regions affected by dust extinction and, somewhat ironically, the relative lack of sensitivity to bluer and/or older stellar populations compared to optical imaging, means that detailed analysis of the nuclear regions of nearby galaxies is more readily achieved. Near-infrared surveys are particularly sensitive to young and intermediate-age AGB stars and also to older RGB populations and provide excellent extra leverage on the age-metallicity-extinction degeneracies that plague optical surveys. Furthermore, most giant-branch stars are significantly redder than the majority foreground Milky Way dwarf population, which allows good discrimination against foreground contaminants. However, there are some notable drawbacks too, mainly caused by the combination of interesting detector characteristics and the bright spatially and temporally varying sky background. This adds considerably to the complexity of data processing although the availability of the excellent all-sky calibration system provided by 2MASS (Skrutskie et al. 2006) goes a long way to compensate.

Data Processing

All of our WFCAM survey data was processed using the Cambridge Astronomical Survey Unit (CASU) pipeline as part of the normal nightly processing operations for WFCAM. Single band data products include stacked images and catalogues with morphological classification of detected objects and, as usual for WFCAM, the astrometric and photometric calibration is provided by 2MASS (see Hodgkin et al. 2009 for more details).

Each set of pawprint-level catalogues in J,H,K-bands were then merged individually at the detector level to form a series of sub-region catalogues. For each galaxy surveyed these sub-regions were then concatenated to form an overall band-merged catalogue. The final step for each galaxy is to isolate and remove duplicate entries based on positional coincidence, always keeping the entry with the best photometric error.

Uniformity of the resulting catalogues is crucial for this type of survey and mainly relies on two factors: the excellent imaging quality of UKIRT + WFCAM; and the accuracy of the overall astrometric and photometric calibration. The examples shown in the next section provide a good illustration of this.

A Selection of WFCAM Survey Results

We have now completed WFCAM JHK surveys of 20 square degrees of M31 covering all of the central regions and outer disk substructures previously detected in our optical surveys (e.g. Ferguson et al. 2002; Ibata et al. 2007). Figure 21.1 shows

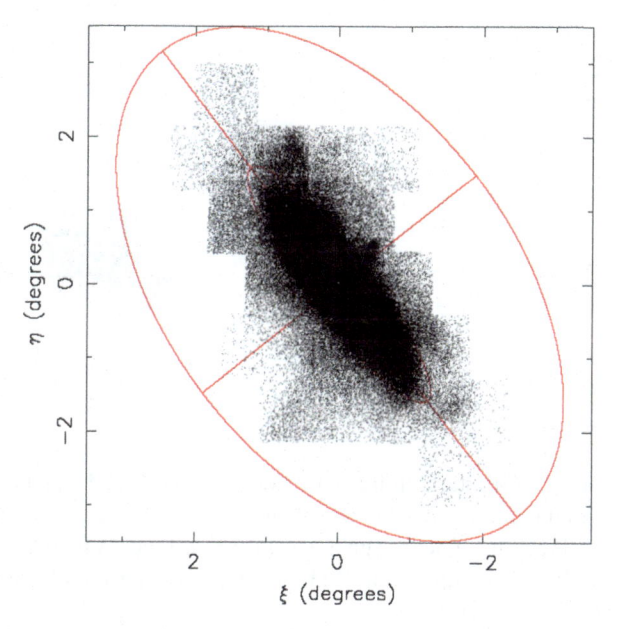

Fig. 21.1 The distribution of AGB stars from our M31 region survey. The full extent of the classical M31 disk, to a radius of 27 kpc, is shown by the barely visible inner ellipse, while the outer ellipse denotes an ellipsoidal halo with c/a = 0.6 and 55 kpc radius. Note the large extent of this intermediate-age AGB population and the prominent clumps and stream-like features clearly separated from the main disk. These features are also seen in our optical surveys of this system and demonstrate unambiguously the chaotic nature of recent events in the growth of M31

Fig. 21.2 The near-infrared J,K colour-magnitude diagram for NGC147 provides a good illustration of the power of near-infrared surveys to discriminate intermediate age AGB populations, cool C-stars and M-giants, from contaminating foreground Milky Way stars. The boundaries indicated on the accompanying two-colour J,H,K diagram demonstrate the clear separation between foreground Milky Way dwarf stars which lie to the left of the vertical division and the distinction between M-giants below the horizontal line with the C-stars above it

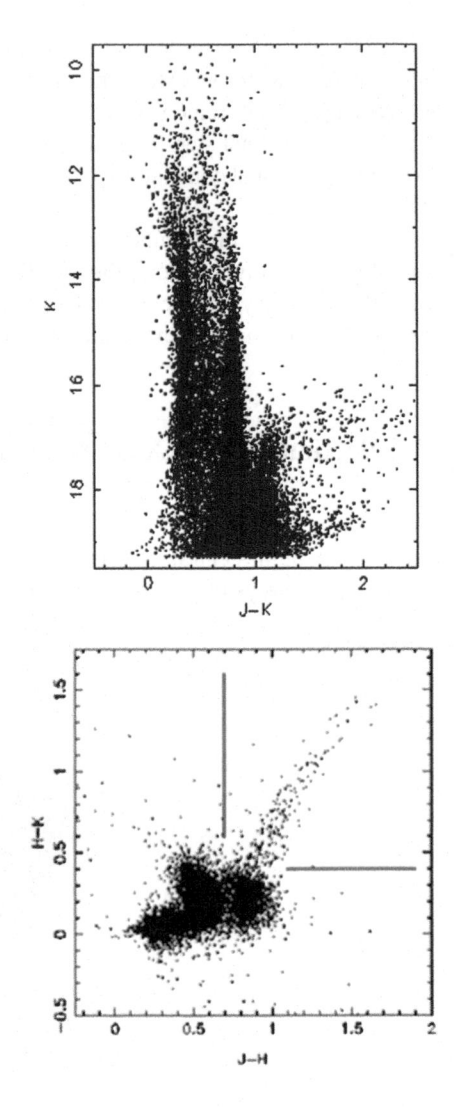

the distribution of AGB stars, in this case cool C-stars and M-giants, over the full survey region and demonstrates the advantage of near-infrared surveys for probing the dense central regions of nearby galaxies. In contrast, our WFCAM survey results for a 3 square-degree region around M33 show a seemingly undisturbed classical disk galaxy (Cioni et al. 2008).

The nearby dwarf and irregular galaxies also provide excellent laboratories to study stellar evolution. In particular the luminous stellar populations of the AGB phase are direct tracers of the structural, chemical and kinematic evolution of the intermediate-age components of these galaxies. However, with the exception of a few small, restricted fields, large scale observations of nearby galaxies in the near-infrared have until now been missing.

Fig. 21.3 The distribution of AGB stars around NGC147 and NGC185 as seen in WFCAM surveys. NGC147 shows the twisted isophotes typical of strong tidal interaction, presumably with M31, whereas NGC185 shows no obvious signs of any tidal disturbance

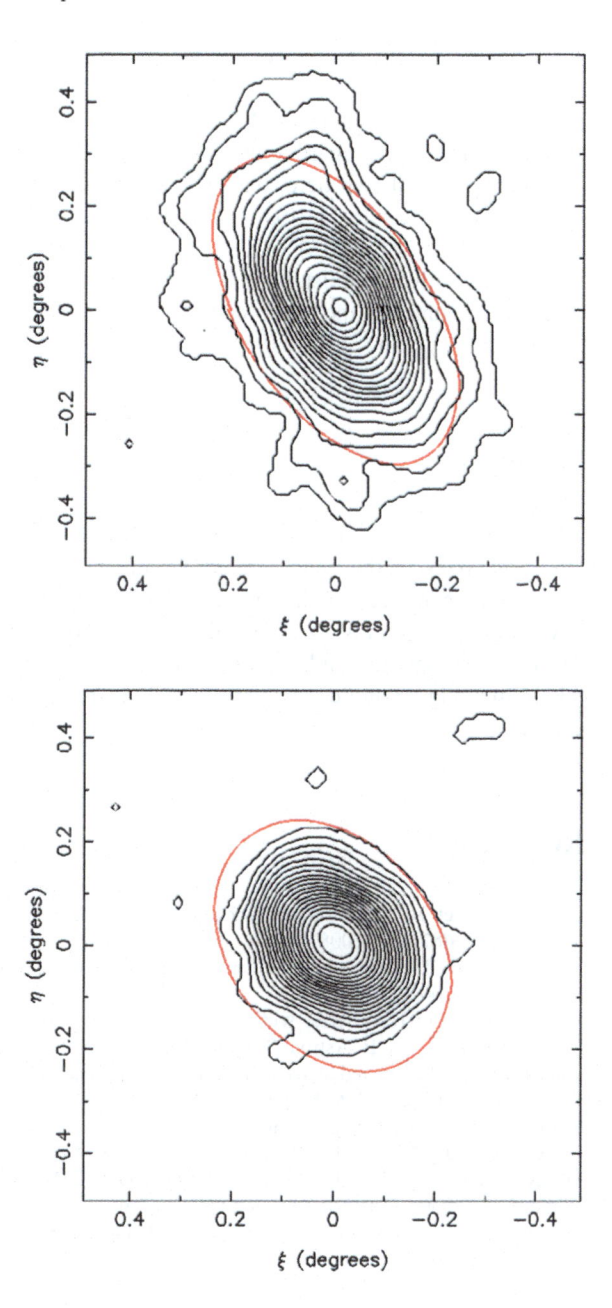

We illustrate the power of WFCAM JHK surveys using data on NGC147, a dwarf satellite of M31 as an example. As noted previously, discrimination against foreground stars is excellent and this is illustrated in Fig. 21.2, which shows colour-magnitude and two-colour diagrams for NGC147. With a clean selection of dwarf galaxy member stars it is then straightforward to analyse their spatial distribution and look for signs, for example, of tidal disturbances. Interestingly the postulated binary pair of M31 satellites, NGC148 and NGC185 show completely different characteristics in this regard as shown in Fig. 21.3.

Although not shown here, the same overarching tidal distortion is also clearly seen in the AGB star distributions around NGC205 and M32 which have been covered as part of the M31 survey. In these cases other strong evidence for tidal interactions with M31 exist, but the clear visibility of these effects in their AGB star distributions supports the interpretation of the effects of strong tidal interactions in the NGC147 system.

Summary

The field-of-view and sensitivity of WFCAM is well-matched to surveying Local Group galaxies in the near-infrared. Complete maps of the cool C-star and M-giant components across the entirety of M31, M33 and their dwarf satellites are now available enabling direct studies of the properties of their intermediate-age AGB populations. In combination with optical imaging and spectroscopic surveys this is allowing an unprecedented census and study of their stellar populations and evolution.

References

Cioni, M.-R., et al.: AGB stars as tracers of metallicity and mean age across M33. Astron. Astrophys. **487**, 131 (2008)
Ferguson, A., et al.: Evidence for Stellar substructure in the halo and outer disk of M31. Astron. J. **124**, 1452 (2002)
Hodgkin, S., et al.: The UKIRT wide field camera ZYJHK photometric system: calibration from 2MASS. Mon. Not. R. Astron. Soc. **394**, 675 (2009a)
Ibata, R., et al.: The haunted halos of Andromeda and Triangulum: a panorama of galaxy formation in action. Astrophys. J. **671**, 1591 (2007)
Searle, L., Zinn, R.: Compositions of halo clusters and the formation of the galactic halo. Astrophys. J. **225**, 357 (1978)
Skrutskie, M.F., et al.: The Two Micron All Sky Survey (2MASS). Astron. J. **131**, 1163 (2006a)
White, S., Rees, M.: Core condensation in heavy halos – a two-stage theory for galaxy formation and clustering. Mon. Not. R. Astron. Soc. **183**, 341 (1978)

Chapter 22
HiZELS: The High Redshift Emission Line Survey with UKIRT

Philip Best, Ian Smail, David Sobral, Jim Geach, Tim Garn, Rob Ivison, Jaron Kurk, Gavin Dalton, Michele Cirasuolo, and Mark Casali

Abstract We report on HiZELS, the High-z Emission Line Survey, our successful panoramic narrow-band campaign survey using WFCAM on UKIRT to detect and study emission line galaxies at $z \sim 1$–9. HiZELS employs the $H_2(S1)$ narrow-band filter together with custom-made narrow-band filters in the J and H-bands, with the primary aim of delivering large identically-selected samples of Hα emitting galaxies at redshifts of 0.84, 1.47 and 2.23. Comparisons between the luminosity function, the host galaxy properties, the clustering, and the variation with environment of these Hα-selected samples are yielding unique constraints on the nature and evolution of star-forming galaxies, across the peak epoch of star formation activity in the Universe. We provide a summary of the project status, and detail the main

---

P. Best (✉) • D. Sobral • T. Garn
SUPA, Institute for Astronomy, Royal Observatory, Blackford Hill, Edinburgh EH9 3HJ, UK
e-mail: pnb@roe.ac.uk

I. Smail • J. Geach
Institute of Computational Cosmology, Durham University, South Road, Durham DH1 3LE, UK

R. Ivison
Astronomy Technology Centre, Royal Observatory, Blackford Hill, Edinburgh EH9 3HJ, UK

J. Kurk
Max-Planck-Institut für Astronomie, D-69117 Heidelberg, Germany

G. Dalton
Department of Physics, University of Oxford, Keble Road, Oxford OX1 3RH, UK

Space Science and Technology, Rutherford Appleton Laboratory, Didcot OX11 0QX, UK

M. Cirasuolo
SUPA, Institute for Astronomy, Royal Observatory, Blackford Hill, Edinburgh EH9 3HJ, UK

Astronomy Technology Centre, Royal Observatory, Blackford Hill, Edinburgh EH9 3HJ, UK

M. Casali
European Southern Observatory, Karl-Schwarzschild-Straße 2, D-85748 Garching bei München, Germany

scientific results obtained so far: the measurement of the evolution of the cosmic star formation rate density out to $z > 2$ using a single star formation indicator, determination of the morphologies, environments and dust content of the star-forming galaxies, and a detailed investigation of the evolution of their clustering properties. We also summarise the on-going work and future goals of the project.

Introduction

The fundamental observables required to understand the basic features of galaxy formation and evolution are the volume-averaged star formation rate as a function of epoch, its distribution function within the galaxy population and the variation with environment. Surveys of the star-formation rate as a function of epoch suggest that the star-formation rate density rises as $\sim(1+z)^4$ out to at least $z \sim 1$ (e.g. Lilly et al. 1996), and then flattens, with the bulk of stars seen in galaxies today having been formed between $z \sim 1$–3. Determining the precise redshift where the star-formation rate peaked is more difficult, however, with different star-formation indicators giving widely different measures of the integrated star-formation rate density (see Hopkins and Beacom 2006). These problems are exacerbated by the effects of cosmic variance in the current samples, which are typically based on small-field surveys.

The Hα emission line is a very well-calibrated measure of star-formation rate in the nearby Universe (e.g. Kennicutt 1998; Moustakas et al. 2006). As it redshifts through the optical and near-IR bands, it offers a single star-formation indicator which can be studied from $z = 0$ to $z \sim 3$, right through the peak star-formation epoch in the Universe. It is relatively immune to dust extinction, and has sufficient sensitivity that estimates of the integrated star-formation rate do not require large extrapolations for faint sources below the sensitivity limit: surveys with a sensitivity of ~ 10 M$_{sun}$/year can be undertaken in Hα at $z \sim 2$ with current instrumentation, compared to limits of ~ 100–1,000 M$_{sun}$/year for other dust-independent tracers such as radio, far-infrared and sub-mm luminosities (Ivison et al. 2007).

The Hα emission line has been widely used as a method of tracing the evolution of star formation, both through spectroscopic surveys and via imaging surveys exploiting narrow-band filters (e.g. Gallego et al. 1995; Yan et al. 1999; Tresse et al. 2002; Doherty et al. 2006). Narrow-band surveys offer a sensitive and unbiased method of detecting emission-line objects lying in large well-defined volumes; the sources are identified on the strength of their emission line and thus crudely represent a star-formation rate-selected sample, and they must lie in a narrow range in redshift. Before the advent of large-area near-IR detectors such as WFCAM, narrow-band Hα surveys in the near-IR (i.e. at $z > 0.7$) were limited to very small areas and sample sizes (the largest at $z \sim 2$ had just ~ 10 candidate sources; Moorwood et al. 2000). The primary goal of HiZELS is to overcome this, with wide-area surveys using narrow-band filters in the J, H and K-bands to detect of order a thousand star-forming galaxies in Hα at each of three redshifts: 0.84, 1.47

and 2.23. These large samples, selected with a uniform selection function, can be used to determine the Hα luminosity function (LF) at each epoch, investigate any strong changes in its shape, and provide the first reliable estimate of the change in the global star-formation rate of the Universe between z = 0 and 2.2 using a single tracer of star formation.

It is not only the global average star formation rate which is important for our understanding of galaxy formation and evolution, but more crucially the nature and distribution of the star-forming galaxies at high redshifts. Galaxies form and evolve within the hierarchically growing dark-matter haloes of a ΛCDM Universe, but the details of the galaxy formation process depend upon the complicated gas dynamics of star formation and feedback, and these processes are poorly understood. A surprising result of many recent studies is that the stellar populations of the most massive galaxies formed earlier than those of less massive galaxies – a process often referred to as "downsizing" (e.g. Cowie et al. 1996). Massive galaxies must therefore form stars rapidly at an early epoch, and then have their star formation truncated, for example by feedback from AGN (e.g. Bower et al. 2006; Best et al. 2006) – but the epoch at which this occurred is still uncertain. In the local Universe, star formation is also suppressed in dense environments (e.g. Lewis et al. 2002; Best 2004); this effect diminishes with increasing redshift, with hints that it disappears altogether at $z \sim 2$ (Kodama et al. 2007). But where precisely, in terms of epoch and environment, does this environmental influence begin to become important, and to what extent is the build-up of galaxies into groups and clusters since $z \sim 1$ responsible for the sharp decline of the cosmic star formation rate density since $z \sim 1$?

HiZELS aims to tackle all of these issues by obtaining large samples of star forming galaxies in representative volumes at three epochs across the peak epoch of star formation in the Universe. Coupled with lower redshift studies, we are determining how the characteristic stellar mass of Hα-selected galaxies declines with redshift between z = 2.2 and z = 0, investigating the physical processes involved in the downsizing activity. We are investigating changes in the Hα luminosity function as a function of environment at each epoch and between different epochs: sky areas of a few square degrees are required to probe the full range of galaxy environments at these redshifts. The samples are also large enough to give a robust measurement the clustering properties of the Hα emitters, split into sub-populations, providing important insights into their properties, including information about the relative masses of their dark matter haloes. Combining all of this information, HiZELS will provide a strong test of theoretical models of galaxy evolution (e.g. Benson et al. 2000; Baugh et al. 2005; Bower et al. 2006) and a direct input into these models.

In these proceedings we outline the current status and future plans of HiZELS. In section "Observations and sample selection", we describe the observational strategy. In section "Scientific results from HiZELS", we show the constraints obtained on the Hα luminosity function and cosmic star formation rate, and discuss the other scientific results to date. We outline our on-going scientific work using HiZELS, and discuss future plans, in section "On-going work and future plans". Section "Conclusions" presents brief conclusions.

Observations and Sample Selection

Observation Strategy and Fields

HiZELS uses observations through narrow-band filters in the J, H and K-bands (NB$_J$, NB$_H$, H$_2$(S1)), with central wavelengths of 1.211, 1.619 and 2.121μm respectively), using WFCAM on UKIRT. Coupled with broad-band filter observations, these are used to detect star-forming galaxies in Hα at redshifts 0.84, 1.47 and 2.23, over several degree-scale regions of the extra-galactic sky. Of course, narrow-band surveys are not sensitive to only one emission line, but will detect many different emission lines at different redshifts, redshifted into the filter. To achieve many of our goals it is necessary to identify which of the emission-line objects are indeed Hα emitters. HiZELS is achieving this by targeting the best-studied regions of the extragalactic sky, in which a wealth of multi-wavelength data already exists. Photometric redshifts and colour-selections are being backed up by statistical analysis of the contamination rates derived from follow-up spectroscopy of sub-samples of emitters, as described below.

The presence of other emission line samples within the filters is actually one of the strengths of HiZELS. The custom-made NB$_H$ and NB$_J$ filters were specially designed such that the [OIII] 5007 and [OII] 3727 lines would fall into those filters for galaxies at $z = 2.23$. This provides both a confirmation of the redshifts for a subset of the $z = 2.23$ Hα-selected sample, as well as allowing a first investigation of the emission line properties of these sources. Other emission lines of interest for these filters include the possibility of detecting Lyα emission from galaxies at $z = 8.90$ in the NB$_J$ filter.

HiZELS was awarded 22 nights on UKIRT over Semesters 07B to 09B for its first phase (of which roughly one-third has been lost to bad weather), and has a provisional allocation of 23 clear nights for a second phase, during semesters 10A to 12A (subject to UKIRT remaining operational). The original survey strategy involved observations of each field in each of the 3 filters for a total on-sky observing time of ∼20 ks/pix, in order to obtain uniform coverage down to a line flux of 10^{-16} erg s^{-1} cm^{-2} in each filter. In addition, a single deeper paw-print (i.e. 0.2 sq. deg., to 65 ks/pix depth) using the H$_2$(S1) filter has been taken to probe further down the Hα luminosity function and assess the completeness of the shallower but wider survey. As the survey has progressed, the observing strategy has been slightly modified based on the survey results, to an overall aim of observing as close as possible to 1,000 Hα emitters at each redshift, split roughly half-and-half above and below the break of the luminosity function at each redshift. To achieve this, additional deeper paw-print observations are proposed using the H$_2$(S1) and NB$_H$ filters, whilst the depth of the standard survey observations using the NB$_J$ and NB$_H$ filters, beyond the first two fields, has been reduced. The full set of proposed observations for HiZELS, along with the current observation status, is provided in Table 22.1.

Table 22.1 HiZELS target fields, proposed exposure times, and current observational status

Field name	Area (sq. deg.)	Target exposure time (ks/pix) NB$_J$	NB$_H$	H$_2$(S1)	Completion (%, Dec 2009) NB$_J$	NB$_H$	H$_2$(S1)
UKIIDSS UDS	0.8	20.0	20.0	20.0	100	100	100
COSMOS-1[a]	0.8	20.0	20.0	20.0	100	50	100
COSMOS-2[a]	0.8	3.0	14.0	20.0	0	0	0
ELAIS N1	0.8	3.0	14.0	20.0	0	38	100
SA 22	0.8	3.0	14.0	20.0	0	0	50
Boötes	0.8	3.0	14.0	20.0	0	0	0
Lockman Hole	0.8	3.0	14.0	20.0	0	0	0
COSMOS-DeepK	0.4			65.0			50
COSMOS-DeepH	0.2		50.0			0	

[a]Two WFCAM pointings will be placed inside the 2 sq. deg. COSMOS field

Selection of Narrow-Band Hα Emitters

Combining the narrow-band observations with broad-band observations of the same fields (either our own dedicated observations, or archival data from the UKIDSS survey), narrow-band emitters are selected according to the following criteria: (i) the object must be robustly detected on the narrow-band image, with a signal-to-noise (S/N) above 3, in a 3-arcsec diameter aperture; (ii) it must present a 'broad-band minus narrow-band' colour excess with a significance $\Sigma \geq 2.5$; (iii) the line emission must have equivalent width above 50 Å; (iv) the object must be visually confirmed as reliable, and not associated with any cross-talk artefact. The top panel of Fig. 22.1 shows these selection criteria for the NB$_J$ observations of the COSMOS-1 pointing. For more details on these selections, see Geach et al. (2008; hereafter G08) or Sobral et al. (2009a; hereafter S09a). S09a have shown that these criteria are very robust.

Our observations identify approximately 800, 350 and 300 narrow-band emitters per pointing (0.8 sq. deg.) in the NB$_J$, NB$_H$, and H$_2$(S1) observations, respectively. Photometric redshifts and colour selections are then used to identify which of the emission line galaxies are Hα. We are able to successfully recover relatively clean Hα samples, since strong contaminating emission lines are sufficiently well spread in wavelength from Hα that photometric redshifts do not have to be very precise: $\Delta z/z \sim 0.5$ – more details can be found in G08, S09a, Sobral et al. (2012), and Geach et al. (2010). In the NB$_J$ observations, photometric redshifts indicate that over half of the detected emitters are indeed Hα emitters at $z = 0.84$, with a significant fraction of the remainder being Hβ or [OIII] emitters at $z \sim 1.4$ (see lower left panel of Fig. 22.1). Archival spectroscopic redshifts for over 100 of the emitters confirm the high completeness and reliability of the photometric selection (Fig. 22.1, lower right panel; see also S09a). In the COSMOS and UDS fields, photometric redshifts provide a similar level of accuracy for selecting Hα emitters at $z = 1.47$ from the NB$_H$ observations (see Sobral et al. 2012), and approximately half of the narrow-band emitters are associated with Hα. The H$_2$(S1) observations suffer considerably more contamination from lower redshift emitters (e.g. Paschen and Brackett series; see G08), but are producing around 90 candidate $z = 2.23$ sources per field.

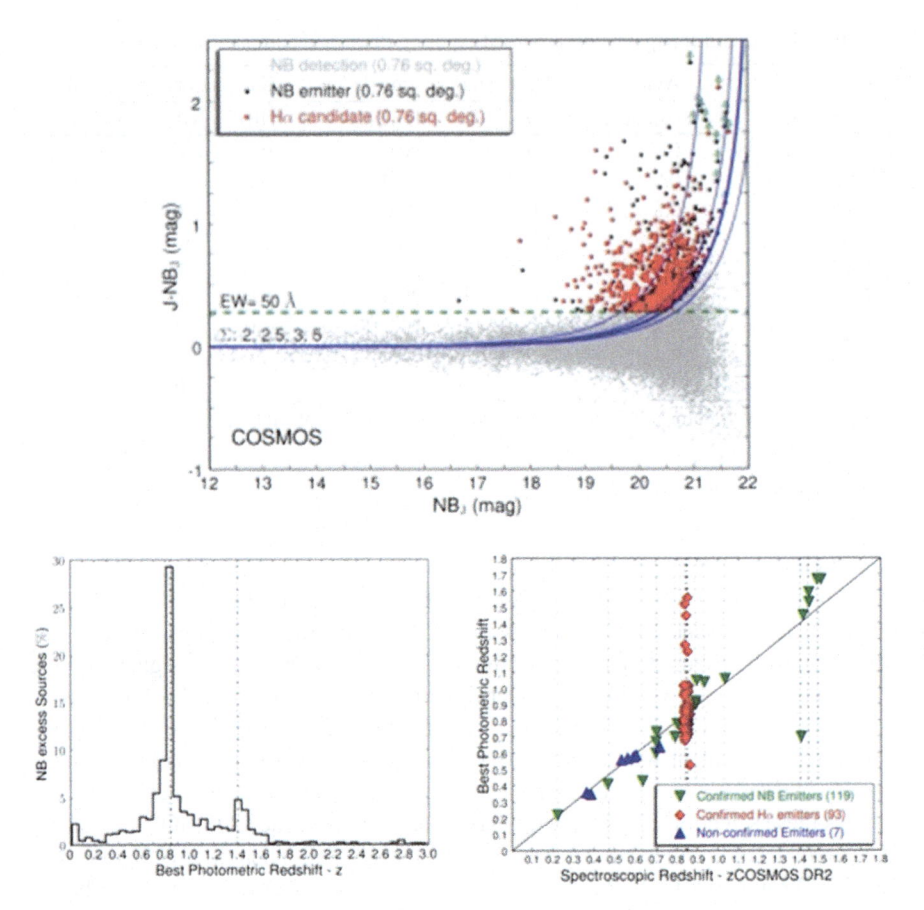

Fig. 22.1 *Top*: A colour-magnitude plot demonstrating the selection of narrow-band excess sources (adopted from S09a). All >3σ detections in the NB$_J$ image are plotted and the curves represent Σ significances of 5, 3, 2.5 and 2, respectively. The dashed line represents an equivalent width cut of 50 Å. All selected narrow-band emitters are plotted in *black*, while candidate Hα emitters (selected using photometric redshifts) are plotted in *red*. *Bottom left*: The distribution of photometric redshifts of the NB$_J$ excess sources, showing clear peaks for Hα at z = 0.84 and Hβ or [OIII] at z ∼ 1.4. *Bottom right*: a comparison between photometric and archival spectroscopic redshifts, demonstrating the reliability of the sample (From Sobral et al. 2009a)

Scientific Results from HiZELS

The Hα Luminosity Function and the Cosmic Star Formation Rate Density

HiZELS has already resulted in by far the largest and deepest survey of emission line selected star-forming galaxies at each of the three targeted redshifts, and has

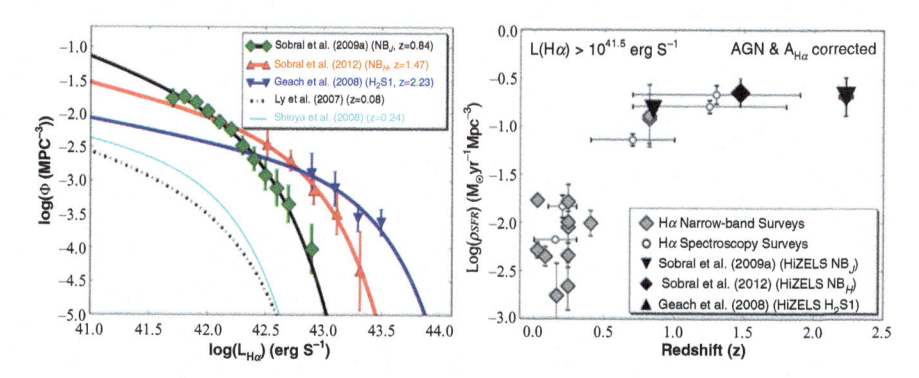

Fig. 22.2 *Left*: The z = 0.84, z = 1.47 and z = 2.23 Hα luminosity functions from HiZELS (corrected for [NII] contamination, completeness, extinction and filter profile biases) with the best-fit Schechter functions overlaid. Other luminosity functions from Hα surveys at different redshifts are presented for comparison, showing a clear evolution of the LF out to z > 2. *Right*: The evolution of ρ_{SFR} as a function of redshift based on Hα (down to the HiZELS limit). This shows a rise in ρ_{SFR} up to at least z ~ 1, slightly steeper than the canonical $(1+z)^4$, followed by a flattening out to at least z ~ 2.2

greatly improved determinations of the Hα luminosity function. It has produced the first reliable Hα LF at z = 2.23 (G08; Geach et al. 2010), as well as providing the first statistically significant samples at redshifts 0.84 (S09a) and 1.47 (Sobral et al. 2012). The luminosity functions are derived after correcting the observations for: (i) contamination of the emission line flux by the nearby [NII] line (using the relation between the flux ratio $f_{[NII]}/f_{H\alpha}$ and the total measured equivalent width, derived by Villar et al. 2008); (ii) extinction of the Hα emission line, taken to be the canonical value of 1 magnitude (but see section "The dust extinction properties of high-z star-forming galaxies" for more details on this); (iii) the detection completeness of faint galaxies, and the selection completeness for detected galaxies with faint emission lines (evaluated through Monte-Carlo simulations); (iv) filter profile effects, due to the filter not being a perfect top-hat (again, evaluated through Monte-Carlo simulations). For more details see G08 and S09a.

The derived luminosity functions show very strong evolution from redshift zero right out to z = 2.23 (G08, S09a; see left panel of Fig. 22.2). At all redshifts the Hα LF is found to be well-fitted by a Schechter function, but the form of the LF undergoes dramatic evolution through the redshift range probed by HiZELS. φ^* and L^* both evolve strongly from the local Universe out to at least z ~ 1, but beyond that L^* continues to rise up to z ~ 2 whilst φ^* peaks at z ~ 1 and then decreases at higher redshifts. The integrated luminosity function is used to estimate the cosmic star formation rate density (ρ_{SFR}) at each redshift; using a single star formation tracer (Hα) from z = 0 to z = 2.23, ρ_{SFR} is found to rise strongly up to z ~ 1 and then appears to flatten out to z ~ 2.2 (right panel of Fig. 22.2).

Fig. 22.3 The merger fraction as a function of Hα luminosity at z = 0.84. This shows a clear dependency, with mergers dominating above L* and passive quiescent galaxies dominating below that (From Sobral et al. 2009a)

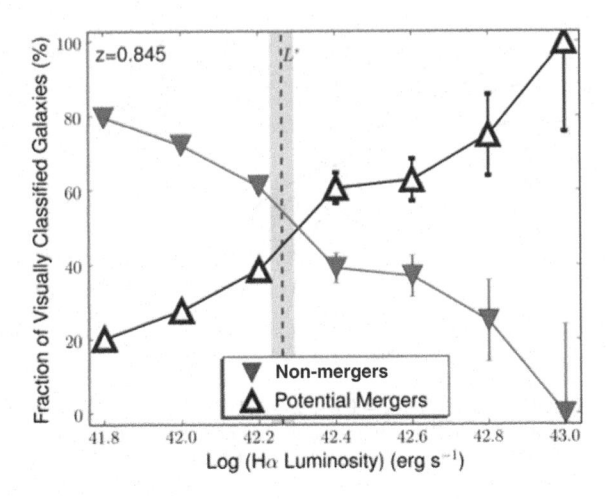

The Morphologies of the Hα Emitters

At z = 0.84, the Hα emitters are mostly morphologically classed as disks, with irregulars and mergers forming a much smaller fraction of the sample (Fig. 22.3; S09a). A strong relation is found between morphology and Hα luminosity, however, with the fraction of irregulars/mergers rising steadily with luminosity and the fraction of quiescent disks falling; the break of the luminosity function seems to define a critical switch-over luminosity between the two populations. Out to $z \sim 1$, the integrated ρ_{SFR} is produced predominantly by disk galaxies and it is their evolution which drives the strong increase in the cosmic star formation rate density from the current epoch to redshift one. In contrast, the continued strong evolution of L* between z = 0.84 and z = 2.23 suggests an increasing importance of merger-driven star formation activity beyond $z \sim 1$, as mergers and irregulars dominate the bright end of the luminosity function. Analysis of the first z = 2.23 Hα emitters (G08) indicates that these do show a range of morphologies, and that indeed many show evidence of on-going merger activity. The completed HiZELS survey will provide statistically significant samples of Hα emitters at z = 2.23 and z = 1.47, allowing a direct test of whether the change in the form of the luminosity function at $z \sim 1$ is driven by the different evolutionary behaviour of these two different populations of star-forming galaxies.

Clustering of Star-Forming Galaxies and Environmental Variations

HiZELS is ideal for investigating the clustering of SF galaxies because the narrow-band selection removes almost all of the projection effects which usually degrade

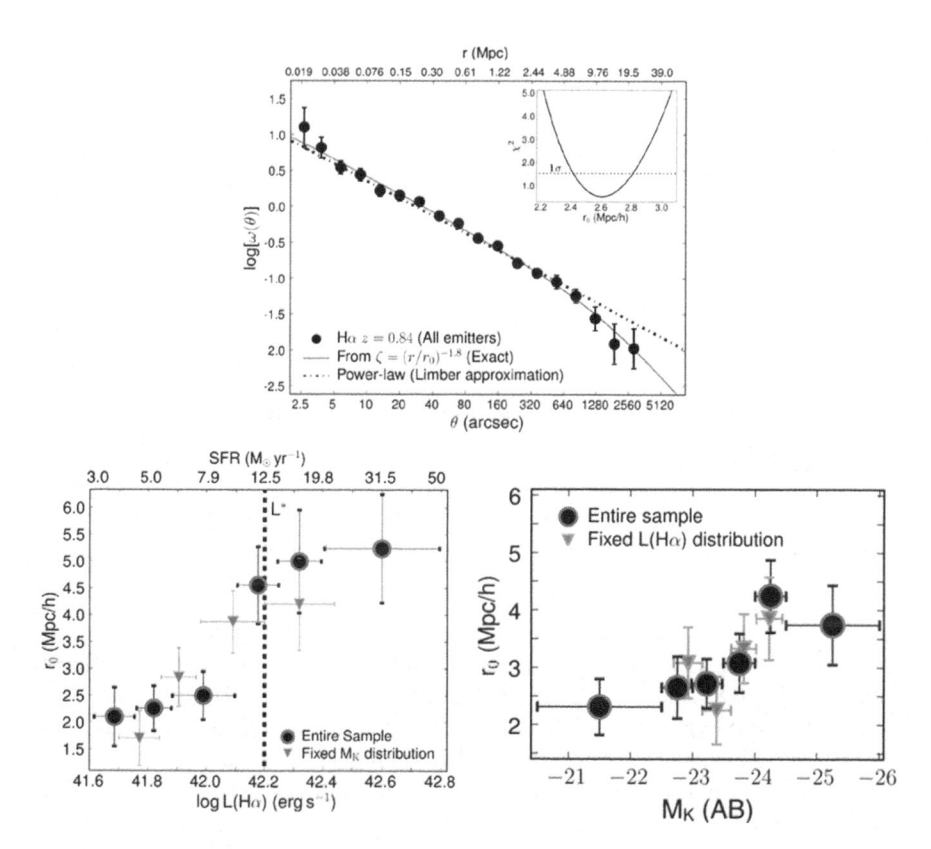

Fig. 22.4 *Top*: The angular cross-correlation function of Hα emitters at redshift z = 0.84 in the COSMOS and UDS fields. *Bottom left, bottom right*: The dependence of the clustering amplitude at this redshift on star formation rate and absolute K-band magnitude (From Sobral et al. 2010)

clustering analysis based on imaging data. The HiZELS emitters are observed to be significantly clustered at all redshifts. At z = 0.84 the characteristic correlation length calculated from the Hα emitters is $r_0 = 2.7 \pm 0.3$ h^{-1} Mpc (Sobral et al. 2010), consistent with them residing in dark matter halos of mass $\sim 10^{12}$ M$_{sun}$ at that epoch, close to that expected for the progenitors of Milky Way galaxies. Using the large sample available at this redshift, the clustering is found to be strongly dependent on Hα luminosity, and also to increase more weakly with near-infrared magnitude (roughly tracing stellar mass). The clustering amplitude is found to be independent of morphology, once the dependencies of morphology on stellar mass and star formation rate have been accounted for. These results, shown in Fig. 22.4, are qualitatively in line with those found in the nearby Universe.

The left-panel of Fig. 22.5 shows the clustering amplitude of Hα emitters measured from narrow-band surveys at different redshifts. However, the strong dependence of clustering amplitude on Hα luminosity implies that considerable care must be taken when attempting to compare these, since they have very different

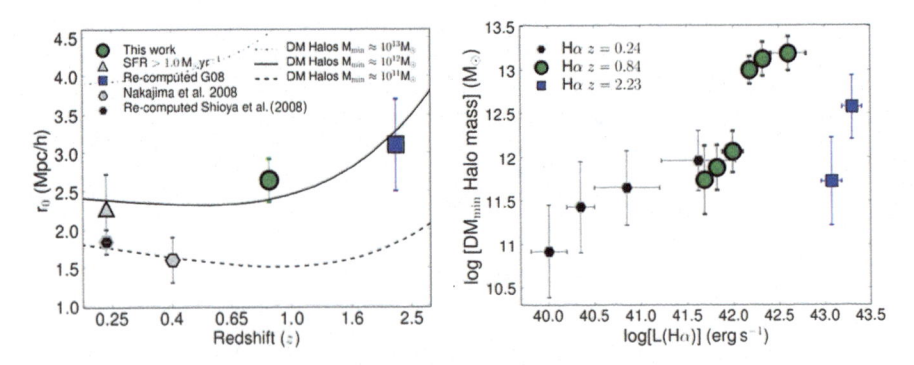

Fig. 22.5 *Left*: The clustering length (r_0) as a function of redshift for Hα emitters selected by narrow-band surveys. The Hα emitters at $z = 0.84$ and $z = 2.23$ studied by HiZELS reside in typical dark matter haloes of $M_{min} \approx 10^{12}$ M_{sun}, consistent with being the progenitors of Milky-Way type galaxies. The lower luminosity Hα emitters found in smaller volumes at $z = 0.24$ and $z = 0.4$ reside in less massive haloes. *Right*: The minimum mass of host dark matter haloes as a function of Hα luminosity at three different redshifts (From Sobral et al. 2010)

luminosity limits. The right-hand panel of Fig. 22.5 compares the clustering amplitude of the Hα emitters, as a function of Hα luminosity, at three different redshifts: $z = 0.84$ and $z = 2.23$ from HiZELS (Sobral et al. 2010, and G08), and $z = 0.24$ from Shioya et al. (2008). It can be seen that at a given halo mass, star formation is much more efficient (higher $L_{H\alpha}$) at higher redshifts.

The increase in $L_{H\alpha}$ at given halo mass with redshift nearly exactly mirrors that of the increase in L^* of the Hα luminosity function. This implies that galaxies in a given dark matter halo mass may form stars at the same fraction of the characteristic star formation rate at that redshift, at all epochs (cf. Sobral et al. 2010). This in turn would suggest a fundamental connection between the strong negative evolution of the Hα L^* since $z \sim 2$ and the quenching of star formation in galaxies within haloes significantly more massive than 10^{12} M_{sun}.

The Masses and Environments of Star-Forming Galaxies

In the nearby Universe, the stellar populations of the most massive galaxies are observed to have formed earlier than those of less massive galaxies. The HiZELS sample allows a direct investigation of the redshift at which this "down-sizing" process began, and a measurement of how the characteristic stellar mass of star-forming galaxies varies with redshift. At $z = 0.84$ we find a strong dependence of the proportion of galaxies which are actively forming stars as a function of mass: for $M \approx 10^{10}$ M_{sun}, around a third of galaxies are detected by HiZELS, but at higher masses ($> 10^{11.5}$ M_{sun}) this fraction falls away to a value consistent with zero (Sobral et al. 2011). This indicates that down-sizing is already in place at $z = 0.84$.

Fig. 22.6 The fraction of galaxies forming stars as a function of local galaxy surface density, for Hα emitters at z = 0.84. The star-forming fraction increases with local surface density in the field, but then decreases in group and cluster environments (From Sobral et al. 2011)

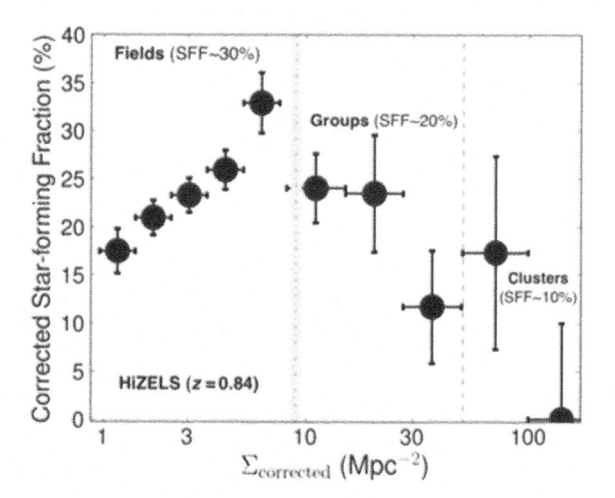

We find that at z = 0.84 the fraction of star-forming galaxies increases with local galaxy surface density at low galaxy surface densities, but then falls in group and cluster environments (Fig. 22.6). The median star formation rate of the Hα emitters increases with local galaxy density: those residing in denser regions are mostly starbursts while the Hα emitters found in less dense regions present much more quiescent star-formation. Our results resolve the apparent contradictions between different studies presented in the literature, which have found the star-forming galaxy fraction to increase (e.g. Elbaz et al. 2007, studying field galaxies) or decrease (e.g. Patel et al. 2009, studying galaxies in the high density regions around a rich cluster) with increasing environmental density: both trends exist, and which is observed depends upon the range of environmental densities studied.

With the full HiZELS sample we will be able to investigate how these relations with mass and environment evolve to higher redshifts, over 80 % of the lifetime of the Universe.

The Dust Extinction Properties of High-z Star-Forming Galaxies

Deep 24 μm data from Spitzer are available in most of the HiZELS fields, and 35 % of the z = 0.84 HiZELS Hα emitters in UDS and COSMOS are detected at 24 μm. Using these detections and stacking analyses, the star formation rate estimates from the HiZELS Hα sample can be compared with those from the mid-infrared to estimate the dust extinction properties of the star-forming galaxies. We find a clear trend for an increase in mean dust extinction with increasing star formation

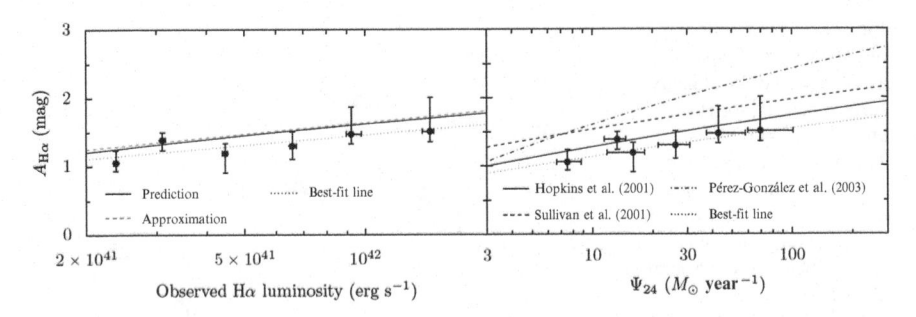

Fig. 22.7 The variation in Hα extinction for Hα emitters binned by their observed Hα flux, as a function of Hα luminosity (*left*) or 24 μm star formation rate (*right*). The best-fit (*dotted line*) shows that, at the 4σ confidence level, extinction increases with star formation rate. The slope and normalisation of the relationship are comparable to those derived in the local Universe using the fit between the Balmer decrement and SFR, by Hopkins et al. (2001) (From Garn et al. 2010)

rate at $z = 0.84$ (Garn et al. 2010; see Fig. 22.7). The relation we determine broadly matches that found in the low redshift Universe by Hopkins et al. (2001), suggesting that there is no significant change in the dust properties of star-forming galaxies with redshift, at least out to $z \sim 1$. We find no variation of the extinction-SFR relation with galaxy morphology, environment or merger status. Carrying out equivalent analyses at the higher HiZELS redshifts will be a key goal when the samples are sufficiently large.

We are exploiting the exquisite multi-wavelength data available for our fields to understand the spectral energy distribution of typical star-forming galaxies across a range of redshifts. We aim to compare additional star-formation tracers (UV continuum, [OII], sub-mm, radio) with our Hα measurements.

Limits on the Space Density of Bright $z = 8.9$ Lyα Emitters

Parallel to our Hα survey using the NB$_J$ filter, we have explored its capabilities to detect bright Lyα emitters at $z = 8.9$, when the Universe was only ∼0.5 Gyr old. Detection of any such objects would have important consequences for our understanding of the early star formation history of the Universe, as well as providing an important probe of the re-ionisation of the Universe. After conducting an exhaustive search for such emitters in the UDS and COSMOS fields, two candidates were isolated, both in the COSMOS field. Follow-up spectroscopy using CGS4 in January 2009, together with follow-up J-band imaging obtained in February 2009 with WFCAM has shown that these cannot be Lyα emitters. These results have allowed us to improve constraints on the bright end of the $z \sim 9$ Lyα luminosity function by three orders of magnitude (Sobral et al. 2009b).

On-Going Work and Future Plans

HiZELS offers a powerful resource for a large number of additional studies. Other on-going work includes the following:

- *Spectroscopic confirmation of emission lines and study of emission line ratios.* We are following-up HiZELS H$_2$(S1) Hα emitters using near-infrared spectroscopic observations, both with VLT/ISAAC and with Gemini. These observations have successfully confirmed the HiZELS emission line detections, and the results will be published soon. We are carrying out optical spectroscopic observations of a large fraction of the emitters selected from all 3 narrow bands (NB$_J$, NB$_H$ & H$_2$(S1)), in the UDS field, using VIMOS. The main aims are to confirm redshifts, to investigate line-ratios and extinction, to identify AGN, and to test the robustness of the selection criteria used for HiZELS. By allowing pre-selection of galaxies with known emission lines in the J- and H-bands, HiZELS will also provide a valuable input sample for early commissioning tests for FMOS, the new near-IR multi-object spectroscope on Subaru; such observations will allow us to confirm emission lines, redshifts, identify AGN, and study sample contaminants.
- *Complementary narrow-band observations.* Using HAWK-I on VLT we have carried out ultra-deep H$_2$(S1) exposures in two small regions in the COSMOS and UDS fields. These will reach a depth of ≈ 3 M$_{sun}$ year^{-1} at z = 2.23, comparable to the HiZELS NB$_J$ sensitivity, and will thus provide an excellent complement to our shallower but much wider HiZELS imaging. This will enable an accurate measurement of the very faint end of the Hα luminosity function at z ∼ 2.23, determining the global star formation history and testing for differential evolution of the most/least active galaxies.
- *Lyα and Hα emission from galaxies at z = 2.23.* We are exploiting existing narrow-band Lyα imaging of the COSMOS field from Nilsson et al. (2009) to compare the Lyα and Hα properties of galaxies at z = 2.23. Lyα emission has been extensively used to identify and study large samples of star-forming galaxies at the highest redshifts, but its reliability as a tracer of complete galaxy populations remains untested at cosmological distances. By combining our Hα sample with Lyα observations we can derive the ratio of Lyα to Hα emission and hence obtain a direct measurement of the escape fraction of Lyα photons with few model assumptions – something which has never been done at high redshift. We find that only a modest fraction, 10 ± 3 %, of Hα emitters are detected in Lyα above a rest-frame equivalent width of 20Å. This detection rate is similar to that seen for z ∼ 2.3 UV-continuum selected galaxies (Reddy and Steidel 2009) and is less than that seen for z ∼ 3 Lyman-break galaxies (Shapley et al. 2004). Stacking the Hα sources we find potential evidence for extended Lyα halos, similar to that seen around z ∼ 3 LBGs by Hayashino et al. (2004).

- *Paschen-series luminosity functions.* In our $H_2(S1)$ imaging we are also detecting large samples of Paα, Paβ and other lines from lower redshift sources. Whilst our photometric selection is very efficient at filtering out these contaminants from our Hα studies, we aim to present a robust analysis of the extent and character of the contamination. This may allow the derivation of Pa-series galaxy luminosity functions, which would provide a new measure of the cosmic star formation rate density, and an estimate of extinction.
- *Kinematics of HiZELS galaxies.* We are using the SINFONI near-IR integral field spectrograph on VLT to map the kinematics of the Hα emission in a sample of 18 Hα emitters at $z = 0.84$, $z = 1.47$ and $z = 2.23$ in the COSMOS field. This project exploits HiZELS panoramic coverage to select sources at each redshift which are matched on Hα-luminosity and, critically, are close to natural AO guide stars. These observations will allow us to map the distribution of Hα emission on $\sim 0.2''$ scales within these galaxies to search for kinematic evidence of rotating discs, major mergers, etc. This detailed follow-up will be a valuable addition to HiZELS, allowing us to study the properties of the progenitors of Milky Way-type galaxies seen at the time when their star formation was at its peak.
- *Comparison with galaxy formation models.* HiZELS provides a large sample of star-forming galaxies at three different redshifts, the selection of which can be cleanly replicated in (and compared with) theoretical semi-analytic models of galaxy formation. An initial comparison of the luminosity function and clustering properties of the Hα emitters at $z = 2.23$ was presented in G08, showing that these were broadly in line with those predicted by one particular "recipe" of galaxy formation, that of Bower et al. (2006). Now with larger datasets available at all three wavelengths, far more detailed investigations are beginning, folding in the variations of the luminosity function and clustering of the HiZELS galaxies with morphology, mass, etc. These studies will offer an invaluable ingredient to these models.

Conclusions

The advent of wide-field imaging cameras in the near-IR, especially WFCAM on UKIRT, has revolutionised the study of high-redshift emission line galaxies. HiZELS aims to detect and study up to a thousand Hα emitting galaxies at each of three redshifts, 0.84, 1.47 and 2.23, spanning the peak epoch of star formation in the Universe. This goal is already nearly completed at $z = 0.84$, and has resulted in a plethora of results related to the luminosity function, masses, morphologies, environments, clustering, dust extinction, and other properties of the star-forming galaxies at this redshift. Results at the higher redshifts remain sparser to date, because the current samples are smaller due to the lower sky density of Hα emitters at these higher redshifts: as the survey progresses, we will replicate our $z = 0.84$ studies at the higher redshifts, where arguably our Hα studies will be most unique, and the impact of HiZELS will be highest.

Acknowledgements Observations obtained with the Wide Field CAMera (WFCAM) on the United Kingdom Infrared Telescope (UKIRT). We are indebted to Andy Adamson, Luca Rizzi, Chris Davis, Tim, Thor and Jack for their support at the telescope. PNB is grateful for support from the Leverhulme Trust, DS for support from the FCT, and JEG for support from STFC.

References

Baugh, C.M., et al.: Can the faint submillimetre galaxies be explained in the Λ cold dark matter model? Mon. Not. R. Astron. Soc. **356**, 1191 (2005)

Benson, A.J., Cole, S., Frenk, C.S., Baugh, C.M., Lacey, C.G.: The nature of galaxy bias and clustering. Mon. Not. R. Astron. Soc. **311**, 793 (2000)

Best, P.N.: The environmental dependence of radio-loud AGN activity and star formation in the 2dFGRS. Mon. Not. R. Astron. Soc. **351**, 70 (2004)

Best, P.N., Kaiser, C.R., Heckman, T.M., Kauffmann, G.: AGN-controlled cooling in elliptical galaxies. Mon. Not. R. Astron. Soc. **368**, L67 (2006)

Bower, R., et al.: Breaking the hierarchy of galaxy formation. Mon. Not. R. Astron. Soc. **370**, 645 (2006)

Cowie, L.L., Songaila, A., Hu, E.M., Cohen, J.G.: New insight on galaxy formation and evolution from Keck spectroscopy of the Hawaii deep fields. Astron. J. **112**, 839 (1996)

Doherty, M., Bunker, A., Sharp, R., Dalton, G., Parry, I., Lewis, I.: The star formation rate at redshift one: Hα spectroscopy with CIRPASS. Mon. Not. R. Astron. Soc. **370**, 331 (2006)

Elbaz, D., et al.: The reversal of the star formation-density relation in the distant universe. Mon. Not. R. Astron. Soc. **468**, 33 (2007)

Gallego, J., Zamorano, J., Aragon-Salamanca, A., Rego, M.: The current star formation rate of the local universe. Astrophys. J. Lett. **455**, L1 (1995)

Garn, T., et al.: Obscured star formation at $z = 0.84$ with HiZELS: the relationship between star formation rate and Hα or ultraviolet dust extinction. Mon. Not. R. Astron. Soc. **402**, 2017 (2010)

Geach, J.E., Smail, I., Best, P.N., Kurk, J., Casali, M., Ivison, R.J., Coppin, K.: HiZELS: a high-redshift survey of Hα emitters – I. The cosmic star formation rate and clustering at $z = 2.23$. Mon. Not. R. Astron. Soc. **388**, 1473 (2008)

Geach, J., et al.: Empirical Hα emitter count predictions for dark energy surveys. Mon. Not. R. Astron. Soc. **402**(1330), 407 (2010)

Hayashino, T., et al.: Large-scale structure of emission-line galaxies at z=3.1. Astron. J. **128**, 2073 (2004)

Hopkins, A.M., Beacom, J.F.: On the normalization of the cosmic star formation history. Astrophys. J. **651**, 142 (2006)

Hopkins, A.M., Connolly, A., Haarsma, D.B., Cram, L.E.: Toward a resolution of the discrepancy between different estimators of star formation rate. Astron. J. **122**, 288 (2001)

Ivison, R., et al.: The SCUBA HAlf degree extragalactic survey – III. Identification of radio and mid-infrared counterparts to submillimetre galaxies. Mon. Not. R. Astron. Soc. **380**(199) (2007)

Kennicutt Jr., R.C.: Star formation in galaxies along the hubble sequence. Annu. Rev. Astron. Astrophys. **36**, 189 (1998)

Kodama, T., et al.: The first appearance of the red sequence of galaxies in proto-clusters at $2 \lesssim z \lesssim 3$. Mon. Not. R. Astron. Soc. **377**, 1717 (2007)

Lewis, I., et al.: The 2dF Galaxy Redshift Survey: the environmental dependence of galaxy star formation rates near clusters. Mon. Not. R. Astron. Soc. **334**, 673 (2002)

Lilly, S.J., Le Fevre, O., Hammer, F., Crampton, D.: The Canada-France redshift survey: the luminosity density and star formation history of the universe to Z approximately 1. Astrophys. J. Lett. **460**, L1 (1996)

Ly, C., et al.: The luminosity function and star formation rate between redshifts of 0.07 and 1.47 for narrowband emitters in the Subaru Deep Field. Astrophys. J. **657**, 738 (2007)

Moorwood, A.F.M., van der Werf, P.P., Cuby, J.-G., Oliva, E.: Hα emitting galaxies and the cosmic star formation rate at z =∼ 2.2. Astron. Astrophys. **326**, 9 (2000)

Moustakas, J., Kennicutt Jr., R.C., Tremonti, C.A.: Optical star formation rate indicators. Astrophys. J. **642**, 775 (2006)

Nakajima, A., et al.: Clustering properties of low-luminosity star-forming galaxies at z = 0.24 and 0.40 in the Subaru Deep Field. Pub. Astron. Soc. Jpn **60**, 1249 (2008)

Nilsson, K.K., et al.: Evolution in the properties of Lyman-α emitters from redshifts z ∼ 3 to z ∼ 2. Astron. Astrophys. **498**, 13 (2009)

Patel, S.G., Holden, B.P., Kelson, D.D., Illingworth, G.D., Franx, M.: The dependence of star formation rates on stellar mass and environment at z ∼ 0.8. Astrophys. J. **705**, L67 (2009)

Pérez-Gonzáles, P.G., Zamorano, J., Gallego, J., Aragón-Salamanca, A., Gil de Paz, A.: Spatial analysis of the Hα emission in the local star-forming UCM galaxies. Astrophys. J. **591**, 827 (2003)

Reddy, N.A., Steidel, C.C.: A steep faint-end slope of the UV luminosity function at z ∼ 2–3: implications for the global stellar mass density and star formation in low-mass halos. Astrophys. J. **692**, 778 (2009)

Shapley, A.E., Erb, D.K., Pettini, M., Steidel, C.C., Adelberger, K.L.: Evidence for solar metallicities in massive star-forming galaxies at z ≳ 2. Astrophys. J. **612**, 108 (2004)

Shioya, Y.: The Hα luminosity function and star formation rate at z ∼ 0.24 in the COSMOS 2 square degree field. Astrophys. J. Suppl. **175**, 128 (2008)

Sobral, D., et al.: HiZELS: a high-redshift survey of Hα emitters – II. The nature of star-forming galaxies at z = 0.84. Mon. Not. R. Astron. Soc. **398**, 75 (2009a)

Sobral, D., et al.: Bright Lyα emitters at z ∼ 9: constraints on the LF from HizELS. Mon. Not. R. Astron. Soc. **398**, L68 (2009b)

Sobral, D., Best, P.N., Geach, J.E., Smail, I., Cirasuolo, M., Garn, T., Dalton, G.B., Kurk, J.: The clustering and evolution of Hα emitters at z ∼ 1 from HiZELS. Mon. Not. R. Astron. Soc. **404**, 1551 (2010)

Sobral, D., Best, P.N., Smail, I., Geach, J.E., Cirasuolo, M., Garn, T.S., Dalton, G.: The dependence of star formation activity on environment and stellar mass at z ∼ 1 from the HiZELS-Hα survey. Mon. Not. R. Astron. Soc. **411**, 675 (2011)

Sobral, D., Best, P.N., Matsuda, Y., Smail, I., Geach, J.E., Cirasuolo, M.: Star formation at z=1.47 from HiZELS: an H>α+[O II] double-blind study. Mon. Not. R. Astron. Soc. **420**, 1926 (2012)

Sullivan, M., Mobasher, B., Chan, B., Cram, L., Ellis, R., Treyer, M., Hopkins, A.: Astrophys. J. **558**, 72 (2001)

Tresse, L., Maddox, S.J., Le Fevre, O., Cuby, J.-G.: Mon. Not. R. Astron. Soc. **337**, 369 (2002)

Villar, V., Gallego, J., Pérez-González, P.G., Pascual, S., Noeske, K., Koo, D.C., Barro, G., Zamorano, J.: Astrophys. J. **677**, 169 (2008)

Yan, L., et al.: Astrophys. J. Lett. **519**, L47 (1999)

Chapter 23
The HiZELS/UKIRT Large Survey for Bright Lyα Emitters at z ~ 9

David Sobral, Philip Best, Jim Geach, Ian Smail, Jaron Kurk, Michele Cirasuolo, Mark Casali, Rob Ivison, Kristen Coppin, and Gavin Dalton

Abstract We present the largest area survey to date (1.4 deg^2) for Lyα emitters (LAEs) at z ~ 9, as part of the Hi-z Emission Line Survey (HiZELS). The survey, which primarily targets Hα emitters at z < 3, uses the Wide Field Camera on the United Kingdom Infrared Telescope and a custom narrowband filter in the J band to reach a Lyα luminosity limit of ~$10^{43.8}$ erg s^{-1} over a co-moving volume of 1.12×10^6 Mpc^3 at z = 8.96 ± 0.06. Two candidates were found out of 1,517

In this work, an H_0 = 70 km s^{-1}Mpc^{-1}, $\Omega_M = 0.3$ and $\Omega_\Lambda = 0.7$ cosmology is used and magnitudes are given in the Vega system. For full details, please refer to Sobral et al. (2009b).

D. Sobral (✉) • P. Best
IfA, Edinburgh University, Edinburgh, UK
e-mail: drss@roe.ac.uk

J. Geach • I. Smail • K. Coppin
ICC, Durham University, Durham, UK

J. Kurk
Max-Planck-Institut für Astrophysik, Karl-Schwarzschild Straße 1, D-85741 Garching, Germany

M. Cirasuolo
UK Astronomy Technology Centre, Royal Observatory of Edinburgh, Blackford Hill, Edinburgh EH9 3HJ, Garching, Germany

R. Ivison
SUPA, Institute for Astronomy, Royal Observatory of Edinburgh, Blackford Hill, Edinburgh EH9 3HJ, Garching, Germany

M. Casali
European Southern Observatory, Karl-Schwarzschild-Strasse 2, D-85738 Garching bei München, Germany

G. Dalton
Department of Physics, University of Oxford, Keble Road, Oxford OX1 3RH, UK

Space Science and Technology, Rutherford Appleton Laboratory, Didcot OX11 0QX, UK

line emitters, but those were rejected as LAEs after follow-up observations. This improves the limit on the space density of bright Lyα emitters by three orders of magnitude and is consistent with suppression of the bright end of the Lyα luminosity function beyond $z \sim 6$. Combined with upper limits from smaller but deeper surveys, this rules out some of the most extreme models for high-redshift Lyα emitters. The potential contamination of narrowband Lyα surveys at $z > 7$ by Galactic brown dwarf stars is also examined, leading to the conclusion that such contamination may well be significant for searches at $7.7 < z < 8.0$, $9.1 < z < 9.5$ and $11.7 < z < 12.2$.

Introduction

Understanding how and when the first stars and galaxies formed is one of the most fundamental problems in astronomy. Furthermore, whilst many sophisticated models of early galaxy formation and evolution have been constructed, it is clear that observations of the most distant galaxies are mandatory to really test, refine, or refute such models. Indeed, considerable manpower and telescope time have been invested in such observations, with the detection of a Gamma Ray Burst (GRB) at $z \approx 8.2$ (Tanvir et al. 2009) being one of the most recent highlights of this extraordinary endeavor. However, despite the recent success in using GRBs to find the most distant sources, the current samples of high redshift galaxies have been mostly assembled using two methods: the broadband drop-out technique and narrowband imaging surveys.

The widely used drop-out technique (pioneered at $z \sim 3$ by Steidel et al. 1996) requires very deep broadband imaging, and can identify $z > 7$ galaxies as z-band dropouts (e.g. Bouwens et al. 2008; Richard et al. 2008). Furthermore, the use of this technique, combined with the recent installation of the Wide Field Camera 3 (WFC3) on the Hubble Space Telescope (HST), has led to the identification of roughly 20 $z \approx 7$–8 candidates (e.g. Bouwens et al. 2008; Oesch et al. 2010; McLure et al. 2010; Bunker et al. 2010; Yan et al. 2010). While this is an efficient method for identifying candidates, it still requires detailed spectroscopic follow-up to confirm them, especially to rule out contributions from other populations with large z–J breaks, such as dusty or evolved $z \sim 2$ galaxies and ultra-cool galactic stars (e.g. McLure et al. 2006a, b). Confirming the candidates is actually quite a significant challenge, since the typical $z > 7$ candidates found so far are just too faint for spectroscopic follow-up.

The narrowband imaging technique has the advantage of probing very large volumes looking for Lyα in emission, and whilst it can only detect sources with strong emission lines and still depends on the Lyman-break technique to isolate very high-redshift emitters, it can yield the perfect targets for follow-up spectroscopy with the current instrumentation. Narrowband Lyα searches at $3 < z < 7$ have been extremely successful in detecting and confirming emitters (e.g. Hu et al. 1998) and, so far, this technique has resulted in the spectroscopic confirmation of the highest redshift galaxy ($z = 6.96$: Iye et al. 2006). Even more recently, Hibon et al. (2010)

identified 7 candidate Lyα emitters at z = 7.7. There have been attempts to detect Lyα emitters at z ∼ 9 (e.g. Willis and Courbin 2005; Cuby et al. 2007; Willis et al. 2008), but all such studies have been unsuccessful to date, having surveyed very small areas (a few tens of square arcmin at most).

With the advent of wide-field, near-IR detectors it is now possible to increase the sky area studied by over 2–3 orders of magnitude and reach the regime where one can realistically expect to detect z ∼ 9 objects. This is a key aim of, for example, the narrowband component of the UltraVISTA Survey (c.f. Nilsson et al. 2007). It is also an aim of HiZELS, the High-z Emission Line Survey (c.f. Geach et al. 2008; Sobral et al. 2009a,b; Garn et al. 2010), that we are carrying out using the WFCAM instrument on the 3.8 m UK Infrared Telescope (UKIRT). HiZELS is using a set of existing and custom-made narrowband filters in the J, H and K bands to detect emission lines from galaxies at different redshifts over ∼10 square degrees of extragalactic sky. In particular, the narrowband J filter (hereafter NBJ) is sensitive to Lyα at z = 8.96.

Data, Selection and Candidates

Deep narrowband J (NBJ ≈ 21.6, 3σ, $F_{lim} = 7.6 \times 10^{-17}$ erg s^{-1} cm^{-2}) imaging was obtained across 1.4 deg^2 in the UKIRT Infrared Deep Sky Survey Ultra Deep Survey (UKIDSS UDS; Lawrence et al. 2007) and the Cosmological Evolution Survey (COSMOS; Scoville et al. 2007) fields, both of which have a remarkable set of deep multi-wavelength data available – this resulted in the selection of 1,517 potential line emitters. The NBJ filter ($\lambda = 1.211 \pm 0.015$ μm) is sensitive to Lyα emission at z = 8.96 ± 0.06, probing a co-moving volume of 1.12×10^6 Mpc^3 – by far the largest probed by a narrowband survey at these wavelengths. Details regarding the observations, data reduction and the general selection of NBJ emitters can be found in Sobral et al. (2009a).

For a source to be considered a candidate z ≈ 9 Lyα emitter it is required to: (i) be selected as a narrowband emitter in Sobral et al. (2009a); (ii) have at least one other detection >3σ in the near infrared; (iii) be visually believable in NBJ and the other band(s), avoiding noisy areas; and (iv) be undetected (<3σ and direct visual analysis) in the available visible band imaging (B,V,r,i,z) – SUBARU and ACS/HST.

No candidates were found in the UKIDSS UDS field, with all emitters that passed tests (i) to (iii) being clearly detected in z-band imaging. In COSMOS, however, two candidates were found that satisfied all criteria. The brightest source was followed up spectroscopically using the CGS4 instrument on UKIRT in January 2009 – these data failed to confirm an emission line. Both candidates were then reobserved using WFCAM (further J imaging in February 2009), resulting in the non-detection of both candidates. Further investigation shows that the sources are likely to be artifacts caused by an unfortunate coincidence of a set of slightly hot pixels (not sufficient to be flagged as bad pixels) which, combined with the dither pattern, produced a few σ excess at one location on the combined image.

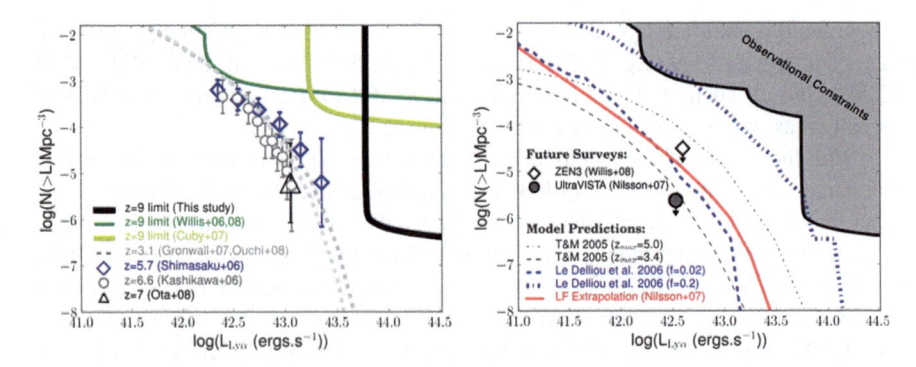

Fig. 23.1 *Left*: Comparison between the measured Lyα luminosity function at z ∼ 3 (*dotted lines*; Gronwall et al. 2007; Ouchi et al. 2008) with data from z ∼ 6–7 (Kashikawa et al. 2006; Shimasaku et al. 2006; Ota et al. 2008). No evidence of significant evolution is found, especially when accounting for cosmic variance. Limits for the z ∼ 9 LF from Willis and Courbin (2005), Cuby et al. (2007) and Willis et al. (2008) are also presented, together with the one presented in this contribution. *Right*: The observational limits on the z ∼ 9 Lyα luminosity function compared to different model predictions and proposed future surveys, showing that the most recent versions of these models are completely consistent with the observations

Lyα Luminosity Function at z ∼ 9

A non-detection of (L > 7.6 × 10^{43} erg s^{-1}) Lyα emitters at z ∼ 9, in a co-moving volume of 1.12 × 10^6 Mpc^3 allows the tightest constraint on the bright end of the z ∼ 9 Lyα luminosity function, as previous surveys (Willis and Courbin 2005; Cuby et al. 2007; Willis et al. 2008) have only covered very small areas (a factor ∼1,000 smaller). However, since those surveys have gone significantly deeper (up to a factor of ∼100), combining all the results from the literature can constrain the Lyα luminosity function across a wide range of luminosities (10^{42} < L < 10^{45} erg s^{-1}) for the first time. The left panel of Fig. 23.1 presents such constraints, indicating the inverse of the volume selection function for each survey. These are compared to the measured Lyα luminosity functions from z ∼ 3 to z ∼ 7, revealing that there is little evolution in the bright end of the luminosity function between z ∼ 3 and z ∼ 5.7. Nevertheless, those bright emitters seem to become much rarer at z = 6.5 (Kashikawa et al. 2006), indicating that L* is not increasing from z ∼ 6 onwards. The results presented here are also consistent with no evolution in L* (Δlog(L*) < 0.5) from z = 5.7 to z ∼ 9.

Comparison with Models and Future Surveys

Many authors have made predictions regarding the Lyα luminosity function at z ∼ 9, either by extrapolating the luminosity function of these emitters from lower redshift, or by using numerical or semi-analytical models. The semi-analytical models discussed here are obtained from GALFORM (Baugh et al. 2005) – these are based on ΛCDM, having been successful in reproducing a wide range of galaxy properties at different redshifts, including Lyα emitters up to $z \sim 6$ (c.f. Baugh et al. 2005; Le Delliou et al. 2006; Orsi et al. 2008). The observational approach, as in Nilsson et al. (2007), extrapolates the Schechter function parameters based on those obtained in the $3.1 < z < 6.5$ redshift range. In practice, this results in little L^* evolution but a significant negative φ^* evolution. Finally, the phenomenological approach in Thommes and Meisenheimer (2005) assumes that Lyα emitters at high redshift are spheroids seen during their formation phase. Each galaxy is assumed to be visible as a Lyα emitter during a starburst phase of fixed duration that occurs at a specific redshift, drawn from a broad distribution (c.f. Thommes and Meisenheimer 2005).

The right panel of Fig. 23.1 presents predictions from GALFORM (Le Delliou et al. 2006), the observational luminosity function extrapolation from Nilsson et al. (2007) and updated phenomenological predictions (Thommes and Meisenheimer 2005) assuming peak redshifts of $z_{max} = 3.4$ and $z_{max} = 5.0$. While most predictions are consistent with the current limits, GALFORM models with high escape fractions are marginally rejected both at faint and bright levels. Earlier phenomenological models (e.g. the $z_{max} = 10$ model of Thommes and Meisenheimer 2005, not shown in Fig. 23.1) are also clearly rejected by our results.

High Redshift Lyα Searches and Cool Galactic Stars

It has become widely realised in recent years that broadband searches for $z > 6$ galaxies using the Lyman-break technique may suffer from significant contamination by cool Galactic L, T, and possibly Y-dwarf stars (e.g. McLure et al. 2006a, b). These low-mass brown dwarfs display extremely red $z - J$ colours reaching as high as $z - J \approx 4$ (e.g. Burningham et al. 2008), coupled with relatively flat $J - K$ colours. Such colours can mimic very closely those expected of a $z > 6$ star forming galaxy with a strong Lyman break.

Since the near-infrared continuum spectra of low-mass, brown dwarfs show considerable structure due to broad molecular absorption features (especially methane and ammonia; e.g. Leggett et al. 2007), as shown in the top panel of Fig. 23.2, it can easily produce a positive broadband minus narrowband (BBNB) colour (see lower panel of Fig. 23.2) if the narrowband filter is located within one of the spectral peaks (this is much less of an issue for surveys which difference two closely-located narrowband filters). Lyα narrowband surveys in the redshift ranges $7.7 < z < 8.0$, $9.1 < z < 9.5$ and $11.7 < z < 12.2$ are therefore prone to contamination by cool Galactic stars – this includes the $z = 7.7$ and $z = 9.4$ atmospheric windows for narrowband searches of Lyα emitters. Narrowband surveys at redshifts $z < 7.7$, or between $8.0 < z < 9.1$ – which include both HiZELS ($z = 8.96$) and the narrowband component of the UltraVISTA Survey ($z = 8.8$; e.g. Nilsson et al. 2007) – will not only be free of such contamination, but can potentially select very cool T dwarf

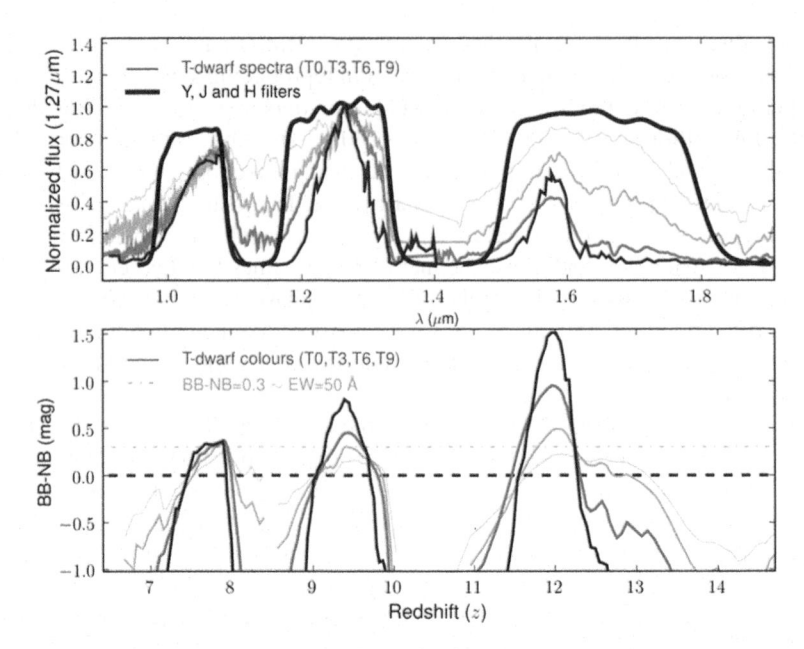

Fig. 23.2 *Top panel*: The near-infrared spectra of T0, T3, T6 and T9 dwarf stars (T0 – lighter, T9 – darker, from Burningham et al. 2008) compared to near-IR broad band filter profiles. *Lower panel*: The consequences for measured broadband minus narrowband (BBNB) colours, clearly demonstrating the redshifts/wavelengths at which searches for Lyα emitters can be significantly contaminated by these very cool stars. For $7.7 < z < 8.0$ and $9.1 < z < 9.5$ searches, these stars can easily mimic Lyα emitters, with strong Y-z or J-z breaks and significant positive BB-NB colours. Searches at higher redshift $11.6 < z < 12.2$ in the H band can detect T9s with BB-NB ~ 1.5, although the lack of strong H-J or H-Y breaks will make it easier to distinguish T dwarfs from Lyα emitters

stars via a narrowband deficit (due to the strong methane absorption feature at these wavelengths). Indeed, motivated by such finding, a T dwarf search was conducted among narrowband deficit sources from Sobral et al. (2009a). The results show that all those sources are galaxies with $z_{photo} \sim 1.4 - 1.5$, probably placing the Hβ and [OIII] emission lines just outside the narrowband coverage, but contributing significantly to the measured J flux, which results in the observed narrowband deficits. No T dwarf candidate was found.

Summary

- Deep narrowband imaging in the J band ($\lambda = 1.211 \pm 0.015$ μm) has been used to search for bright Lyα emitters at $z = 8.96$ over an area of 1.4 deg^2. No Lyα emitter was found brighter than $L \approx 7.6 \times 10^{43}$ erg s^{-1}.
- The Lyα luminosity function constraints at $z \sim 9$ have been significantly improved for $10^{42} < L < 10^{45}$ erg s^{-1}.

- The results rule out significant positive evolution of the Lyα Luminosity Function beyond z ~ 6; they are in line with recent semi-analytic & phenomenological model predictions, rejecting some extreme models.
- It has been shown that for narrowband searches, T dwarfs can mimic Lyα emitters at 7.7 < z < 8.0, 9.1 < z < 9.5 and 11.7 < z < 12.2; they will not contaminate the future UltraVISTA narrowband survey (and can even be identified via a narrowband deficit), but they may contaminate narrowband Lyα searches within the z = 7.7 and z = 9.4 atmospheric windows.

These results show that bright $L > 10^{43.8}$ erg s^{-1} Lyα emitters are extremely rare. Although the area coverage is absolutely important, a depth + area combination is likely to be the best approach for gathering the first sample of these very high-redshift galaxies. In fact, that is the strategy of the narrowband component of the UltraVISTA survey (c.f. Nilsson et al. 2007), using the VISTA telescope, which will map 0.9 deg^2 of the COSMOS field to a planned 5σ luminosity limit of $L = 10^{42.53}$ erg s^{-1} and a surveyed volume of 5.41×10^5 Mpc^3 (see right panel of Fig. 23.1) at z = 8.8. This combination lies below all current predictions for the z ~ 9 Lyα LF and the survey is expected to detect 220 Lyα emitters at $z = 8.8 \pm 0.1$. Furthermore, the continuation of HiZELS on UKIRT and the extension of the narrowband J survey to a wider area might be able to detect one of the brightest Lyα emitters at z ~ 9, perfectly suited for spectroscopic follow-up and potentially enabling the detailed studies which simply won't be possible for much fainter emitters, even if they are detected.

References

Baugh, C.M., et al.: Can the faint submillimetre galaxies be explained in the Λ cold dark matter model? Mon. Not. R. Astron. Soc. **356**, 1191 (2005)

Bouwens, R.J., Illingworth, G.D., Franx, M., Ford, H.: z ~ 7–10 Galaxies in the HUDF and GOODS fields: UV luminosity functions. Astrophys. J. **686**, 230 (2008)

Bouwens, R.J., et al.: Discovery of z ~ 8 galaxies in the Hubble Ultra Deep Field from ultra-deep WFC3/IR observations. Astrophys. J. **709**, L133 (2010)

Bunker, A., et al.: The contribution of high-redshift galaxies to cosmic reionization: new results from deep WFC3 imaging of the Hubble Ultra Deep Field. Mon. Not. R. Astron. Soc. **409**, 855 (2010)

Burningham, B., et al.: Exploring the substellar temperature regime down to ~550K. Mon. Not. R. Astron. Soc. **391**, 320 (2008)

Cuby, J.G., et al.: A narrow-band search for Lyα emitting galaxies at z = 8.8. Am. Assoc. Pediatr. **461**, 911 (2007)

Garn, T., et al.: Obscured star formation at z = 0.84 with HiZELS: the relationship between star formation rate and Hα or ultraviolet dust extinction. Mon. Not. R. Astron. Soc. **402**, 2017 (2010)

Geach, J.E., et al.: HiZELS: a high-redshift survey of Hα emitters – I. The cosmic star formation rate and clustering at z = 2.23. Mon. Not. R. Astron. Soc. **388**, 1473 (2008)

Gronwall, C., et al.: Lyα emission-line galaxies at z = 3.1 in the extended Chandra Deep Field-South. Astrophys. J. **667**, 79 (2007)

Hibon, P., et al.: Limits on the luminosity function of Lyα emitters at z = 7.7. Astron. Astrophys. **515**, A97 (2010)

Hu, E.M., Cowie, L.L., McMahon, R.G.: The density of Ly alpha emitters at very high redshift. Astrophys. J. Lett. **502**, L99C (1998)

Iye, M., et al.: A galaxy at a redshift z = 6.96. Nature **443**, 186 (2006)

Kashikawa, N., et al.: The end of the reionization epoch probed by Lyα emitters at z = 6.5 in the Subaru Deep Field. Astrophys. J. **648**, 7 (2006)

Lawrence, A., et al.: The UKIRT Infrared Deep Sky Survey (UKIDSS). Mon. Not. R. Astron. Soc. **379**, 1599 (2007)

Le Delliou, M., Lacey, C.G., Baugh, C.M., Morris, S.L.: The properties of Lyα emitting galaxies in hierarchical galaxy formation models. Mon. Not. R. Astron. Soc. **365**, 712 (2006)

Leggett, S.K., et al.: Physical and spectral characteristics of the T8 and later type dwarfs. Astrophys. J. **667**, 537 (2007)

McLure, R.J., et al.: On the evolution of the black hole: spheroid mass ratio. Mon. Not. R. Astron. Soc. **368**, 1395 (2006a)

McLure, R.J., et al.: The discovery of a significant sample of massive galaxies at redshifts 5 < z < 6 in the UKIDSS Ultra Deep Survey early data release. Mon. Not. R. Astron. Soc. **372**, 357 (2006b)

McLure, R.J., et al.: Galaxies at z = 6 9 from the WFC3/IR imaging of the Hubble Ultra Deep Field. Mon. Not. R. Astron. Soc. **403**, 960 (2010)

Nilsson, K.K., et al.: Narrow-band surveys for very high redshift Lyman-α emitters. Astron. Astrophys. **474**, 385 (2007)

Oesch, P.A., et al.: Structure and morphologies of z ∼ 7–8 galaxies from ultra-deep WFC3/IR imaging of the Hubble Ultra-deep Field. Astrophys. J. Lett. **709**, 21 (2010)

Orsi, A., Lacey, C.G., Baugh, C.M., Infante, L.: The clustering of Lyα emitters in a ΛCDM Universe. Mon. Not. R. Astron. Soc. **391**, 1589 (2008)

Ota, K., et al.: Reionization and galaxy evolution probed by z = 7 Lyα emitters. Astrophys. J. **677**, 12 (2008)

Ouchi, M., et al.: The Subaru/XMM-Newton Deep Survey (SXDS). IV. Evolution of Lyα emitters from z = 3.1 to 5.7 in the 1 deg^2 field: luminosity functions and AGN. Astrophys. J. Suppl. **176**, 301 (2008)

Richard, J., et al.: A hubble and spitzer space telescope survey for gravitationally lensed galaxies: further evidence for a significant population of low-luminosity galaxies beyond z = 7. Astrophys. J. **685**, 705 (2008)

Scoville, N., et al.: The Cosmic Evolution Survey (COSMOS): overview. Astrophys. J. Suppl. **172**, 1 (2007)

Shimasaku, K., et al.: Lyα emitters at z = 5.7 in the Subaru Deep Field. Publ. Astron. Soc. Jpn. **58**, 313 (2006)

Sobral, D., et al.: HiZELS: a high-redshift survey of Hα emitters – II. The nature of star-forming galaxies at z = 0.84. Mon. Not. R. Astron. Soc. **398**, 75 (2009a)

Sobral, D., et al.: Bright Lyα emitters at z ∼ 9: constraints on the LF from HiZELS. Mon. Not. R. Astron. Soc. **398**, L68 (2009b)

Steidel, C.C., et al.: Spectroscopic confirmation of a population of normal star-forming galaxies at redshifts Z < 3. Astrophys. J. Lett. **462**, L17C (1996)

Tanvir, N., et al.: A γ-ray burst at a redshift of z ∼ 8.2. Nature **461**, 1254 (2009)

Thommes, E., Meisenheimer, K.: The expected abundance of Lyman-α emitting primeval galaxies. I. General model predictions. Astron. Astrophys. **430**, 877 (2005)

Willis, J.P., Courbin, F.: A deep, narrow J-band search for protogalactic Lyα emission at redshifts z ∼ 9. Mon. Not. R. Astron. Soc. **357**, 1348 (2005)

Willis, J.P., Courbin, F., Kneib, J.P., Minniti, D.: ZEN2: a narrow J-band search for z ∼ 9 Lyα emitting galaxies directed towards three lensing clusters. Mon. Not. R. Astron. Soc. **384**, 1039 (2008)

Yan, H., et al.: Galaxy formation in the reionization epoch as hinted by Wide Field Camera 3 observations of the Hubble Ultra Deep Field. Astron. Astrophys. **10**, 867 (2010)

Chapter 24
Observations of Gamma-Ray Bursts at UKIRT

Nial Tanvir

Abstract UKIRT is a powerful facility for GRB follow-up. The infrared capability allows us to search for dusty and distant afterglows and the integrated observing and data reduction systems are well suited to rapid observation and pipeline processing of results. UKIRT has operated a GRB target-of-opportunity programme for a number of years, which has localised and monitored many GRB afterglows; has provided critical observations of GRB-like phenomena, such as the SGR 0501+4516; and recently made the discovery of the afterglow of GRB 090423 at a record-breaking redshift of $z = 8.2$.

Introduction

Although gamma-ray bursts (GRBs) were first discovered a decade before UKIRT was built (Klebesadel et al. 1973), it was not until 1997 that they became a focus for its efforts, following the realisation that GRBs are accompanied by longer-lived afterglows at other wavelengths (van Paradijs et al. 1997). This breakthrough led rapidly to the conclusion that GRBs are generally at cosmological distances and hence are the most luminous sources known. As such it was clear that they could, in principle, be seen at very high redshifts (e.g. Lamb and Reichart 2000), and therefore be natural targets for near-infrared (nIR) follow-up, where the afterglows may still appear bright despite the Lyman-alpha break extinguishing optical flux. Being produced by the deaths of short-lived massive stars also suggests that they might have followed soon after the formation of the first generation of stars. The fact that a small proportion of GRBs were optically very dark (e.g. Groot et al. 1998) encouraged such searches.

N. Tanvir (✉)
Department of Physics and Astronomy, University of Leicester, University Road,
Leicester LE1 7RH, UK
e-mail: nrt3@star.le.ac.uk

UKIRT has a number of advantages as a GRB follow-up machine. In the first instance it occupies, of course, one of the world's premier observing sites and, being a dedicated IR telescope, it is available for nIR observations a large fraction of the year. The observing infrastructure at UKIRT is well developed, with flexible scheduling, an integrated observing and pipeline reduction system, and the eSTAR (Allan et al. 2008) triggering system. Last, but not least, with a $\sim$10 h time difference to the UK, it lends itself to observations during UK daytime when the triggerers are hopefully at their most alert! Undoubtedly a further advantage until recently was that the Cassegrain instrument suite allowed easy switching between imaging and spectroscopy – unfortunately not the case while the telescope is operating in WFCAM only mode.

Until the advent of *Swift*, the sample of well-localised GRBs had a median redshift around $z = 1$, and the highest redshift measured was $z = 4.5$ (Andersen et al. 2000). *Swift* has changed this situation markedly, with the median redshift of *Swift* bursts being about $z = 2.5$ (Jakobsson et al. 2006) and a clear tail of events to much higher redshifts. Various attempts to predict the proportion of very high redshift ($z > 6$) Swift GRBs have concluded it is likely to be a few percent (Bromm and Loeb 2006; Tanvir and Jakobsson 2007), but observationally it remains challenging to rapidly identify these rare events, and then to acquire the requisite deep observations before the afterglow fades.

The Potential of High Redshift GRBs

GRBs are not only sufficiently luminous to be detected at very high redshifts, but in many cases their afterglows are also bright enough to allow nIR spectroscopic observations provided they can be targeted sufficiently early. Spectroscopy not only provides accurate redshifts, but also diagnostics of the state of the interstellar medium (ISM) of the host and intergalactic medium (IGM) along the line-of-sight. Indeed, the power-law continua of GRBs make them almost ideal backlights for such absorption-line spectroscopy. Thus we may, for example, be able to determine the chemical make up and dynamics of the host galaxy even, as is likely, the host itself is extremely faint (e.g. Vreeswijk et al. 2004). Once the afterglow has faded, the host of a high-redshift GRB can be searched for in direct imaging with both present and future technology. This ability to study the evolution with redshift of the distribution of size and degree of chemical enrichment of star-forming galaxies, independently of the luminosity of the galaxies themselves, is highly complementary to traditional studies of flux-limited samples.

Excitingly, GRBs also have huge potential for investigating the signature of the Gunn-Peterson trough, the integrated Lyman-alpha absorption arising from the tenuous neutral hydrogen along the line-of-sight in the intergalactic medium. If this can be quantified then it offers a route to determining directly the degree to which the universe has become re-ionized at the redshift of the burst (Barkana and Loeb 2004; McQuinn et al. 2008). Compared to bright quasars, which have more complex underlying spectra and have a significant "proximity effect" on the surrounding IGM, the underlying power-law SEDs and relatively small, low luminosity hosts of GRBs are a great advantage. However, even with GRBs these observations are challenging. Firstly, at the redshifts of interest ($z > 6$) the Gunn-Peterson trough itself is likely to be essentially saturated and hence one must determine the abundance via the shape of its damping wing. Secondly, absorption by HI gas in the host may dominate that from the IGM (e.g. Fynbo et al. 2009), or at least make an important contribution. In principle the two components can be distinguished, but this requires very good signal-to-noise ratio observations of the damping wing, and is greatly helped by an exact determination of the GRB redshift via metal lines. Finally, even if the neutral fraction could be determined for one GRB, that would only pertain to that particular sight-line, whereas re-ionization is expected to be a clumpy process that takes place at different rates depending on local factors (e.g. Bolton and Haehnelt 2007; Alvarez et al. 2009). Thus, many high-redshift afterglows will need to be observed at high signal-to-noise to give a good picture of the progress of re-ionization as a whole.

Nonetheless, re-ionization is a key phase-change in the Universe, and is thought most likely to represent the global effects of UV radiation from the first stars and galaxies on the neutral gas from the Big Bang. An understanding of this era therefore amounts to an understanding of the formation of the first collapsed structures, and fills in the last major gap in our knowledge of the large-scale evolution of matter. Currently the main constraint we have is from the polarization signal observed in the cosmic microwave background by the WMAP satellite. This indicates that the main phase of re-ionization occurred at $z = 10.9 \pm 1.4$ (Komatsu et al. 2009). Clearly, to make progress we must understand the nature and number of the ionizing sources, protogalaxies and the early stellar populations they hosted, which drove re-ionization. GRBs both provide a route to measuring the evolving neutral fraction and also investigating the nature of the hosts and the stellar populations from which the GRBs derive, which are likely to be representative of the sources driving the re-ionization process. Hence, although the observations are hard, in the long-term it is very likely that GRBs will prove the tool of choice with which to accomplish this.

Highlights of the UKIRT Campaign

UKIRT has contributed to many GRB follow-up campaigns over the years. For instance, providing infrared data points to construct, in tandem with optical observations, spectral energy distributions that can be used to constrain and test relativistic fireball models (e.g. Levan et al. 2006a). Of greatest interest, though, are observations of particularly red sources, and I discuss some of the most striking individual results below.

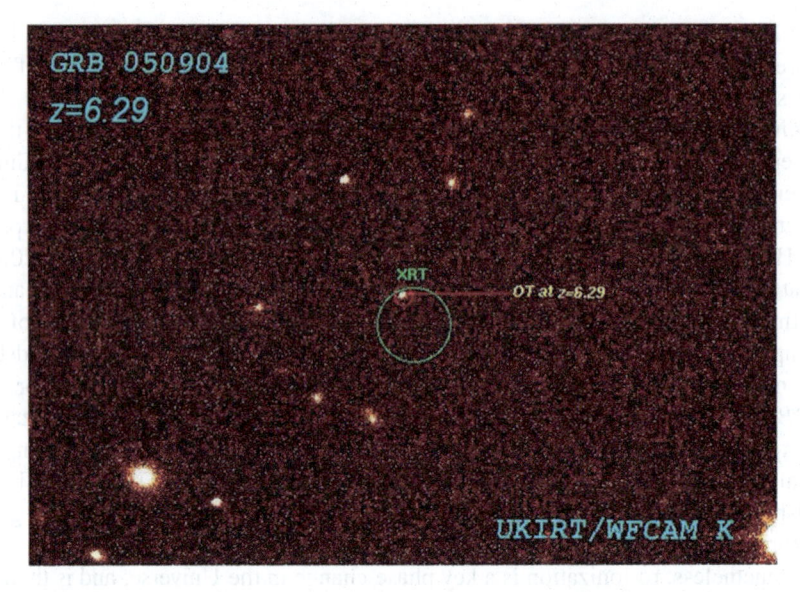

Fig. 24.1 UKIRT/WFCAM image of the afterglow of GRB 050904, which showed it to be IR bright. Subsequent SUBARU spectroscopy showed this to be the first burst detected above redshift 6

GRB 050904

Typically, the first indication that a GRB is at very high redshift is from the absence of an afterglow detection in deep optical imaging. Multiband near-infrared photometry may then show them as dropout sources and, in combination with the optical limits, can provide impressively precise photometric redshifts, thanks to the power-law nature of the underlying afterglow SED. Ideally, of course, the redshift will be confirmed via deep spectroscopy on a large telescope to determine the position of the Lyman-alpha break and/or metal absorption lines.

The first opportunity to observe a very high redshift GRB came with GRB 050904. J-band imaging from the SOAR telescope provided the first indication of an extreme object, which was confirmed by our subsequent UKIRT (Fig. 24.1) and Gemini photometric redshift of $z = 6.39 \pm 0.12$ (Haislip et al. 2006). Ultimately a spectrum was obtained by SUBARU confirming the redshift to be $z = 6.29$ (Kawai et al. 2006). This particular burst was also remarkable in being the intrinsically brightest event witnessed until that time ($J < 13$ at peak), emphasising the potential uniqueness of GRBs as high redshift beacons.

GRB 060923A

Not all optically-dark GRBs turn out to be at high redshift. One of the most extreme events so far detected was GRB 060923A, the infrared afterglow of which

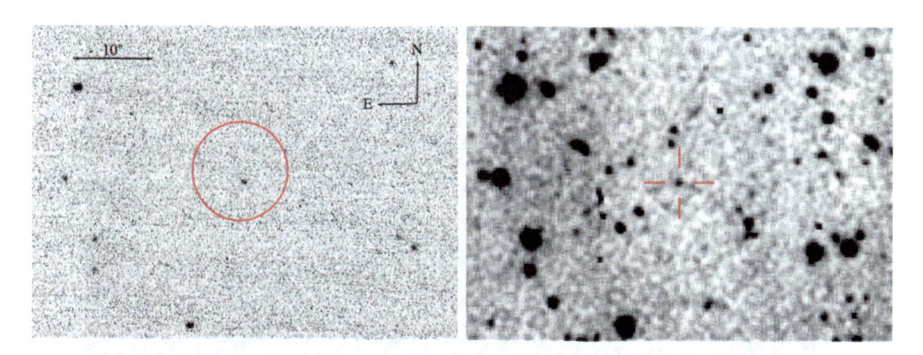

Fig. 24.2 (*Left*) UKIRT/UFTI discovery image of a K-band afterglow of GRB 060923A, which was undetected in J and H imaging. (*Right*) VLT/FORS2 deep late-time image showing a compact, blue host galaxy at exactly the same location (the cross hairs are centred on the GRB position), showing that in this case we are dealing with a high extinction rather than high redshift event

was discovered in UKIRT K-band observations within an hour of the trigger, but was undetected to deep limits in J or H (from UKIRT and Gemini). This made it a candidate ultra-high redshift event. Ultimately, however, our detailed study of this burst revealed that despite the lack of optical emission from the afterglow, the position coincided exactly with a faint galaxy that was detected at least down to 4,600 Å (Fig. 24.2; Tanvir et al. 2008). Most likely this is the host, and indeed it is similar to other GRB hosts found at more typical redshifts e.g. $z = 1-3$. This suggests this event was a case of a particularly dusty sight-line to the burst, despite the host being relatively blue. Although several similar dusty GRBs have now been identified, it remains interesting that these seem to constitute only a small proportion of the whole GRB population, consistent with the typical GRB hosts being comparatively small, unevolved and dust-free systems.

GRB 090423

As of the beginning of 2009, the second most distant spectroscopically known source was a GRB (GRB 080913; Greiner et al. 2009), but the record holder remained the Lyman-alpha emission line galaxy reported by Iye et al. (2006).

Years of effort and preparation finally paid off in April 2009 with the discovery of GRB 090423. Our UKIRT/WFCAM observations, obtained despite high winds at Mauna Kea that night, which ultimately forced the telescope to close, located a K-band afterglow of the GRB about 30 min post-burst. The lack of any emission in the early optical observations again pointed to this burst being a candidate high-z event. Gemini-N/NIRI imaging in YJH at about 90 min post-burst showed the afterglow to be a Y-band dropout, and crucially to have a relatively blue colour in J-H (Fig. 24.3).

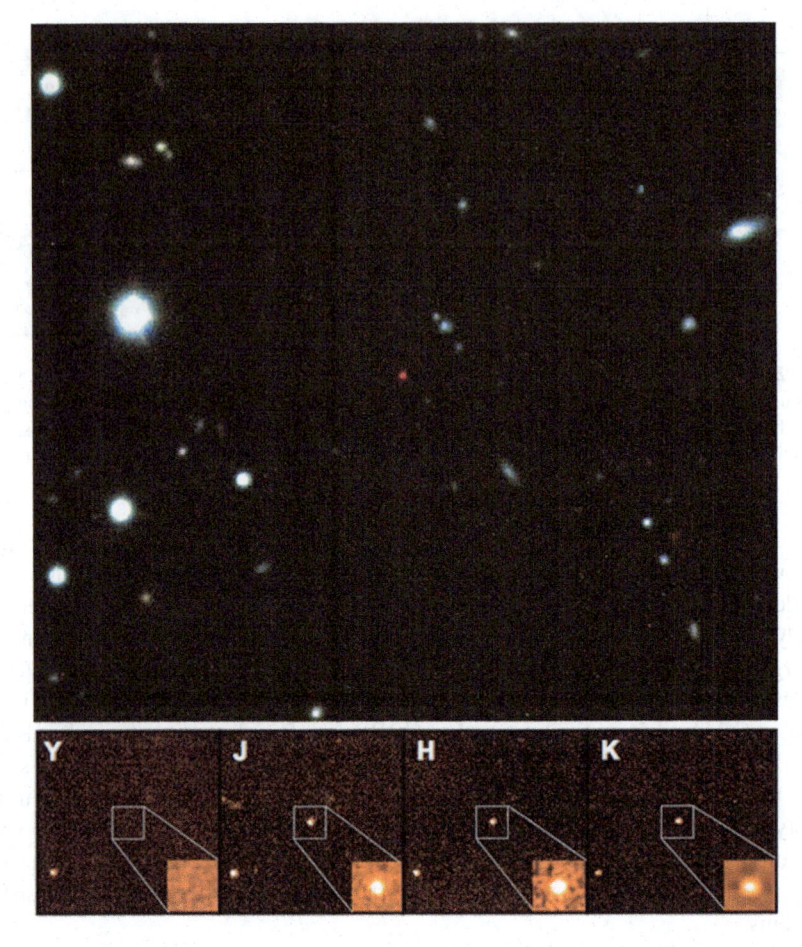

Fig. 24.3 Two ways of illustrating the drop-out nature of GRB 090423. (*Top*) shows a "true colour" representation, constructed from images in z, Y and J, all obtained in Chile. The afterglow being only visible in J is strikingly red in this image. (*Bottom*) shows directly the Y, J, H and K imaging obtained from Hawaii. The blow-ups show the field around the afterglow smoothed and at high contrast to emphasis the lack of flux in Y

This suggested a redshift $z > 8$, which was confirmed by spectroscopy at VLT (Fig. 24.4; Tanvir et al. 2009) and La Palma (Salvaterra et al. 2009) the following night.

Although in this case our spectroscopy was not of sufficient signal-to-noise to provide a clear measurement of the IGM neutral fraction, or to clearly detect metal lines in the host, it does illustrate that GRBs were being produced within a few 100 million years after the Big Bang, and pinpoints its faint host galaxy which can now be targeted for further study.

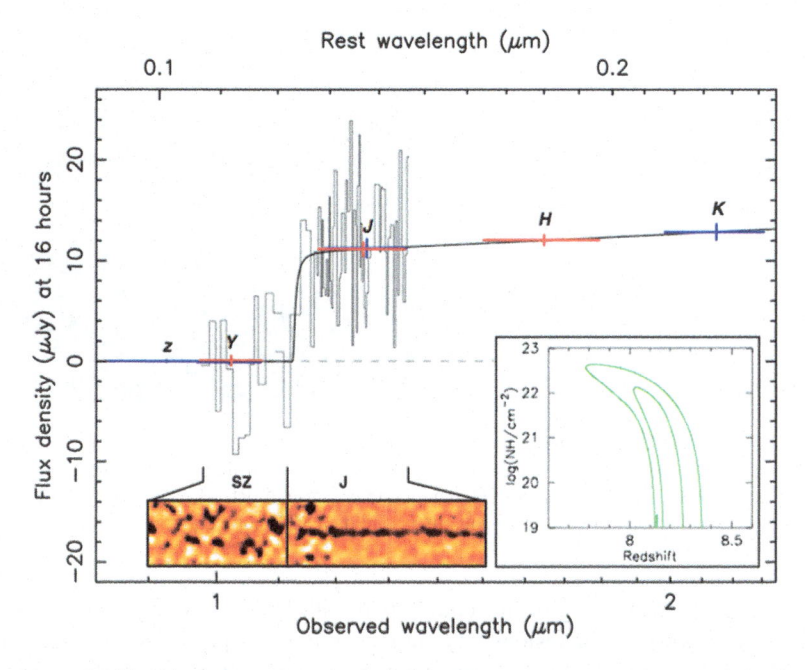

Fig. 24.4 VLT/ISAAC spectrum (both 1D and 2D) of the afterglow, overlayed with photometric data points (2-sigma flux errors shown) and the best-fitting model (*bold line*) showing the Lyman-alpha break at z = 8.2. The inset panel shows confidence contours of redshift versus the neutral hydrogen column in the host galaxy

SGR 0501+4516

Soft gamma-ray repeaters (SGRs), along with their cousins, the anomalous X-ray pulsars (AXPs), are thought to be highly magnetised neutron stars dubbed magnetars (Mereghetti 2008). The characteristic bursts of gamma-rays they occasionally emit (which give rise to the name) are similar to those of classical short-GRBs, although unlike GRBs they do repeat. There are only about half a dozen SGRs definitely identified in the Milky-Way, although it is expected that giant flaring events from SGRs in other nearby galaxies have been recorded as S-GRB events, for example in the BATSE catalogue (Hurley et al. 2005; Palmer et al. 2005; Tanvir et al. 2005).

SGR 0501+4516 was discovered by Swift and announced as a GRB in August 2008. Ongoing observations with Swift detected further flares, which soon identified the source as a SGR (Barthelmy et al. 2008). In the mean-time we obtained observations with UKIRT that identified a faint infra-red point source coincident with the Swift X-ray localisation (Fig. 24.5). The discovery of an isolated counterpart of an SGR in the near-IR is an important event, since they are typically in very crowded, high-extinction fields that makes such identifications at best highly challenging and in most cases unfeasible.

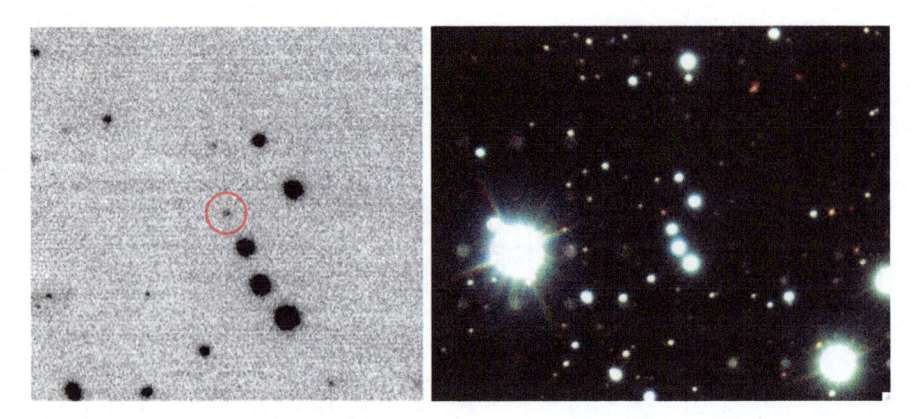

Fig. 24.5 (*Left*) UKIRT/UFTI K-band discovery image of the infrared counterpart of the SGR 0501+4516, the field, by SGR standards, having very little crowding or extinction. (*Right*) is a colour image made from JHK frames obtained with Gemini-N/NIRI. Although a faint source, it is nonetheless significantly brighter than any other candidate SGR counterpart

Subsequent monitoring of this counterpart with UKIRT, WHT and Gemini has led to the following conclusions: (1) the source exhibits periodicity in the optical and infrared on the same 5.8 s timescale as the X-ray (Dhillon et al. 2011); (2) the average IR flux has declined monotonically ever since the initial flaring event, reducing in flux by a factor $\sim$5 over 14 months; and (3) there is no apparent proper motion of the source, which rules out its association with the nearby supernova remnant HB9, or any other known remnant or stellar cluster, assuming the commonly held expectation that the age of the magnetar should be less than about 15,000 years (Levan et al. in prep.).

Quite what the full implications of these results are for our understanding of SGRs is not yet known, but it does look like they must either be older than expected, or possibly have been produced by some different channel than simply in a core-collapse supernova, for example via the accretion induced collapse of a white dwarf (e.g. Levan et al. 2006b). In any event, it is clear that this SGR is the most promising for multi-wavelength observation in the future.

Conclusions

UKIRT remains on the front-line of the global effort to follow-up gamma-ray bursts, providing key observations particularly of red GRB afterglows and GRB-like events. The lesson of Swift has been that very high redshift bursts are being detected, but they are rare. The potential importance of these for our understanding of the early universe and the era of re-ionization means that it is vital that we attempt deep IR and optical observations for as many afterglows as possible, and in particular those without any counterpart in early (but usually shallow) optical imaging.

The ultimate reward for our long-running campaign came in April 2009 with the discovery of GRB 090423 at redshift z = 8.2. As discussed in this contribution, the importance is not simply in breaking a record, but in opening a new route to the detailed study of the re-ionization era. To fully exploit this opportunity requires continued commitment of resources to build up significant samples of bursts at such distances, and to search for and characterise their hosts and any neighbouring structures.

References

Allan, A., et al.: Autonomous software: myth or magic? Astronomische Nachrichten **329**, 266 (2008)
Alvarez, M.A., et al.: Connecting reionization to the local universe. Astrophys. J. **703**, L167 (2009)
Andersen, M.I., et al.: VLT identification of the optical afterglow of the gamma-ray burst GRB 000131 at z=4.50. Astron. Astrophys. **364**, L54 (2000)
Barkana, R., Loeb, A.: Gamma-ray bursts versus quasars: Lyα signatures of reionization versus cosmological infall. Astrophys. J. **601**, 64 (2004)
Barthelmy, S.D., et al.: New soft gamma repeater SGR 0501+4516. Astron. Telegr. **1676**, 1 (2008)
Bolton, J.S., Haehnelt, M.G.: The observed ionization rate of the intergalactic medium and the ionizing emissivity at z $>=$ 5: evidence for a photon-starved and extended epoch of reionization. Mon. Not. R. Astron. Soc. **382**, 325 (2007)
Bromm, V., Loeb, A.: High-redshift gamma-ray bursts from population III progenitors. Astrophys. J. **642**, 382 (2006)
Dhillon, V.S., et al.: The first observation of optical pulsations from a soft gamma repeater: SGR 0501+4516. Mon. Not. R. Astron. Soc. **416**, L16 (2011)
Fynbo, J.P.U., et al.: Low-resolution spectroscopy of gamma-ray burst optical afterglows: biases in the swift sample and characterization of the absorbers. Astrophys. J. Suppl. **185**, 526 (2009)
Greiner, J., et al.: GRB 080913 at Redshift 6.7. Astrophys. J **693**, 1610 (2009)
Groot, P.J., et al.: A search for optical afterglow from GRB 970828. Astrophys. J **493**, L27 (1998)
Haislip, J.B., et al.: A photometric redshift of z = 6.39 ± 0.12 for GRB 050904. Nature **440**, 181 (2006)
Hurley, K., et al.: An exceptionally bright flare from SGR 1806–20 and the origins of short-duration γ-ray bursts. Nature **434**, 1098 (2005)
Iye, M., et al.: A galaxy at a redshift z = 6.96. Nature **443**, 186 (2006)
Jakobsson, P., et al.: A mean redshift of 2.8 for Swift gamma-ray bursts. Astron. Astrophys **447**, 897 (2006)
Kawai, N., et al.: An optical spectrum of the afterglow of a γ-ray burst at a redshift of z = 6.295. Nature **440**, 184 (2006)
Klebesadel, R., et al.: Observations of gamma-ray bursts of cosmic origin. Astrophys. J. **182**, L85 (1973)
Komatsu, E., et al.: Five-year wilkinson microwave anisotropy probe observations: cosmological interpretation. Astrophys. J. Suppl. **180**, 330 (2009)
Lamb, D.Q., Reichart, D.E.: Gamma-ray bursts as a probe of the very high redshift universe. Astrophys. J. **536**, 1 (2009)
Levan, A.J., et al.: The first swift X-ray flash: the faint afterglow of XRF 050215B. Astrophys. J. **648**, 1132 (2006a)
Levan, A.J., et al.: Short gamma-ray bursts in old populations: magnetars from white dwarf-white dwarf mergers. Mon. Not. R. Astron. Soc. **368**, L1 (2006b)
Levan, A. J., et al.: Constraints on the nature of SGR 0501+4526 (2013, in prep)

McQuinn, M., et al.: Probing the neutral fraction of the IGM with GRBs during the epoch of reionization. Mon. Not. R. Astron. Soc. **388**, 1101 (2008)

Mereghetti, S.: The strongest cosmic magnets: soft gamma-ray repeaters and anomalous X-ray pulsars. Astron. Astrophys. Rev. **15**, 225 (2008)

Palmer, D.M., et al.: A giant γ-ray flare from the magnetar SGR 1806–20. Nature **434**, 1107 (2005)

Salvaterra, R., et al.: GRB090423 at a redshift of z 8.1. Nature **461**, 1258 (2009)

Tanvir, N.R., et al.: An origin in the local Universe for some short γ-ray bursts. Nature **438**, 991 (2005)

Tanvir, N.R., Jakobsson, P.: Observations of GRBs at high redshift. R. Soc. Lond. Philos. Trans. A **365**, 1377 (2007)

Tanvir, N.R., et al.: The extreme, red afterglow of GRB 060923A: distance or dust? Mon. Not. R. Astron. Soc. **388**, 1743 (2008)

Tanvir, N.R., et al.: A γ-ray burst at a redshift of z 8.2. Nature **461**, 1254 (2009)

van Paradijs, J., et al.: Transient optical emission from the error box of the γ-ray burst of 28 February 1997. Nature **386**, 686 (1997)

Vreeswijk, P.M., et al.: The host of GRB 030323 at z=3.372: a very high column density DLA system with a low metallicity. Astron. Astrophys. **419**, 927 (2004)

Part III
UKIDSS and the Future

Chapter 25
The UKIRT Infrared Deep Sky Survey (UKIDSS): Origins and Highlights

Andy Lawrence

Abstract UKIDSS is now roughly half finished, covering several thousand square degrees, and expected to complete by the middle of 2012. So far 13.5 billion rows of data have been downloaded by several hundred users, resulting in 154 publications concerning high-z quasars, the nearest and coolest brown dwarfs, the evolution of galaxies, the sub-stellar mass function, and many other topics. I summarise the origins and status of UKIDSS, and briefly describe some scientific highlights.

The UKIRT Infrared Deep Sky Survey (UKIDSS)

UKIDSS is currently the major activity of UKIRT. It is the successor to the Two Micron All Sky Survey (2MASS; Skrutskie et al. 2006), but many times deeper, and so is the IR equivalent of the Sloan Digital Sky Survey (SDSS; York et al. 2000). Unlike 2MASS it does not cover the whole sky. In fact it is really a portfolio of five sub-surveys. Three of these are "shallow" surveys – the Large Area Survey (LAS), the Galactic Plane Survey (GPS), and the Galactic Clusters Survey (GCS) – designed to cover almost 7,000 sq.deg. in several bands to a depth of $K \sim 18$ (I will be using Vega magnitudes throughout.) Two of the surveys are smaller and deeper – the Deep Extragalactic Survey (DXS) aiming to cover 35 sq.deg. in four fields to $K \sim 21$, and the Ultra Deep Survey (UDS) covering a single 0.8 sq.deg. field to a target depth of $K \sim 23$. More details can be found in other talks at this meeting, in the UKIDSS core reference (Lawrence et al. 2007), in papers describing the calibration, data access, and processing (Hewett et al. 2006; Hodgkin et al. 2009;

A. Lawrence (✉)
Institute for Astronomy, University of Edinburgh, Royal Observatory, Blackford Hill, Edinburgh EH9 3HJ, UK
e-mail: al@roe.ac.uk

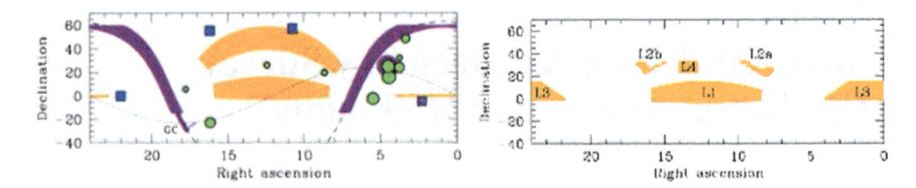

Fig. 25.1 *Left*: original UKIDSS survey design from 2001. *Orange* is the Large Area Survey (*LAS*); *purple* is the Galactic Plane Survey (*GPS*); *green* is the Galactic Clusters Survey (*GCS*); *blue* is the Deep Extragalactic Survey (*DXS*); *pink* is the the Ultra Deep Survey (*UDS*). *Right*: revised footprint for LAS from the September 2009 completion plan approved by the UKIRT Board. Other footprint changes since 2001 are that the DXS XMM-LSS field is now abutting the UDS rather than surrounding it; the GPS is not surveying Taurus-Auriga-Perseus; and the GCS is now not surveying the Hyades

Hambly et al. 2008; Irwin et al. in preparation), and at the UKIDSS web page.[1] The survey is implemented using UKIRT's Wide Field Camera (WFCAM; Casali et al. 2007)

Figure 25.1 shows the UKIDSS survey design, together with the revised footprint for the LAS approved by the UKIRT Board in November 2009, for completion by mid-2012. The survey plan maximises multi-wavelength coverage – in particular the UDS is the same field as the Subaru-XMM Deep Survey, and the LAS has a large commonality with SDSS, making a unique large area, well matched ugrizYJHK data set.

As well as providing a rich legacy database, UKIDSS aims are to find the nearest and faintest substellar objects; to discover Population II brown dwarfs, if they exist; to determine the substellar mass function; to break the $z = 7$ quasar barrier; to determine the epoch of re-ionization; to measure the growth of structure from $z = 3$ to the present day; to determine the epoch of spheroid formation; and to map the Milky Way through the dust, to several kpc.

The data is public but released in two steps for each release slice – first to registered European astronomers, and then to the world 18 months later. All data are accessible through a flexible query interface at the WFCAM Science Archive (WSA), which also hosts all other WFCAM data.[2]

Origins

UKIDSS is an unusual community based project, mixing elements of the public and the private. It is therefore interesting to see how it developed. The idea of an ambitious survey programme was developed and proposed along with concept of

---

[1] http://www.ukidss.org

[2] http://surveys.roe.ac.uk/wsa

the camera itself, and was then developed as a "wedding cake" programme and proposed to PPARC by a relatively small number of people (Lawrence et al. 1998, 1999). The reaction of the Board was to make an Announcement of Opportunity to build on this concept and involve more people. The result was the formation of a large consortium with a single purpose, but much improved scientific design. The consortium included almost all astronomers in the UK likely to be actively interested in such a survey. When the UK joined ESO, the consortium expanded again to include astronomers from across Europe. Finally a single proposal was put to the UKIRT Board in March 2001. The Board has approved UKIDSS in a series of provisional stages, each peer reviewed, with significant updates in Nov 2001, Nov 2006, and Sept 2009.

Although UKIDSS has distinct scientific and practical components, it is designed and implemented as a single whole, with frequent discussions between working groups, including many trade-offs and compromises to optimise the whole programme, occasional full consortium meetings, a single web site, and a single method of data access. Most such large coherent projects have in the past been conducted by public bodies (e.g. the UK Schmidt Unit) on behalf of the community. Alternatively, telescopes run as facilities tend to allocate time to many small groups. Increasingly such facilities have offered "key programme" time, and sometimes proposing consortia are obliged to make their data public in due course, but they are still essentially private projects. UKIDSS represents perhaps the first time a community has come together to propose use of the *majority* of a public facility. VISTA is in some ways a second example, but in that case the distinct surveys are run as separate projects. The fact that UKIDSS acts a single community project may perhaps be a lucky accident due to growing organically from a seed, rather than from a facility requesting proposals to fill a space.

Although I am stressing that UKIDSS is a community project, this is an oversimplification. In fact part of the reason for its success is that it operates within a larger UKIRT ecosphere, each component of which is professionally conducted. First and foremost, UKIDSS relies on the existence of UKIRT as a mature and efficient operating observatory. Second, it relies on the existence of WFCAM, built by the UKATC, and expressly designed for UKIDSS-like surveys (as well of course as regular PATT PI time, and other smaller "campaign" proposals). Third, the pipeline processing and science archive development has also been run as a professional funded project by Cambridge, Edinburgh and QMUL, with a transition to VISTA anticipated (the VISTA Data Flow System – VDFS). The community could therefore concentrate on what it does best – scientific design and implementation, and input of requirements to UKIRT, ATC, and VDFS. The relationship between the UKIDSS consortium and the VDFS team was particularly close and detailed.

Along with UKIRT, UKATC, VDFS, and the UKIDSS consortium, the Fifth Estate is the wider community which exploits the data and writes papers. The UKIDSS consortium has no proprietary rights over the data. Its purpose is to design and implement the survey. Of course people in the UKIDSS working groups have been thinking about the survey for many years, and so are typically in a strong

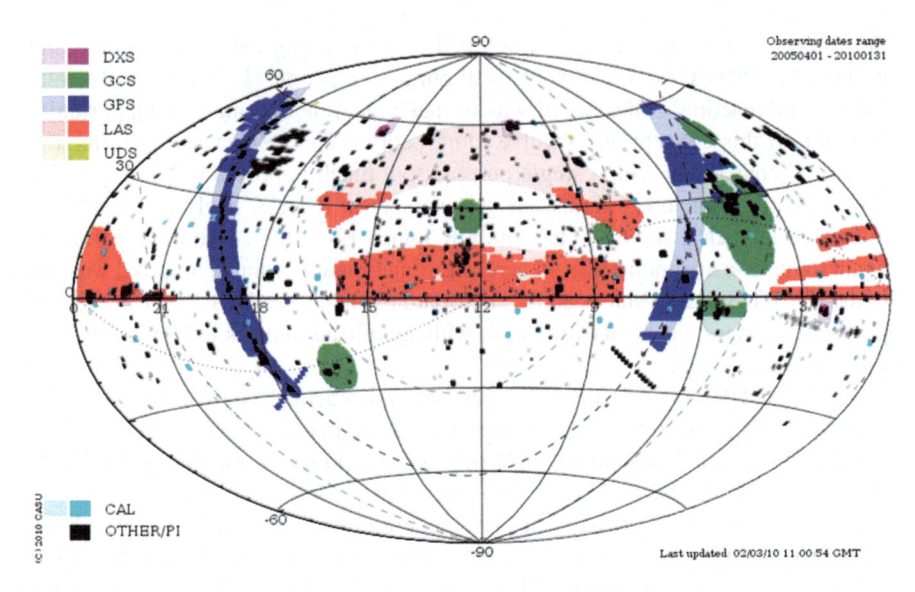

Fig. 25.2 Status of WFCAM observations as of 1-Feb-2010. The *light coloured areas* show the original design footprint. The *darker coloured areas* show areas with observations in at least one filter. *Black squares* show non-UKIDSS WFCAM observations

position to exploit it; but they have no structural privileges. Interestingly, unlike some other large private programmes I am aware of, UKIDSS has had little trouble staffing its observing runs (Fig. 25.2).

Status

At the time of writing (April 2010), each of the UKIDSS surveys is already the largest of its kind and will remain so until VISTA has been operating for a few years. In the latest data release (DR7, February 2010), UKIDSS is approximately 40 % complete compared to the 2001 goals, and 50 % complete compared to the 2009 revised plan. The UDS has so far accumulated a depth $K \sim 22.3$. The LAS currently covers 2,468 sq.deg. in any filter, and 1,818 sq.deg. in the complete YJHK set, out of a planned total of 3,792 sq.deg. The total number of detections in the database is well over 10^8, dominated by the GPS (See the DR7 release page at the WSA for more details).

The rate of completing the survey has been a little slower than hoped, but the quality of the data collected is very close to that planned. For example, the median seeing is $0.82''$; the median stellar ellipticity is 0.07; the astrometric accuracy is 50–100 mas depending on latitude; and the absolute photometric accuracy is 2 %. The 5σ depths reached in the shallow surveys are close to expectation (e.g. for LAS $Y = 20.16$, $J = 19.56$, $H = 18.81$, $K = 18.19$). In the deep (stacked) surveys the rate of improvement of depth with accumulated exposure falls a little short of the naive $t^{1/2}$ expectation, but is coming closer at each data release due to processing improvements (e.g. dealing with structured background). For the GPS, the effective depth reached is very sensitive to source crowding.

The UKIDSS data has been heavily used. As of Sept 2009, there were 890 registered users for the initial Europe-only data releases, and of course an unknown number of anonymous users of the subsequent world wide releases. The number of SQL queries run as of Sept 2009 was 805,000, extracting 13.5 billion rows of data. The publication rate for papers wholly or partly based on UKIDSS data is increasing; 7 in 2006, 24 in 2007, 35 in 2008, 60 in 2009, and 26 in the first 3 months of 2010.

The completion of UKIDSS depends of course on the status of UKIRT itself, but all being well we expect to complete the project by the middle of 2012.

Highlights

Other papers from this meeting present scientific progress from the UKIDSS sub-surveys. Here I present a personal selection of highlights.

The LAS is particularly well suited to finding rare but important objects. Figure 25.3 shows the optical spectra of four new $z \sim 6$ quasars (Mortlock et al. 2009 and Patel et al. in preparation). Around $z \sim 6$ UKIDSS finds the quasars already seen

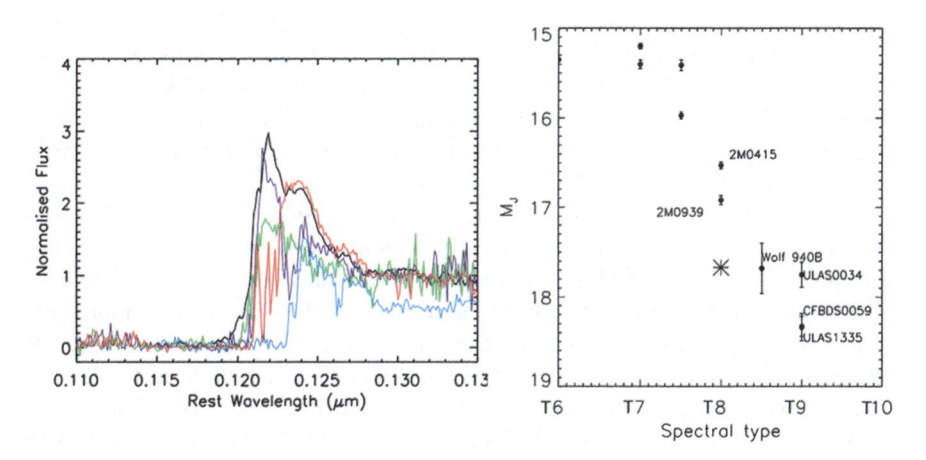

Fig. 25.3 *Left*: four new UKIDSS-LAS-selected $z = 6$ quasars from Patel et al. (in preparation), compared to the mean spectrum of $z = 6$ quasars from SDSS (*black curve*, Fan et al. 2004). UKIDSS is finding weaker lined objects missed by SDSS. *Right*: Absolute magnitude M_J for *cool brown* dwarfs as a function of spectral type. The four LAS brown dwarfs with parallaxes are the T8.5 dwarf Wolf 940B, and the three T9 dwarfs ULAS0034, CFBDS0059, and ULAS1335 (Burningham et al. 2009; Warren et al. 2007). They are all substantially less luminous than 2MASS J0415−09, the lowest luminosity brown dwarf known at the start of the survey

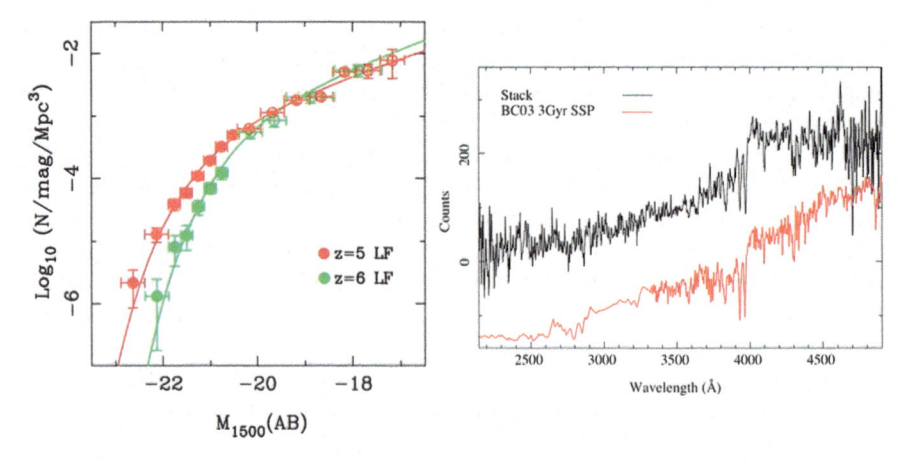

Fig. 25.4 *Left*: evolution of the luminosity function of UKIDSS-UDS UV-selected galaxies at z > 5 (From McLure et al. 2009). *Right*: Mean spectrum of 38 distant Luminous Red Galaxies selected from UKIDSS-DXS, showing strong Hδ absorption indicating a significant contribution by young (<2 Gyr) stars

by SDSS, and around the same number again not found by SDSS. However, LAS has not yet found quasars at z > 6.4, as predicted from extrapolation of the Fan et al. (2004) luminosity function evolution. This is now becoming highly significant, and indicates a steep rise in quasar numbers at just the time of re-ionisation.

At the opposite end of the distance scale, Fig. 25.3 shows the properties of very cool brown dwarfs. UKIDSS has now found the majority of known T dwarfs, including a succession of "coldest star" record holders (Warren et al. 2007; Burningham et al. 2009; Lucas et al. 2010). Along with new 2MASS and CFHT work, UKIDSS has extended the brown dwarf sequence by nearly 2 magnitudes. We have not yet however definitively identified the expected new class of Y dwarfs, where Ammonia absorption should become apparent.

The deeper surveys (DXS and UDS) are especially effective at statistical studies of the higher redshift universe. An early highlight was the identification of a supercluster structure 30 Mpc across at $z \sim 1$ (Swinbank et al. 2007). Combining UDS with Subaru data to obtain photometric redshifts has led to some impressive results, including the evolution of the K-band luminosity function from $z \sim 4$ to the present day (Cirasuolo et al. 2010), and the detection of evolution of the Lyman Break Galaxy (LBG) population between $z = 5$ and $z = 6$ (McLure et al. 2009). Figure 25.4 shows the LBG result, and also the mean spectrum of 38 luminous Distant Red Galaxies from selected from DXS, showing strong Hδ absorption, indicating a significant contribution from young stars.

The prime aim of the GCS is to measure the substellar luminosity function and its dependence on environmental factors. Figure 25.5 shows the measured mass functions for three large Galactic Clusters (Lodieu et al. 2009). All three are consistent with a log Gaussian form, which therefore shows that the star formation

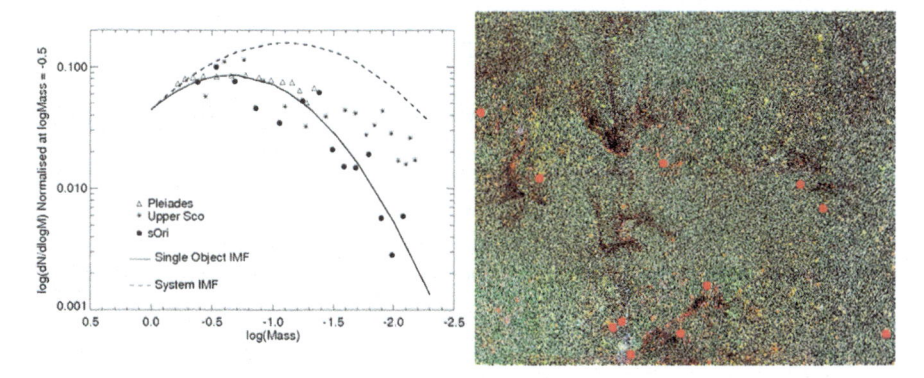

Fig. 25.5 *Left*: Mass functions (number of stars per unit logarithmic mass interval) derived from GCS data. The figure compares Upper Sco, the Pleiades and the sigma-Orionis clusters (*solid circles, open triangles* and *asterisks* respectively plotted without error bars for clarity – for details see Lodieu et al. 2009 and references therein). Also plotted is the log-normal parameterised field IMF assuming 50 % binarity; the *dashed line* is the corresponding underlying single-star IMF. *Right*: Location of high amplitude variables found in the Serpens OB2 complex. These were found in an automated search of 31 sq.deg. with two epoch K-band data, locating objects where $\Delta K > 1.0$

process has a characteristic mass, which is very close to the nuclear burning threshold. The GPS has located ~400 open clusters in the Galactic Plane, only half of which were previously known, leading to a large variety of follow-on projects. More surprising has been the discovery of a number of new high-amplitude variables within a relatively small area (DK > 1 mag; see Fig. 25.5). The nature of these is currently unknown, but they are likely to be eruptive pre-main sequence stars, or FU Orionis stars (Lucas 2010). The GPS therefore holds the possibility of finding large numbers of such objects.

The Future

UKIDSS is expected to complete in mid-2012, UK politics permitting. The mantle for large near-IR surveys now passes to VISTA, which is an ESO 4 m telescope with an even larger IR array, dedicated to survey programmes. It will take a few years for VISTA to overtake UKIDSS in volume. At longer wavelengths, WISE is as I write surveying the sky in the mid-IR, at a sensitivity well matched to UKIDSS-LAS for a variety of object types. At shorter wavelengths, PanSTARRS-1 is just starting a survey that covers all the Northern sky and some of the Southern sky. Eventually we will have LSST producing a very deep southern sky survey, and possibly PanSTARRS-4 producing a northern sky survey that is almost as deep. Until that time, in the near future, the southern hemisphere will be better served in the IR, and the northern hemisphere better served in visible light. With this prospect in mind, in 2006 the UKIDSS consortium proposed that UKIRT should follow on from UKIDSS by conducting a UKIRT Hemisphere Survey (UHS), which together with the VISTA Hemisphere Survey (VHS) would make an (almost) complete NIR sky survey.

The current default is that the UHS will not now happen, as UKIRT is planned for closure after completion of UKIDSS. However, if new UKIRT funding partners emerge, there is still a possibility that UHS could be part of a continuing UKIRT programme. Alternatively, the idea of a relatively modest space-based NIR sky survey obviously has appeal. Several manifestations of such a mission have come close to happening, but not quite made the cut. The international community should keep trying!

References

Burningham, B., et al.: The discovery of an M4+T8.5 binary system. Mon. Not. R. Astron. Soc. **395**, 1237 (2009)
Casali, M.M., et al.: The UKIRT wide-field camera. Astron. Astrophys. **467**, 777 (2007)
Cirasuolo, M., et al.: A new measurement of the evolving near-infrared galaxy luminosity function out to z = 4: a continuing challenge to theoretical models of galaxy formation. Mon. Not. R. Astron. Soc. **401**, 1166 (2010)
Fan, X., et al.: A Survey of z>5.7 quasars in the Sloan Digital Sky Survey. III. Discovery of five additional quasars. Astron. J. **128**, 515 (2004)
Hambly, N.C., et al.: The WFCAM science archive. Mon. Not. R. Astron. Soc. **384**, 637 (2008)
Hewett, P.C., Warren, S.J., Leggett, S.K., Hodgkin, S.L.: The UKIRT Infrared Deep Sky Survey ZY JHK photometric system: passbands and synthetic colours. Mon. Not. R. Astron. Soc. **367**, 454–468 (2006)
Hodgkin, S.L., Irwin, M.J., Hewett, P.C., Warren, S.J.: The UKIRT wide field camera ZYJHK photometric system: calibration from 2MASS. Mon. Not. R. Astron. Soc. **394**, 675 (2009b)
Lawrence, A., Adamson, A., Hawarden, T., Peacock, J., Saunders, W., Hambly, N., Casali, M.: Paper to PPARC Ground Based Facilities Committee, 22 Sept 1998
Lawrence, A., Hambly, N., Peacock, J., Saunders, W., Adamson, A., Hawarden, T., McMahon, R., Irwin, M., Casali, M., Warren, S., Ward, M., Eales, S., Sekiguchi, K.: Proposal to UKIRT Board, 1 Nov 1999
Lawrence, A., et al.: The UKIRT Infrared Deep Sky Survey (UKIDSS). Mon. Not. R. Astron. Soc. **379**, 1599 (2007)
Lodieu, N., et al.: A census of very-low-mass stars and brown dwarfs in the σ Orionis cluster. Astron. Astrophys. **505**, 1115 (2009a)
Lucas, P.W., et al.: Thirty Years of Astronomical Discovery with UKIRT. Astrophysics and Space Science Proceedings, vol. 37. Springer, New York/Berlin
Lucas, P.W., et al.: The discovery of a very cool, very nearby brown dwarf in the Galactic plane. Mon. Not. R. Astron. Soc. **408**, 56 (2010). arXiv:1004.0317 (2010)
McLure, R.J., et al.: The luminosity function, halo masses and stellar masses of luminous Lyman-break galaxies at redshifts $5 < z < 6$. Mon. Not. R. Astron. Soc. **395**, 2196 (2009)
Mortlock, D., et al.: Discovery of a redshift 6.13 quasar in the UKIRT infrared deep sky survey. Astron. Astrophys. **505**, 97 (2009)
Skrutskie, M.F., et al.: The Two Micron All Sky Survey (2MASS). Astron. J. **131**, 1163 (2006b)
Swinbank, A.M., et al.: The discovery of a massive supercluster at z = 0.9 in the UKIDSS Deep eXtragalactic Survey. Mon. Not. R. Astron. Soc. **379**, 1343 (2007)
Warren, S.J., et al.: A very cool brown dwarf in UKIDSS DR1. Mon. Not. R. Astron. Soc. **381**, 1400 (2007)
York, D.G., et al.: The Sloan Digital Sky Survey: technical summary. Astron. J. **120**, 1579 (2000)

Chapter 26
A Billion Stars: The Near-IR View of the Galaxy with the UKIDSS Galactic Plane Survey

P.W. Lucas, D. Samuel, A. Adamson, R. Bandyopadhyay, C. Davis, J. Drew,
D. Froebrich, M. Gallaway, A. Gosling, R. de Grijs, M.G. Hoare,
A. Longmore, T. Maccarone, V. McBride, A. Schroeder, M. Smith, J. Stead,
and M.A. Thompson

Abstract The scope and status of the UKIDSS Galactic Plane Survey is described. Some issues regarding automatic processing of crowded fields and comparisons with other surveys are identified and some early results are presented. The detection of more than 100 new clusters and 16 new high amplitude variables are reported.

P.W. Lucas (✉) • D. Samuel • J. Drew • M. Gallaway • M.A. Thompson
Centre for Astrophysics Research, University of Hertfordshire, College Lane, Hatfield AL10 9AB, Hertfordshire, UK
e-mail: p.w.lucas@herts.ac.uk

A. Adamson • C. Davis
Joint Astronomy Centre, 660 N. A'Ohoku Place, Hilo, HI, USA

R. Bandyopadhyay
Department of Astronomy, 211 Bryant Space Science Center, Gainesville, FL 32611-2055, USA

D. Froebrich • M. Smith
Centre for Astrophysics and Planetary Science, University of Kent, Canterbury CT2 7NH, UK

A. Gosling
Department of Astrophysics, University of Oxford, Keble Road, Oxford OX1 3RH, UK

R. de Grijs
Kavli Institute for Astronomy and Astrophysics, Peking University, Beijing 100871, China

M.G. Hoare • J. Stead
School of Physics and Astronomy, University of Leeds, Leeds, LS2 9JT, UK

A. Longmore
UK Astronomy Technology Centre, Royal Observatory, Blackford Hill, Edinburgh EH9 3HJ, UK

T. Maccarone • V. McBride
School of Physics and Astronomy, University of Southampton, Hampshire SO17 1BJ, UK

A. Schroeder
Hartebeesthoek Radio Astronomy Observatory, Johannesburg, South Africa

Introduction to the GPS

The UKIDSS Galactic Plane Survey (GPS) is one of the five UKIDSS surveys with the Wide Field Camera (WFCAM) (Casali et al. 2007) on UKIRT. It began in 2005 and will continue until 2012. The survey aims, some demonstration science results and a guide to the GPS data are described in Lawrence et al. (2007) and in Lucas et al.(2008). Briefly, the GPS comprises data in the J, H and K bandpasses, with on-source integration times of 80 s, 80 s and 40 s respectively. Images in these three filters are taken near simultaneously (within a 20 min interval) unless one of the images fails quality control and is repeated later. There is also a 2nd epoch of K band imaging (again with 40 s integration time) taken at least 2 years before or after the three colour imaging. The first results from the 2nd epoch data are described in section "New extreme infrared variable stars and the stellar birthline". The survey depth varies due to source confusion in the 1st Galactic quadrant but reaches $K \sim 18.0$ (5σ, Vega system) throughout the 2nd and 3rd quadrants. In general the GPS is 3 magnitudes deeper than the 2MASS survey, due to greater sensitivity in uncrowded fields and higher spatial resolution in fields where source confusion is significant.

The main survey area is a 10° wide band in the northern and equatorial sky. In detail, the area is a narrow strip at longitudes $l = 358°$ to $l = 15°$ at latitudes $|b| < 2°$, and two broader bands at $l = 15-107°$ at $|b| < 5°$ and $l = 142-230°$ at $|b| < 5$. The northern limit of UKIRT is Dec $= +60°$, which puts the $l = 107-142°$ region off limits. The southern limit of the survey is at Dec $= -15°$ except for the narrow southern strip between $l = 358-15°$. The GPS also includes a survey of a 200 deg^2 area located off the plane in the Taurus-Auriga-Perseus molecular cloud complex, observed in the J, H, and K filters plus a narrow band 2.12 μm filter sensitive to H_2 line emission (e.g. Davis et al. 2008).

The aim of the GPS is to produce a legacy database of $\sim 10^9$ stars that will be useful for all areas of Galactic astronomy. Some particular areas where the large number of stars and the panoramic coverage are expected to lead to scientific advances are star formation, Galactic structure and the Galactic X-ray population. It is also hoped that the 2 epoch K band data will lead to the detection of high amplitude variable stars that probe brief, rarely observed phases of stellar evolution, e.g. FU Orionis stars and AGB stars with unstable helium shell burning.

Progress

At the end of semester 2009A (31 July 2009) the GPS is approximately 35 % complete relative to the original 2001 goals. Top priority was given to the region that overlaps with the Spitzer-GLIMPSE and GLIMPSE-II surveys (Benjamin et al. 2003). This region and in fact the whole section of the mid-plane from $l = 358$ to 107, $|b| < 1.3$, have now been completed. In total ~ 800 deg^2 of the GPS region has been observed in the J, H and K filters and the "2nd epoch" K band data have been taken for a further ~500 deg². Progress has been slower in the section from l = 141–230, owing to a combination of relatively poor winter weather and greater competition with the other UKIDSS surveys and Cassegrain observation blocks. For the future, we have prioritised the mid-plane in the outer Galaxy, where the mid-plane is being defined in a manner that follows the northern warp in the outer galaxy.

Cross Matches with Other Galactic Surveys

The GPS is one of many panoramic surveys that are surveying the Milky Way at wavelengths from the ultraviolet to the radio, e.g. IPHAS, UVEX, VPHAS+, VISTA VVV, GLIMPSE I-III, Cygnus-X, GLIMPSE-360, MIPSGAL, HERSCHEL Hi-Gal, Scuba2 JPS, Scuba2 SASSy, BOLOCAM and VLA CORNISH. Naturally, most types of Galactic science will benefit from drawing on data from two or more of these surveys. For stellar sources, this means that tools for combining the source catalogues of the optical, near IR and mid-IR surveys are essential.

Fortunately, the WFCAM Science Archive or WSA (Hambly et al. 2008) includes copies of many of the desired catalogues, e.g. the GLIMPSE I and GLIMPSE II version 2.0 catalogues (both the "highly reliable catalogues" and the "more complete archives"). It also includes the 2MASS catalogues. These catalogues have been cross-matched by the archive team to create neighbour tables: i.e. a table that lists one or more near neighbours in the appropriate catalogue (typically with a maximum radius of 10 arcsec). These tables facilitate rapid SQL queries by users who wish to cross match the GPS with these other catalogues to detect sources with particular characteristics.

We have performed GPS/GLIMPSE and GPS/IPHAS cross matches (the latter using Astrogrid[1] since no neighbour table yet exists in the WSA). The GPS/IPHAS search was straightforward (see Lucas et al. 2008) and shows that the (i′–J) vs. (J–H) two colour diagram has potential to aid spectrophotometric typing of field stars, breaking the degeneracy that exists between spectral type and extinction when using the GPS dataset alone.

A GPS/GLIMPSE cross match using GPS DR4 data was performed to search for Young Stellar Objects (YSOs) and other objects with hot circumstellar dust using the dereddened [K-4.5] colour, where the 4.5 μm magnitude comes from Spitzer/IRAC band 2. We find that in the crowded fields of the GLIMPSE survey this cross match produces many candidates, a large fraction of which are likely to be false positives.

Overdensities of candidates are detected in the vicinities of some known star formation regions, e.g. the young cluster BDSB 126 (Bica et al. 2003) which is associated with the HII region G29.96-0.02. This indicates that GPS/GLIMPSE

---

[1] http://www.astrogrid.org/wiki/Help/UsageExamples/Datasets/Iphas

cross matches may be useful to discover very young clusters. However, the cross match also shows an absence of detections in both catalogues in the vicinity of star formation regions with extensive bright nebulosity, e.g. W43, M17, BDSB 121.

This problem has been addressed in the DR5 release of the GPS on 28 August 2009. The data reduction team at the Cambridge Astronomical Survey Unit (CASU) have reprocessed all GPS fields that contain nebulosity using a new median filter algorithm that measures the sky background on much finer spatial scales than the standard UKIDSS pipeline. The fields to be reprocessed were identified by visual inspection by Lucas during quality control. The GPS catalogue is now much more complete than hitherto in bright nebulous fields, though it should be noted that photometry of faint stars embedded in bright, spatially structured nebulosity is always less reliable than in most other fields. Similar reprocessing of the GLIMPSE data in nebulous star formation regions is desirable if resources can be found to do it.

Searches for New Star Clusters

The GPS has the potential to advance star formation science a statistical analysis of the properties of a large number of star formation regions. This requires that we first assemble the sample, which should naturally include pre-main sequence clusters that have not previously been detected due to their large distances and high extinction at optical wavelengths.

We have used two main methods to search for new clusters: a Bayesian search of the GPS point source catalogue (*gpsSource*) and the visible inspection of the GPS jpeg images during quality control. Both methods are difficult because of the high density of field stars in the 1st Galactic quadrant, which is where many young clusters are expected to lie because of the higher star formation rate in the inner Galaxy.

The Bayesian search (Samuel, M.Sc. thesis, in prep) was performed at Hertfordshire on a local copy of the DR4 version of *gpsSource* that contains only a small subset of the data columns in the WSA. Each 13.7 arcminute WFCAM array was searched separately. First, poor quality data at the edges of the array are removed (an area smaller than the overlap region of the arrays in UKIDSS observations). Then we filtered out all the point sources in the catalogue, leaving only sources classified as extended (mergedClass $= +1$). This raises the contrast of small clusters in rich fields because clusters tend to contain many pairs of stars that are only marginally resolved in the images, and therefore appear as single extended sources in the catalogue. Furthermore, bright nebulosity in star formation regions is often mis-identified as a group of faint extended sources by the UKIDSS source detection algorithm (in data releases prior to DR5). This also aids the detection of pre-main sequence clusters.

The data were then filtered by K magnitude, excluding sources with $K > K_{max}$, where K_{max} was set at 13.0, 14.0, 15.0, 16.0 and 17.0 in different searches. Next the array was split into a grid of $n \times n$ sub-windows with an average of $\sim$500 stars each. This increases the contribution of any small cluster in the field to the model fit for the sub-window and increases the accuracy of the description of the field stars as a Poisson background with a surface density that does not vary within the sub-window. Then some edge effects are removed by wrapping an artificially generated Poisson background of stars around the outskirts of the sub-window. This greatly reduces the number of false positive detections.

The model adopted for the cluster search is a circularly symmetric Gaussian cluster profile added to a Poisson background of stars. The Bayesian search is then performed using the Expectation Maximisation (EM) Method as follows. Initial cluster parameters are first randomly generated within the sub-window and the likelihood of the data, given the model, is then calculated (the E-step). Then the data are used to update the model parameters (the M-step), weighting the contribution of each star by the probability that it is a cluster member. The EM method always shifts the model parameters in the direction of increased likelihood, though it could in principle be misled by small local peaks in the probability distribution for the model parameter space. The E- and M-steps are then repeated for a total of 30 iterations. This was found to be a sufficient number for convergence to occur.

A preliminary cluster candidate is identified if the cluster + Poission field model is preferred over a simple Poisson field model. The preference is determined by calculating the Bayesian Information Criterion (BIC) for the cluster model. This is a function that is based on the log of the likelihood of the model and the number of free parameters in the model. Additional parameters improve the likelihood of the fit so they are represented as penalty terms in the BIC in order to avoid a bias towards more complicated models. The BIC is defined in such a way that it is zero for the Poisson field model. The cluster model is then preferred if the BIC is negative. The search is then repeated with $(n+1) \times (n+1)$ sub-windows per array, and with $(n-1) \times (n-1)$ sub-windows, to remove the incompleteness due to the location of clusters at the edges of the sub-windows.

The EM iteration is then repeated for fields with a preliminary cluster candidate after including the point sources (mergedClass $= -1$) and sources with an uncertain classification (mergedClass $= -2$ or -3). If the cluster remains more significant than the simple Poisson field model then its parameters are logged as a candidate worthy of visual examination. Visual inspection is then used to eliminate false positives, which are typically associated with bright stars (frequently resolved by the source detection algorithm as false clusters if the image profile is saturated over an area much larger than the seeing disc) or with image artefacts of various kinds.

This method has successfully identified 316 clusters in DR4 that pass visual inspection. These are a mixture of pre-main sequence clusters (identified by bright nebulosity or dark clouds) and older clusters. JHK three colour images of four of the new clusters are shown in Fig. 26.1. The 316 clusters include a handful of candidate clusters that require further analysis to confirm that that are not merely chance overdensities. Slightly under half of the 316 are previously unknown. In future data releases a copy of the version of gpsSource produced by the standard data reduction pipeline will be retained so that the filtering based on mergedClass $= +1$ can still be employed for nebulous regions, even though the public version of gpsSource will comprise only the improved catalogue described in section "Cross matches with other Galactic surveys", above.

Fig. 26.1 Four new pre-main sequence (*PMS*) clusters discovered in the GPS by the Bayesian search (see section "Searches for new star clusters"). These JHK colour images are 2–3 arcmin on a side

Results of the visual search are still being collated but at present it appears that it is increasing the number of clusters detected by ∼25 %. It is therefore useful even though there is significant overlap with the results of the Bayesian search.

New Extreme Infrared Variable Stars and the Stellar Birthline

An exciting recent result from DR5 has been the detection of 16 new high amplitude variable stars ($\Delta K > 1$ mag) in the two epochs of GPS K band data. They were found in a search for stars that had changed by $\Delta K > 1$ mag and were brighter than K = 16

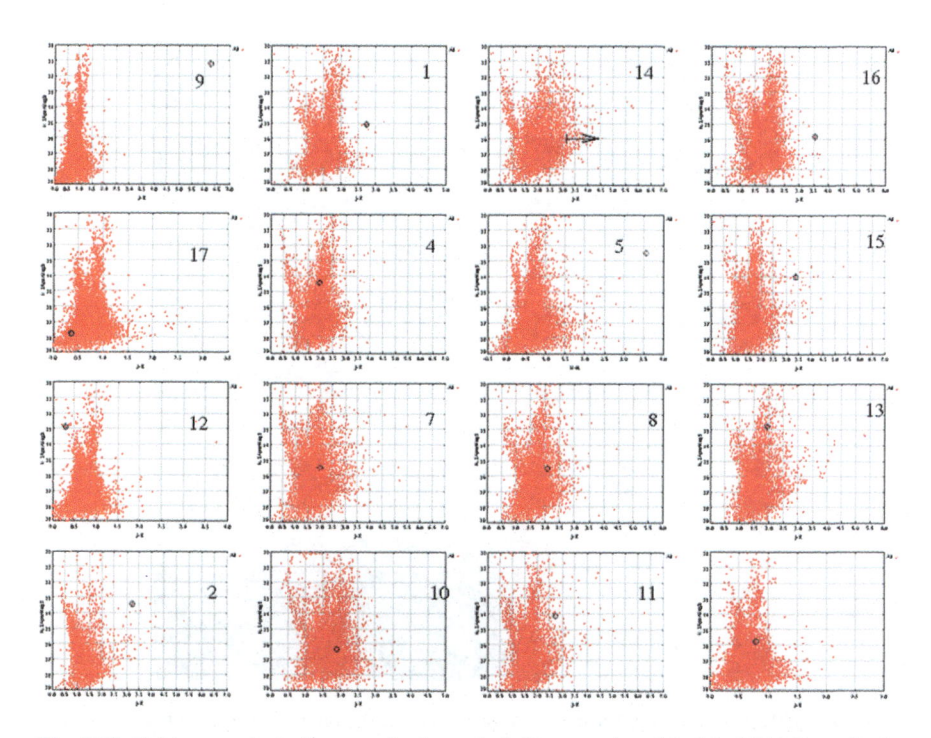

Fig. 26.2 Colour magnitude diagrams for 6 arcmin fields centred on 16 of the 17 high amplitude variables. Most are either *extremely red* or are projected against the giant branch. Most plots show (J–K) vs. K. The exception is the plot for the source labelled no. 5. For this the (H–K) colour is plotted because it is a J band drop out

in at least one epoch. Seventeen stars were found out of five million stars that had K < 16 in at least one epoch in the 31 deg^2 of sky with two epoch coverage. DR5 is the first GPS data release with 2 epoch data. Only one of these variables is previously known in the literature (Nova Sct 2003, now a faint blue object). The maximum flux difference is 3.75 mag. Eight have 2MASS K band fluxes that provide a third epoch at low S/N, apparently confirming the variability. Fourteen of the sixteen new discoveries are red objects, eight of them much redder than giant stars in the same field (see Fig. 26.2). These stars are all much too faint to be R Cor Bor stars or Asymptotic Giant Branch variables such as Miras. Pre-main sequence (PMS) stars are usually variable for a variety of reasons but a large study by Carpenter et al. (2001) indicated that r.m.s. variations are always <0.5 mag at K, in the absence of eruptive behaviour, which is rarely observed.

The colours, magnitudes and locations of the 16 new variables indicate that many of them are likely to be eruptive (PMS) variable stars, either FU Orionis stars (FUORs) or their even younger equivalents. In Fig. 26.2 we show the colour magnitude diagrams for 6 arcmin fields centred on 16 of the 17 variables. All but two of the sources (Nova Sct 2003 and one brighter blue source) are either projected against the giant branch or are much redder than the giant branch. Two are J band

Fig. 26.3 JHK three colour image of a ~ 1 deg^2 field in the region of the Serpens OB2 association. The 11 high amplitude variables are marked with *red circles*

drop-outs, one of which is also undetected in the H band. Eleven of the variables are located in a 1 deg^2 area centred on the Serpens OB2 association (Forbes 2000), a massive 5 Myr old association in which star formation is still ongoing. These 11 are mostly located at the outskirts of molecular clouds in the complex (see Fig. 26.3). All 11 have red colours and evidence for K band excess emission due to hot circumstellar dust, based on their location in the (J–H) vs. (H–K) diagram, see Fig. 26.4. Detection of FUORs was one of the original GPS science goals. Only 10–15 FUORs are known but it is thought that they may represent a ubiquitous phase of PMS evolution, in which the accretion rate increases by up to three orders of magnitude (e.g. Hartmann and Kenyon 1996). Baraffe et al. (2009) have recently proposed that this can explain much of the scatter that is generally observed in HR diagrams of PMS clusters (e.g. Mayne and Naylor 2008; Weights et al. 2009) as well as the luminosity problem (Kenyon and Hartmann 1990). The luminosity problem is that YSOs in nearby low mass star formation regions tend to have lower luminosities than would be expected if they are low mass stars presently located on the Hayashi track, above the main sequence.

If this hypothesis is correct, the unfortunate consequence is that the stellar birth line in the HR diagram is essentially a meaningless concept and the ages and masses of PMS stars and brown dwarfs derived from existing theoretical isochrones

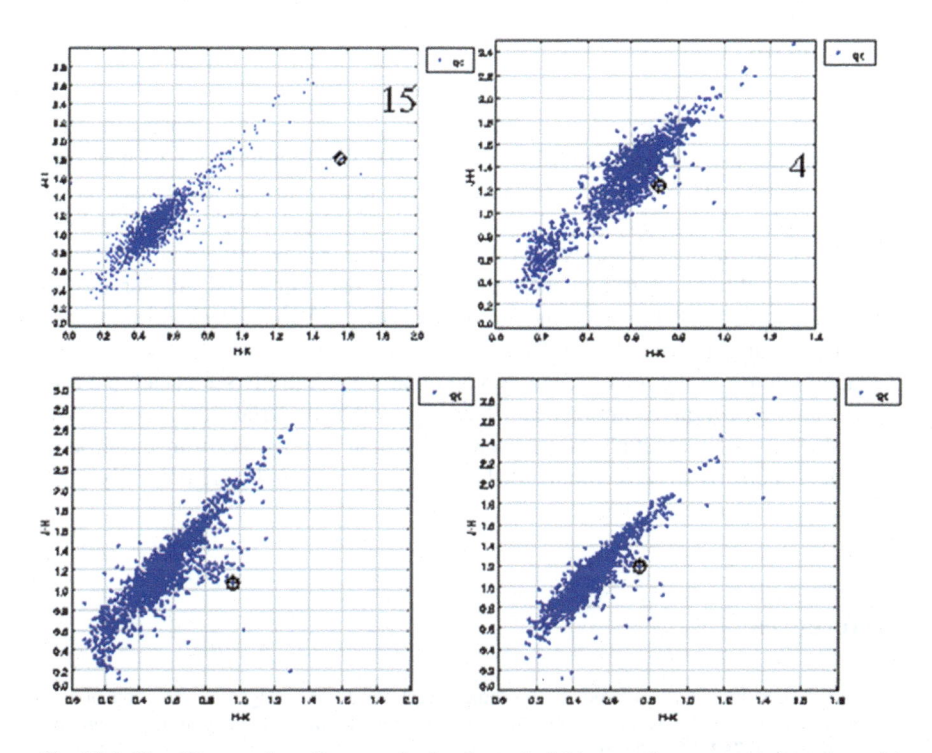

Fig. 26.4 Near IR two colour diagrams for 6 × 6 arcmin fields around a representative subset of 4 of the 11 variables in the Serpens OB2 association. We see that all four sources (*circles*) are located at the *right hand side* of the *reddening* sequences, indicating the presence of a K band excess due to hot dust. This supports the identification of these as FUORs

are often likely to be wrong. Future GPS 2 epoch data (aided by spectroscopic confirmation of their distinctive near IR absorption features) will permit the FUOR phenomenon to be quantified for the first time by counting the number of FUORs in a sample of several hundred PMS clusters and associations (identified by the searches described in section "Cross matches with others Galactic surveys").

The nature of the five new variables that are not located in the environs of the Serpens OB2 association is unclear. One possibility is that they are low luminosity Symbiotic stars (Brent Miszalski, private communication) Whatever their nature, these early results of the two epoch K band photometry show that the growing GPS catalogue has great promise for finding rare variables. It should find >1,000 high amplitude variables at the end of the 2nd epoch, provided that the survey is fully completed. The VISTA VVV survey will be able to undertake similar science in a portion of the southern Galactic plane and the Bulge. VVV will have far higher cadence but since it is optimised for short period variables (RR Lyrae stars and Cepheids), and has a smaller area than the GPS, it does not have a clear advantage for the detection of long period variables such as FUORs.

Conclusion

In summary, the UKIDSS Galactic Plane Survey is proceeding as planned. Here we have reported the detection of more than 100 new clusters and 16 new high amplitude variables. Much work remains to be done to follow up these discoveries.

The exploitation of this huge dataset is just beginning, aided by similarly large surveys that are underway in several other wavebands. Many of these large Galactic surveys are led by the astronomers in the UK. The panoramic Galactic surveys are now being joined by large area synoptic surveys such as VISTA VVV (Minniti et al. 2010) and Pan-STARRS 4. The Gaia mission promises further great advances in Galactic science in the next decade. Beyond that we can dimly see ahead to a time when panoramic high resolution near IR surveys will be undertaken from space with several orders of magnitude greater sensitivity than is possible from the ground. This field is clearly to be recommended to young scientists who are presently at the beginning of their research careers.

References

Baraffe, I., Chabrier, G., Gallardo, J.: Episodic accretion at early stages of evolution of low-mass stars and brown dwarfs: a solution for the observed luminosity spread in H-R diagrams? Astrophys. J. **702**, L27 (2009)

Benjamin, R., Churchwell, E., et al.: GLIMPSE. I. An SIRTF legacy project to map the inner galaxy. Publ. Astron. Soc. Pac. **115**, 953 (2003)

Bica, E., Dutra, C.M., Soares, J., Barbuy, B.: New infrared star clusters in the Northern and Equatorial Milky Way with 2MASS. Astron. Astrophys. **404**, 223 (2003)

Carpenter, J.M., Hillenbrand, L.A., Skrutskie, M.F.: Near-infrared photometric variability of stars toward the Orion a molecular cloud. Astron. J. **121**, 3160 (2001)

Casali, M., Adamson, A., et al.: The UKIRT wide-field camera. Astron. Astrophys **467**, 777 (2007)

Davis, C.J., Scholz, P., Lucas, P., Smith, M.D., Adamson, A.: A shallow though extensive H_2 2.122-μm imaging survey of Taurus-Auriga-Perseus – I. NGC 1333, L1455, L1448 and B1. Mon. Not. R. Astron. Soc. **387**, 954 (2008)

Forbes, D.: The serpens OB2 association and its thermal "Chimney". Astron. J. **120**, 2594F (2000)

Hambly, N.C., et al.: The WFCAM science archive. Mon. Not. R. Astron. Soc. **384**, 637 (2008)

Hartmann, L., Kenyon, S.J.: The FU Orionis phenomenon. Annu. Rev. Astron. Astrophys. **34**, 207 (1996)

Kenyon, S.J., Hartmann, L.W.: On the apparent positions of T Tauri stars in the H-R diagram. Astrophys. J. **349**, 197 (1990)

Lawrence, A., Warren, S., Almaini, O., Edge, A.C., Hambly, N.C., Jameson, R.F., Lucas, P., Casali, M., et al.: The UKIRT Infrared Deep Sky Survey (UKIDSS). Mon. Not. R. Astron. Soc. **379**, 2599 (2007)

Lucas, P.W., Hoare, M.G., Longmore, A., Schoeder, A.C., Adamson, A., Davis, C., Bandyopadhyay, R.M., et al.: The UKIDSS Galactic plane survey. Mon. Not. R. Astron. Soc. **391**, 136 (2008)

Mayne, N., Naylor, T.: Fitting the young main-sequence: distances, ages and age spreads. Mon. Not. R. Astron. Soc. **386**, 261 (2008)

Minniti D., Lucas P.W., Emerson J.P., Saito J.P., Hempel M., et al.: New Astron., submitted (2010) VISTA Variables in the Via Lactea (VVV): The public ESO near-IR variability survey of the Milky Way And reference is New Astron., 15,433 (2010)

Weights, D., Lucas, P.W., Roche, P.F., Pinfield, D.J., Riddick, F.: Infrared spectroscopy and analysis of brown dwarf and planetary mass objects in the Orion nebula cluster. Mon. Not. R. Astron. Soc. **392**, 817 (2009)

Chapter 27
The UKIDSS Galactic Clusters Survey

Sarah Casewell and Nigel Hambly

Abstract The UKIDSS Galactic Cluster Survey (GCS) was suggested as a way of providing homogenous data on a number of open star clusters, star forming regions, and associations to aid in determining the shape of the initial mass function (IMF) in the substellar regime. We describe the goals, design and some highlights of the GCS.

Science Goals

The IMF is the distribution of stellar masses that formed in the galaxy and as such is an essential tool in understanding the formation of stellar and substellar objects. There is, however, no direct way of measuring it although we are able to measure the cluster luminosity function and, when combined with evolutionary models, the mass function of open star clusters and associations. As the cluster members are gravitationally bound, the cluster mass function is generally considered a proxy for the IMF although in some cases, in particular the older clusters, the dynamical evolution of the cluster must be taken into account.

The stellar IMF is well characterised down to 0.1 solar masses (Chabrier 2003), however below this mass it is still undefined due to sparse data and small number statistics. The questions still to be answered are: What is the functional form of the IMF? Is it smoothly varying? Is there a characteristic mass? What is the significance of brown dwarfs on the IMF? Is it universal?

S. Casewell (✉)
Department of Physics and Astronomy, University of Leicester, Leicester LE1 7RH, UK
e-mail: slc25@le.ac.uk

N. Hambly
Scottish Universities Physics Alliance (SUPA), Institute for Astronomy,
School of Physics and Astronomy, University of Edinburgh, Royal Observatory,
Blackford Hill, Edinburgh EH9 3HJ, UK

To date there have been many studies of open star clusters searching for the lowest mass members with which to determine the form of the substellar IMF. However, the majority of these surveys have been performed with a variety of instruments, often with small fields of view, resulting in data of different depths, in different filter sets, and with poor or incomplete areal coverage. Ideally, to constrain the IMF the data should be consistent across a range of young star forming regions (e.g. Taurus, Orion $\sim$1 Myr) and more mature star clusters (e.g. Hyades, 625 Myr).

The Survey

The UKIDSS GCS is imaging 10 open star clusters, associations and star forming regions with ages ranging from 1 Myr to 600 Gyr. The imaging will be completed in all 5 filter sets used by WFCAM: Z, Y, J, H and K to depths of Z = 20.4, Y = 20.3, J = 19.5, H = 18.6, K = 18.2 (Vega magnitudes, 5 sigma detections) with complete areal coverage, in some cases as much as 300 sq.deg. Additionally, we will observe a second epoch of data taken in the K band to allow proper motions to be measured over a baseline of at least 5 years to an accuracy of $\sim$10 mas/year. This second epoch is vital to the survey as proper motions are an excellent tool for distinguishing cluster objects from non-member contamination.

A summary of the targets can be found in Table 27.1. The target clusters and associations span a range of ages, and consequently, as substellar objects are brighter when younger, a range of masses; from 40 to 10 Jupiter masses, well into the planetary mass regime (M < 13 Jupiter masses). This range of ages is important to study the evolution of the clusters and their dynamical evolution, as well as providing information on the universality of any resultant IMF.

Table 27.1 Summary of targets observed by the GCS

Priority/Name	Type	RA (2000)	Dec. (2000)	Area (sq.deg.)	Age (Myr)	Minimum mass ($M_\odot$)
(1) IC 4665	Open cluster	17 46	+05 43	3.1	40	0.020
(2) Pleiades	Open cluster	03 47	+24 07	79	100	0.024
(3) Alpha Per	Open cluster	03 22	+48 37	50	90	0.025
(4) Praesepe	Open cluster	08 40	+19 40	28	400	0.046
(5) Taurus-Auriga	SF association	04 30	+25 00	218	1	0.010
(6) Orion	SF association	05 29	−02 36	154	1	0.014
(7) Sco	SF association	16 10	−23 00	154	5	0.010
(8) Per-OB2	SF association	03 45	+32 17	12.6	1	0.011
(9) Hyades	Open cluster	04 27	+15 52	291	600	0.041
(10) Coma-Ber	Open cluster	12 25	+26 06	79	500	0.043

Highlights from the Survey

Upper Sco

For the science verification data release of UKIDSS 6.5 sq.deg. of Upper Scorpius were surveyed in all five UKIDSS GCS colours. Using a variety of colour-magnitude diagrams clearly showed a cluster sequence, well separated from field stars. Figure 27.1 shows the Z, Z–J colour magnitude diagram (Fig. 27.1). Using these diagrams and the evolutionary models of the Lyon group (Chabrier et al. 2000; Baraffe et al. 1998), 164 candidate cluster members were recovered. The data for the brighter objects were then cross-correlated with the 2 Micron All Sky Survey (2MASS) and proper motions were measured between the two datasets which are separated by 5 years. The result was that 116 low mass stars and brown dwarfs were discovered to have proper motions consistent with that of the cluster. Additionally 12 new brown dwarfs with masses of less than 20 Jupiter masses were discovered, and although they are too faint to be confirmed by proper motions, they are the lowest mass brown dwarf members of Upper Scorpius discovered to date.

Sigma Orionis

Using Data Release 4, an area within 30 arcmin of the centre of the Sigma Orionis cluster was surveyed in all 5 broadband colours. As for Upper Scorpius,

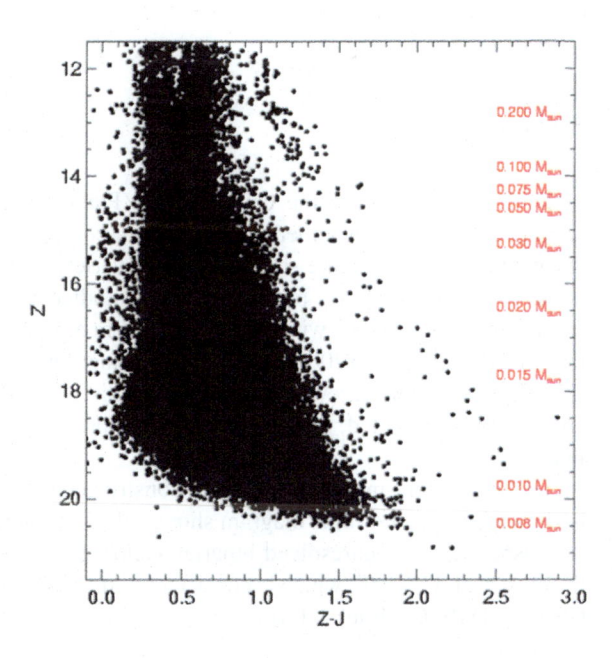

Fig. 27.1 The Z, Z–J colour magnitude diagram for the 6.5 sq.deg. covered in Upper Scorpius using the science verification data (Lodieu et al. 2007a)

Fig. 27.2 Z, Z–J colour magnitude diagram for 30″ of Sigma Orionis. The masses are from the DUSTY (Chabrier et al. 2000) and NextGen models (Baraffe et al. 1998) assuming a distance of 352 pc and an age of 3 Myr (Lodieu et al. 2009a)

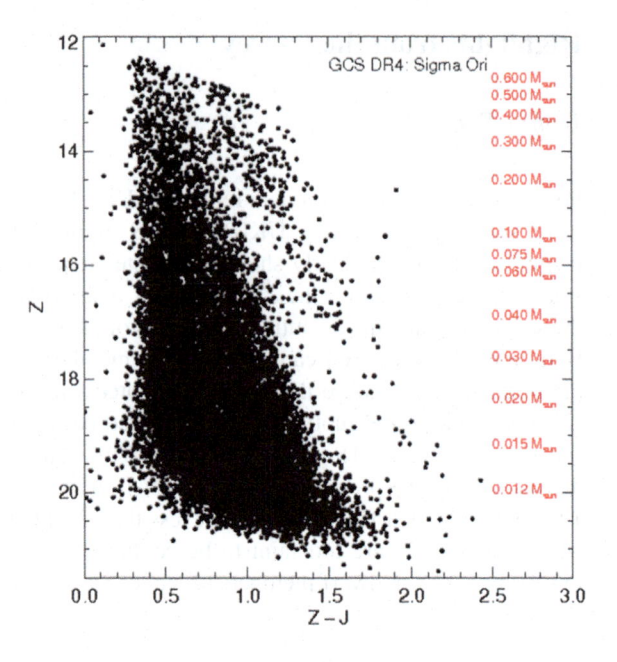

proper motions were measured for candidates that were also bright enough to be detected in 2MASS (down to 30 Jupiter masses) and 261 low mass stars and brown dwarf candidate cluster members were recovered. 2MASS was also used in comparison with UKIDSS to determine if any of the brown dwarfs were variable (Fig. 27.2).

Pleiades

In DR1, 12 sq. deg. of the centre of the Pleiades had been surveyed in all 5 broadband colours (since then more has been completed). As for the other clusters discussed, candidates were selected using a variety of colour magnitude diagrams, however the Pleiades has been extensively studied and there were data available from many other studies with which to perform a proper motion analysis including those of Dobbie et al. (2002) and the new reduction of the data from Moraux et al. (2003) presented in Casewell et al. (2007). This allowed an in depth proper motion analysis to be executed, for even the faintest candidates and the results can be seen in Fig. 27.3.

The final colour-magnitude diagram constructed after the proper motion selection is shown in Fig. 27.4. This diagram shows all 63 substellar candidates, of which 23 are suspected to be unresolved binaries yielding a photometric binary fraction for the cluster of 33–40 %, which is consistent with results in the literature, and also the Chabrier IMF (Chabrier 2003).

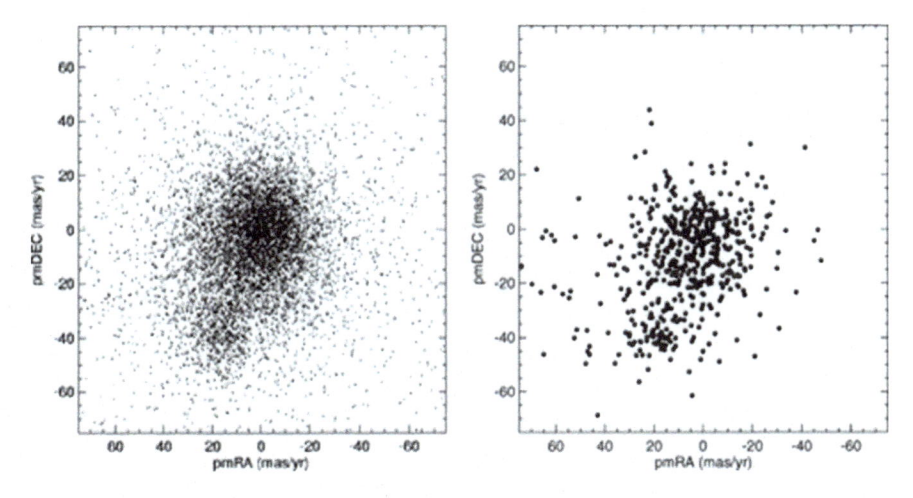

Fig. 27.3 Proper motion vector diagrams for the Pleiades showing brighter candidates (proper motion from 2MASS vs. GCS) in the *left hand panel* and fainter candidates (proper motion from GCS vs. INT or CFHT) in the *right hand panel* (Lodieu et al. 2007b)

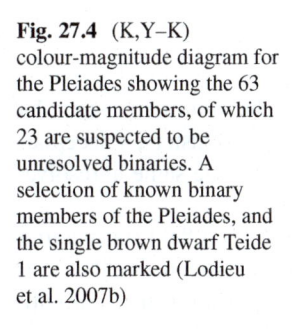

Fig. 27.4 (K,Y–K) colour-magnitude diagram for the Pleiades showing the 63 candidate members, of which 23 are suspected to be unresolved binaries. A selection of known binary members of the Pleiades, and the single brown dwarf Teide 1 are also marked (Lodieu et al. 2007b)

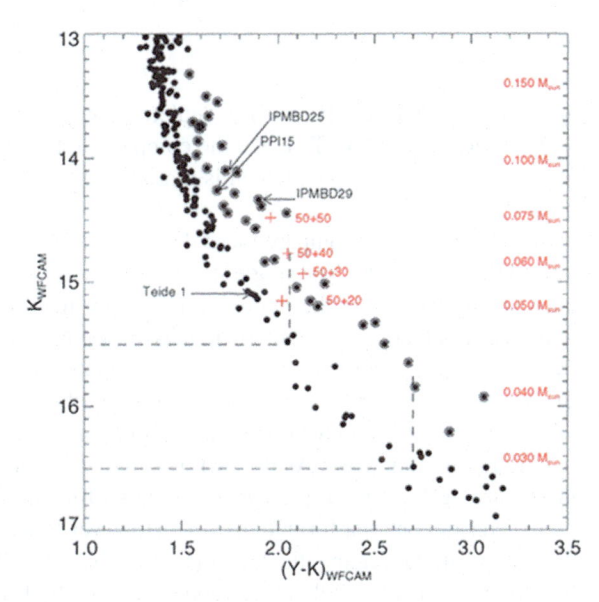

Implications for the IMF

The GCS provides a consistent set of data on open star clusters and formation regions, allowing consistent mass functions for clusters of various ages to be created and compared (Fig. 27.5). These mass functions clearly follow the same trend down to the lower masses, and show the number of brown dwarfs in the intermediate

Fig. 27.5 The cluster mass functions for Upper Scorpius (*light blue*), Sigma Orionis (*black*), the Pleiades (*red*) and IC4665 (*blue*). The *dot–dashed line* represents the completeness limit for the Sigma Orionis data

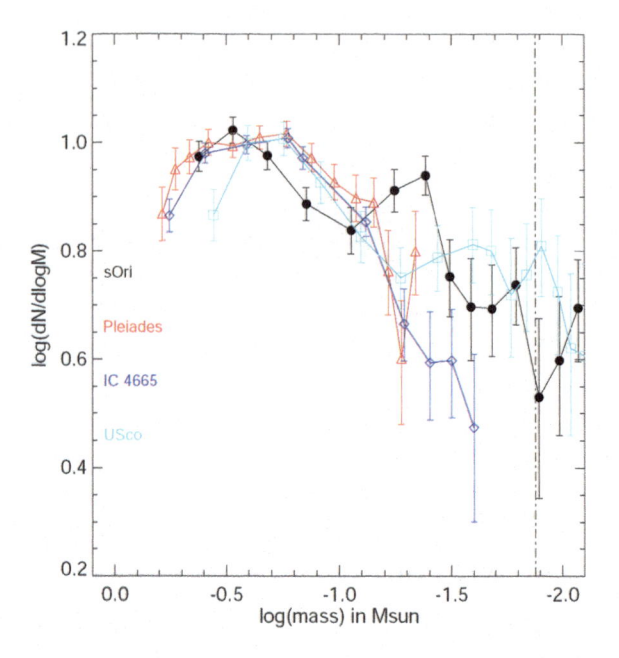

age clusters (Pleiades, IC4665) is depleted compared to the younger clusters due to dynamical evolution. To make the comparison between the clusters, a necessary assumption must be made, which is that all the clusters have the same unresolved binary fraction.

It is interesting to compare these data to the Chabrier (2003) field star parameterised MF (Fig. 27.6). The single star field MF has a characteristic mass of 0.1 solar masses, approximately the hydrogen minimum mass burning limit. It is currently unknown whether this characteristic mass is a coincidence, or whether it is caused by an as yet unquantified effect which is fundamental to the star forming process. There also appears to be no lower mass limit with the GCS IMFs continuous down to planetary masses, indicating a star formation process that has a very low or no required minimum mass. As the IMF is symmetrical about the hydrogen minimum mass burning limit there are equal numbers of stellar and substellar mass objects produced from one star formation event. However, since the IMF is clearly not rising into the substellar regime, the total mass that the substellar objects contribute is less than 10 % of the total stellar mass contributed to the IMF from this event.

Other Uses for the GCS

The GCS is also being used, as is the LAS, to search for foreground field T and Y dwarfs (Lodieu et al. 2009b), and is being combined with 2MASS to provide a large area proper motion survey to search for high proper motion objects

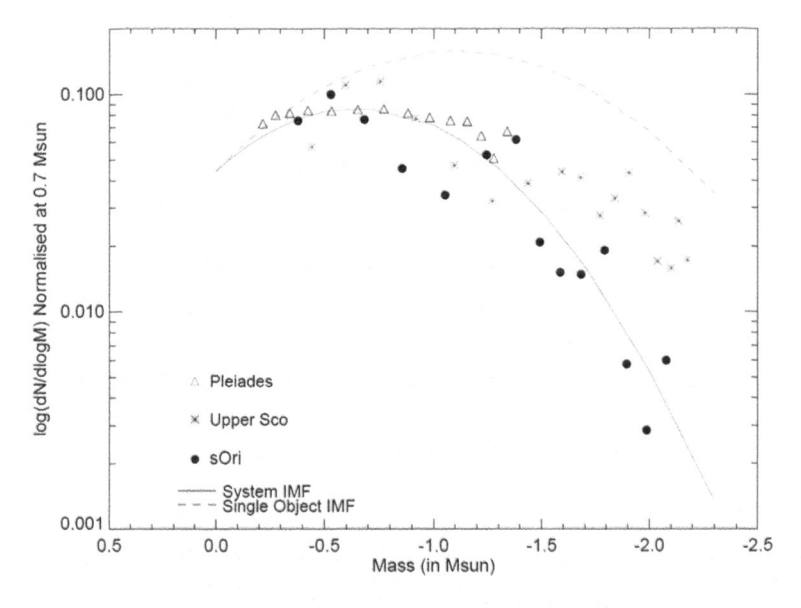

Fig. 27.6 The normalised mass function for the Pleiades (*open triangles*), Upper Scorpius (*crosses*) and Sigma Orionis (*filled circles*). For comparison, the Chabrier (2003) field star "system" MF (i.e. that uncorrected for unresolved binarity) is plotted as a *solid line* and the corresponding single object field star MF as a *dashed line* – these comparisons illustrate the effects of unresolved binarity (as is undoubtedly present in the GCS IMFs) on the MF. We note that it is currently unknown whether Upper Scorpius has a different mass function from the other clusters or whether the level of contamination has been underestimated in this region

(Deacon et al. 2009). The GCS fields are also being searched for background extra-galactic objects such as high redshift QSOs.

In the GCS clusters themselves multi-colour imaging has been used to identify new stellar clumps and disk candidates in Taurus and Orion that will be followed up in the mid-infrared.

Additionally, L dwarf members of various clusters (Pleiades, Upper Sco, Hyades, Alpha Per) have been used to define a colour-age relationship for L dwarfs which could be used to measure the ages of field dwarfs (Jameson et al. 2008). Broadband photometry of cluster white dwarfs can also be used to search for low mass companions and disks (Casewell et al. 2011, 2012) as is also being done with the LAS (Steele et al. 2009).

Acknowledgments The authors thank Niall Deacon and Nicolas Lodieu.

References

Baraffe, I., Chabrier, G., Allard, F., Hauschildt, P.: Evolutionary models for solar metallicity low-mass stars: mass-magnitude relationships and color-magnitude diagrams. Astron. Astrophys. **337**, 403 (1998)

Casewell, S.L., Dobbie, P.D., Hodgkin, S.T., Moraux, E., Jameson, R.F., Hambly, N.C., Irwin, J., Lodieu, N.: Proper motion L and T dwarf candidate members of the Pleiades. Mon. Not. R. Astron. Soc. **378**, 1131 (2007)

Casewell, S.L., Burleigh, M.R., Dobbie, P.D., Napiwotzki, R.: Evolved solar systems in Praesepe. AIPC Planetary systems beyond the main sequence. **1331**, 292 (2011)

Casewell S.L., et al., WD0837+185: The formation and evolution of an extreme mass-ratio white dwarf-brown dwarf binary. ApJ. **759**, L34 (2012)

Chabrier, G.: Galactic stellar and substellar initial mass function. Publ. Astron. Soc. Pac. **115**, 763 (2003)

Chabrier, G., Baraffe, I., Allard, F., Hauschildt, P.: Evolutionary models for very low mass stars and brown dwarfs with dusty atmospheres. Astrophys. J. **542**, 464 (2000)

Deacon, N.R., Hambly, N.C., King, R.R., McCaughrean, M.J.: The UKIDSS-2MASS proper motion survey -I. Ultracool dwarfs in UKIDSS DR4. Mon. Not. R. Astron. Soc. **394**, 857 (2009)

Dobbie, P.D., Kenyon, F., Jameson, R.F., Hodgkin, S.T., Pinfield, D.J., Osborne, S.L.: A deep IZ survey of 1.1deg^2 of the Pleiades cluster: three candidate members with M<0.04 Msolar. Mon. Not. R. Astron. Soc. **335**, 687 (2002)

Jameson, R.F., Lodieu, N., Casewell, S.L., Bannister, N.P., Dobbie, P.D.: The ages of L dwarfs. Mon. Not. R. Astron. Soc. **385**, 1771 (2008)

Lodieu, N., Hambly, N.C., Jameson, R.F., Hodgkin, S.T., Carraro, G., Kendall, T.R.: New brown dwarfs in Upper Sco using UKIDSS Galactic Cluster Survey science verification data. Mon. Not. R. Astron. Soc. **374**, 372 (2007a)

Lodieu, N., Dobbie, P.D., Deacon, N.R., Hodgkin, S.T., Hambly, N.C., Jameson, R.F.: A wide deep infrared look at the Pleiades with UKIDSS: new constraints on the substellar binary fraction and the low-mass initial mass function. Mon. Not. R. Astron. Soc. **380**, 712 (2007b)

Lodieu, N., Zapatero Osorio, M.R., Rebolo, R., Martín, E.L., Hambly, N.C.: A census of very-low mass stars and brown dwarfs in the sigma Orionis cluster. Astron. Astrophys. **505**, 1115 (2009b)

Lodieu, N., Burningham, B., Hambly, N.C., Pinfield, D.J.: Identifying nearby field T dwarfs in the UKIDSS Galactic Clusters Survey. Mon. Not. R. Astron. Soc. **397**, 258 (2009c)

Moraux, E., Bouvier, J., Stauffer, J.R., Cuillandre, J.-C.: Brown dwarfs in the Pleiades cluster: clues to the substellar mass function. Astron. Astrophys. **400**, 891 (2003)

Steele, P., Burleigh, M.R., Barstow, M.A., Jameson, R.F., Dobbie, P.D.: White dwarfs with unresolved substellar companions and debris disks in the UKIDSS survey. J. Phys. Conf. Ser. **172**, 12058 (2009)

Chapter 28
The UKIDSS Deep eXtra-Galactic Survey

A.M. Swinbank

Abstract The UKIDSS Deep eXtra-Galactic (DXS) will image an area of 35 sq.deg. at high Galactic latitudes in the J and K-band filters to a depth K = 21.0 (with 5 sq.deg. also imaged in H-band). When completed in 2012, the DXS will have used 118 nights to survey 4 of the best-studied extra-galactic fields: XMM-LSS, the Lockman Hole, SSA22 and ELAIS-N1. The principal goals of the DXS are to identify and measure the abundance of galaxy clusters at $0.8 < z < 1.5$; to measure galaxy clustering in well defined samples at $z > 1$ (and hence measure the evolution of bias, a key test of hierarchical models); and to provide a multi-wavelength census of the luminosity density in star formation and AGN. Here, we describe some of the recent science highlights. Specifically, the discovery of a supercluster at $z = 0.9$ which spans >30 Mpc and has properties similar to local super-clusters (such as Hercules) and some of the first wide-field ($\sim$3-sq.deg.) clustering measurements in EROs and DRGs.

Introduction

The UKIDSS-DXS is the central component of the UKIDSS survey, with the aim of imaging an area of 35 sq.deg. at high Galactic latitudes in the J and K-band filters to a depth K = 21.0, with 5 sq.deg. also imaged in H-band. The DXS will use 118 nights of UKIRT time over 7 years, and has at his writing completed JK imaging for 12 sq.deg. The survey includes some of the best studied extra-galactic survey fields, including the XMM-LSS, the Lockman Hole and ELAIS-N1.

The goal of the UKIDSS DXS is to compare the properties of galaxies at $1.0 < z < 1.5$ with the properties of galaxies today. Locally, the properties of galaxies

---

A.M. Swinbank (✉)
Department of Physics, Durham University, Durham DH1 3LE, UK
e-mail: a.m.swinbank@durham.ac.uk

have been well quantified using (e.g.) 2dF, SDSS, and IRAS at optical and far-infrared wavelengths. The DXS in combination with other imaging surveys (XMM, GALEX, MEGACAM, SIRTF, and SZ surveys), and including near-infrared spectroscopy with FMOS, aims to produce a comparable map at high redshift. Thus, the purpose of the DXS is to quantify the evolution of the properties of the Universe over a time span greater than half the current age of the Universe. This requires a survey at infrared wavelengths in order to sample the rest-frame optical. By comparing the properties of the Universe at these widely separated cosmic times we can perform critical tests of the current ΛCDM paradigm for the origin and growth of structure. For example the evolution of the abundance of galaxy clusters from $z=0$ to $1<z<1.5$ is very sensitive to cosmological parameters. The DXS will examine the evolution of bias for galaxies of different luminosity and type. The DXS will also provide the crucial near-infrared imaging for identification of reddened populations such as starburst galaxies and obscured AGN, that will allow a census of the individual contributions of stars and AGN to the global energy budget from X-ray to mm wavelengths, as well as providing legacy survey imaging required to identify and interpret other multi-wavelength surveys (e.g. with SCUBA2).

Goals

There are four primary goals of the UKIDSS DXS:

1. To detect a large sample of high-redshift galaxy clusters $0.8<z<1.5$ in order to measure the evolution of the cluster mass function $N(M, z)$.
2. The clustering of galaxies, by type and luminosity, at $z=1$. To obtain a representative view of the properties of galaxies at $z>1$, near-infrared surveys must cover a comparable luminosity range to that probed by current surveys of the local Universe (i.e. Lunimosities down to $\sim L^* + 2$). At $z=1$ this corresponds to a depth of $K=21$. For reference, in the redshift range $1<z<1.5$ a survey over 35 sq.deg. covers the same volume as the 2dF survey out to $z=0.2$.
3. Measure the star-formation rate at $z>1$ from ultraviolet to far-infrared wavelengths. In the local Universe about one-third of the energy from galaxies emerges at far-infrared wavelengths, but at higher redshift this proportion probably increases, and may dominate, as indicated by observations of luminous sub-mm galaxies and of the cosmic optical and far-infrared background radiation. The JHK-band DXS imaging is vital for the measurement of photometric redshifts of galaxies, and the identification of highly reddened objects, detected at longer wavelengths, in order to establish their redshift distribution.
4. Measure the contribution of AGN to the cosmic energy budget. While the cosmic background radiation is dominated by star formation at optical and far-infrared to sub-mm wavelengths, in the mid-infrared AGN make an important contribution, while at X-ray wavelengths they dominate. An important question is thus what is the contribution, and its evolution, of AGN to the total cosmic

energy budget? This is generally believed to be about 10 %, but has proved difficult to quantify because the bulk of the hard X-ray background is attributed to AGN that are heavily obscured at optical wavelengths. A multi-wavelength study covering X-ray (XMM), near-infrared (DXS), and mid-IR (SWIRE) can answer this question. The mid-infrared and X-ray observations are complementary, as these are the wavelengths at which obscured AGN are most visible. Near-infrared imaging is required for identification, to allow spectroscopic observations, and for quantifying the extinction.

Here, we highlight two of the recent results from two of these science questions.

High-z Galaxy Clusters

One of the primary aims of the DXS is to detect a large sample of high-redshift galaxy clusters in the redshift range $0.8 < z < 1.5$ to measure the evolution of the cluster mass function $N(M, z)$. It sets the scope of the survey and the case forms a major part of the DXS proposal. The measurement of the evolution of the cluster mass function $N(M, z)$ can be used to obtain constraints on cosmological parameters (e.g. Viana and Liddle 1996; Eke et al. 1998). These constraints will be complementary to existing and future cosmic microwave background and supernova measures, and can remove degeneracies.

The structure and evolution of clusters of galaxies and their constituent substructures provides a powerful test of our understanding of both the growth of large scale structure and dark matter in the Universe. In the current hierarchical paradigm of structure formation, massive galaxy clusters arise from the extreme tail in the distribution of density fluctuations, so their number density depends critically on cosmological parameters. One of the remarkable successes of the ΛCDM paradigm is the match to the number density, mass and evolution of clusters of galaxies out to $z \sim 0.5$–1 (Jenkins et al. 2001; Evrard et al. 2002). However, due to the very limited number of clusters known at $z > 1$ (where the number density is most sensitive to the assumed cosmology), the full power of the comparisons to theoretical simulations has not yet been exploited.

The paucity of clusters known at $z > 1$ stems from the limitations of current survey methods. For instance, optical colour-selection of clusters (Couch et al. 1991; Gladders and Yee 2000), which relies on isolating the 4,000 Å break in the spectral energy distributions of passive, red early-type galaxies (the dominant population in local clusters) becomes much less effective at $z > 0.7$ where this feature falls in or beyond the i-band, in a region of declining sensitivity of silicon-based detectors. Recent progress has been made in identifying clusters using X-ray selection with Chandra and XMM-Newton (Romer et al. 2001; Rosati et al. 2002; Mullis et al. 2005) and these studies have identified galaxies clusters out to $z \sim 1.5$ (Stanford et al. 2006; Bremer et al. 2006). However, the X-ray gas in these clusters appears more compact than for comparable systems at lower redshifts and hence there are

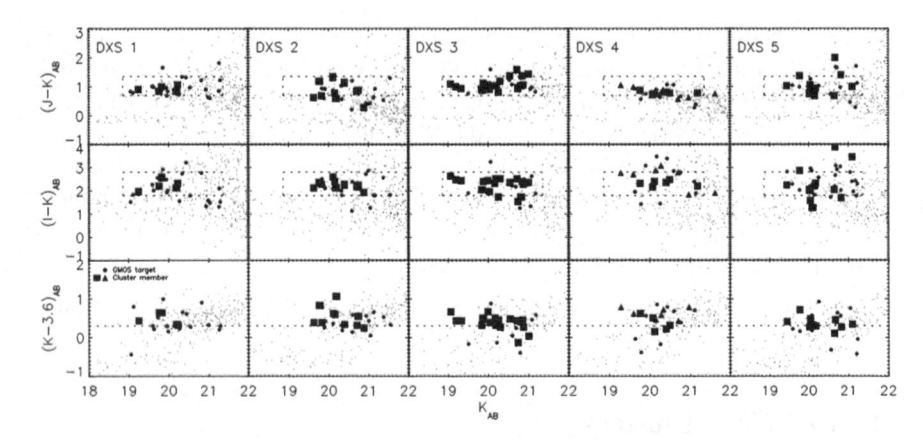

Fig. 28.1 J−K, I−K and K−3.6 μm colour-magnitude plots for each cluster identified in ELAIS-N1 from DXS/WFCAM observations. The colour-magnitude diagrams represent a 1.5 arcmin region around each cluster candidate. The *dashed lines* show the limits used to select candidate cluster members via the *red* sequence. We identify the spectroscopically confirmed z = 0.89 cluster members (*filled squares*), as well as the foreground/background galaxies (*filled circles*) and highlight those sources which are detected at 24 μm. In the DXS4 field we identify separately the members of the higher redshift structure in this field (at z = 1.09). Note also that some spectra were taken for objects outside the colour-magnitude limits (marked as *dashed boxes*) to fully populate the GMOS masks

concerns that the accurate comparison of cluster properties with redshift required to constrain cosmological parameters could be subject to potential systematic effects related to the thermal history of the intracluster medium (Fig. 28.1).

One solution to this problem is to extend the efficient optical colour-selection method beyond $z > 0.7$ using near-infrared detectors (Hirst et al. 2006). This approach has been impressively demonstrated by Stanford et al. (2006) who find a $z = 1.45$ cluster in the NOAO-DW survey selected from optical-near-infrared colours. To further this study, in Swinbank et al. (2007) we utilise the UKIDSS-DXS early data release for the ELAIS-N1 region which covers a contiguous area of $0.86° \times 0.86°$ centred on $\alpha = 16\ 11\ 14.400$; $\delta = +54\ 38\ 31.20$ (J2000). The survey data products for this region reach 5-σ point source limits of $J_{AB} = 22.8$–23.0 and $K_{AB} = 22.9$–23.1. To complement these observations we exploit deep I-band imaging obtained with Suprime-Cam on Subaru Telescope, and reach a 5-σ point-source limit of $I_{AB} = 26.2$. As part of the Spitzer Wide-area InfraRed Extragalactic (SWIRE) survey (Lonsdale et al. 2003), the ELAIS-N1 region was also imaged in the IRAC (3.6, 4.5, 5.8 and 8.0 μm) bands as well as at 24 μm with MIPS.

As this was a pilot study, we identified candidate high-redshift galaxy clusters in three ways: first, we searched for the sequence of passive red galaxies in high-redshift clusters by selecting galaxies from photometric catalogue in the (J−K)–K and (I−K)–K colour-magnitude space. Each resulting spatial surface density plot for a colour slice was then tested for an over-density using consecutively larger apertures from 0.01° to 0.05° (corresponding to approximately 250 kpc to 1 Mpc

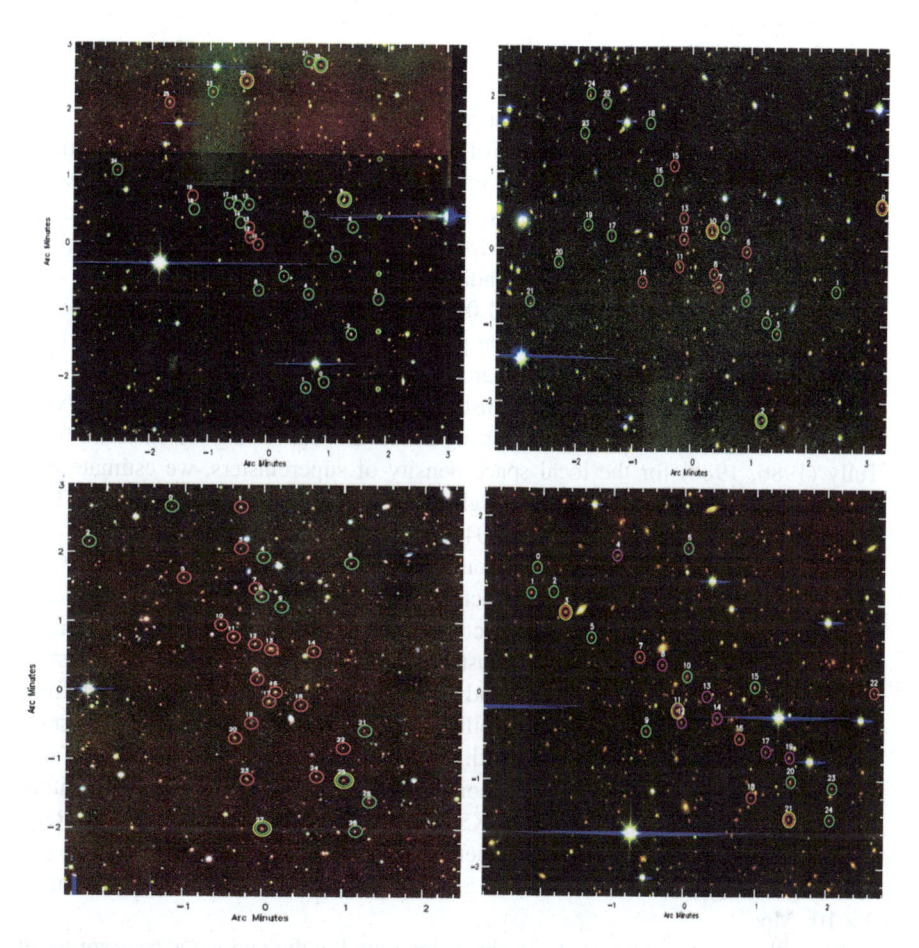

Fig. 28.2 True colour IJK-band imaging of four of the five DXS clusters in the ELAIS N1 field from Swinbank et al. (2007). *Circles* denote galaxies which were targetted with GMOS. *Red circles* mark cluster members at z = 0.90; *green circles* mark non-cluster members. *Yellow circles* mark 24 μm detections. In all four fields

at z = 1). If the over-density in the central aperture was $\geq 3\,\sigma$ above the background and the density decreased with increasing aperture radius then a region was marked as a candidate cluster (Fig. 28.2).

A similar procedure was carried out using the (I−K)−K colour-magnitude space to refine the selection of cluster candidates. Independently, we identified cluster candidates by identifying peaks in the surface density in $K_{AB} - 3.6$ μm colour space (using an approximate colour cut of $K_{AB} - 3.6$ μm > 0.3 which should be efficient at selecting ellipticals at $z \sim 1$). Having defined these cluster candidates, we checked that each of these met the selection criterion recently used by van Breukelen et al. (2006) (which is based on a projected friends of friends algorithm). Using these three selection criteria we identified fifteen candidates, of which eight were identified using all three criteria. Five of the most promising eight cluster candidates (which showed the tightest colour-magnitude sequences and a clear over-density of red objects), were then targeted for spectroscopic follow-up.

Our Gemini/GMOS spectroscopy of five cluster candidates robustly identifies six galaxy clusters. One of these structures lies at $z = 1.1$, however, and somewhat surprisingly, the other five structures lie within $\sim$3,000 kms^{-1} of each other at $z = 0.89$ and spanning at least 30 Mpc in projection. The velocity dispersions and physical sizes of each of these individual clusters is $\sim$500–100 km s^{-1}. The discovery of five clusters within $\sim$1,000 km s^{-1} suggest we have identified a high-redshift super-cluster. Indeed, the properties of each of the five clusters bear similarities to (well studied) local superclusters.

The discovery of a massive supercluster in the first DXS survey field of nearly 50 is surprising given their low space density below $z = 0.1$. Using the statistics from Tully (1986, 1988) for the local space density of superclusters, we estimate that there are five superclusters over the high galactic latitude sky within $z = 0.1$. This corresponds to one supercluster per 0.04 Gpc^3.

The total volume sampled in the complete DXS survey between redshifts 0.7–1.4 (the farthest we can efficiently select galaxy clusters from the DXS and reliably recover redshifts from optical spectroscopy) will be 0.27 Gpc^3 (comoving). Scaling from the local space density of superclusters, we expect a total of seven superclusters in the complete DXS survey. To find one such system in the first field from the DXS is fortunate ($\sim$15 % probability) but not so unlikely for us to question the validity of our interpretation of this system as a rich supercluster. We also measure the mass of the cluster by comparing with N-body simulations. Within the survey volume covered by this DXS field, the halo mass functions from Reed et al. (2005) suggest that at $z = 0.9$ there should be $\sim$60, 25, 2 and 0.05 halos of mass $\log(M/M_\odot) = 13, 13.5, 14.0$ and 15.0 respectively within our survey volume of 3×10^6 Mpc^3.

Thus, the number density of massive halos found in the region we have surveyed is consistent with halos of mass $\sim 10^{13.5-14.0}$ M$_\odot$, suggesting a total mass of the supercluster of order $> 10^{15}$ M$_\odot$.

Galaxy Clustering

The 2dF and SDSS galaxy redshift surveys are providing accurate measures of the clustering of galaxies in the nearby Universe out to the largest scales ($>$100 Mpc). These results provide constraints on cosmological parameters (from the shape of the galaxy power spectrum on large scales) and allow tests of the predictions of theories of galaxy formation, exploring the relation between light and mass through the measurement of clustering as a function of galaxy type, luminosity, and star-formation history. These surveys use catalogues of galaxies selected at optical wavelengths. As a consequence, the measurement of the evolution from a comparable galaxy sample at substantial look-back time, $z > 1$, can only be achieved with deep wide-field imaging at near-infrared wavelengths.

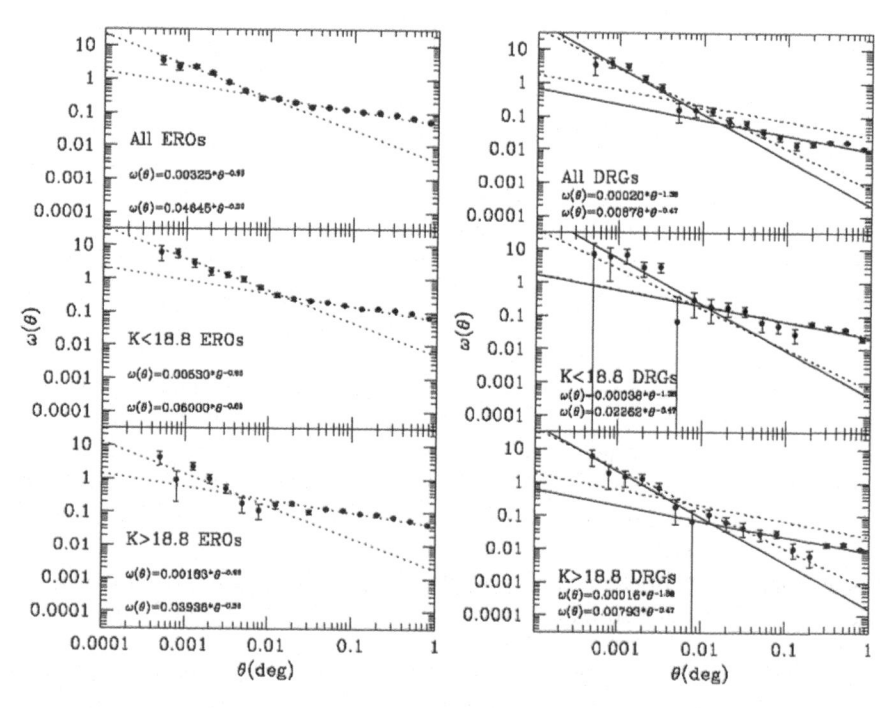

Fig. 28.3 *Left*: The bias-corrected angular two-point correlation functions of all (*top*), K < 18.8 (*middle*) and K > 18.8 (*bottom*) EROs. *Right*: The bias-corrected angular two-point correlation functions of all (*top*), K < 18.8 (*middle*) and K > 18.8 (*bottom*) DRGs. In both panels *dotted lines* and equations in each panel show power law (Aω, $\theta\delta$) fitting results at small and large scale

Colour selection in the near-infrared has become a successful tool for isolating high-redshift galaxy populations, such as extremely red objects (EROs) and distant red galaxies (DRGs) for which these comparisons can be made. EROs exhibit a large colour difference between optical and infrared wave-lengths (I−K > 4). EROs are predominantly at z > 1, and contain some of the most massive populations at this epoch. DRGs are also red galaxies, but defined in the near-infrared wavelengths, with J−K > 2.3. This colour efficiently selected the redshifted 4,000 Å break at z > 2. Both EROs and DRGs select both passive and dusty star-forming galaxy populations, both of which are known to be highly clustered, although (until now) samples have been limited to small fields (~few arcminutes).

Kim et al. (2011) exploit the UKIDSS DXS to study the clustering of these two populations using 3 sq.deg. of contiguous WFCAM JK-band imaging in the SA22 field. Using colour selection to identify ~5,500 EROs and ~3,500 DRGs, Kim et al. construct the angular two-point correlation functions of each population, as well as the real space correlation length. Both populations show strong clustering properties. EROs are described by a double power law with inflection at ~0.6–1.2′ and assuming a power law, $\omega(\theta) = A_\omega(\theta^{-\delta})$, ($A_\omega$, δ) of K < 18.8 and i−K > 4.5 EROs are (0.005, 0.96) and (0.06, 0.39) for small and large scales respectively (Fig. 28.3). The correlation function of EROs also shows a double power law with the inflection at $\sim 0.6'-1.2'$. The bright EROs show higher amplitudes, larger correlation lengths at small scales, but none of these properties are significantly variable on larger scales. Moreover, the correlation functions of bright EROs are steeper than fainter EROs, although redder EROs are more clustered without the slope variability.

The correlation function of DRGs is also well described by a double power law, with $\delta = 1.38$ and 0.47 for small and large scales respectively (Fig. 28.3). However, the DRGs shows slightly different trends with redshift from $2 < z < 3$ DRGs, likely due to the contamination of $1 < z < 2$ DRGs.

The results from this analysis illustrate the importance of sampling the widest possible fields in the near-infrared in order to recover representative clustering properties of distant galaxies. In the near future the combination of UKIDSS and VISTA surveys will cover more than an order of magnitude larger area to comparable depth. Our ability to extract the clustering of EROs in these areas is limited only by the depth of comparable optical imaging. Future studies in different fields and using z–K or Y–K selection will be important to reduce the effects of Cosmic variance and select higher redshift galaxies. Finally, ERO samples are now of sufficient size to offer direct tests to galaxy formation models in terms of number density and clustering so future comparisons to semi-analytic simulations will be more powerful.

The Future

At completion, the greatest achievement of the UKIDSS DXS will be the legacy value of wide-field near-infrared imaging in some of the best-studied extra-galactic survey fields. In particular, the PanSTARRS survey has recently begun observing the DXS fields, which will provide a reliable, optical multi-band (grizY) imaging. In the same fields, the Spitzer Space Telescope warm mission (SERVS) has started observing three DXS fields, providing further (longer-wavelength) imaging. With the commissioning of FMOS on Subaru, and the first multi-object near-infrared spectroscopy on >5′ fields, the DXS will be an important source of targets. Moreover, the roll-out of LOFAR stations over 2010 will allow the first <300 MHz radio surveys to start, and will include DXS fields.

It is also worth noting that Herschel and SCUBA-2 observations are now underway, and there will be a great deal of multi-wavelength data that requires deep J and K-band imaging for identification of counterparts. As the DXS reaches a factor of ten larger area than the UDS, then the general statistical tests and source selections (ERO, DRG, BzK, etc) become more powerful. In short, the DXS is progressing steadily and with DR5 is reaching the size to start addressing issues beyond those that can be answered using the UDS.

References

Bremer, M.N., Valtchanov, I., Willis, J., Altieri, B., Andreon, S., Duc, P.A., Fang, F., Jean, C., et al.: XMM-LSS discovery of a z = 1.22 galaxy cluster. Mon. Not. R. Astron. Soc. **371**, 1427 (2006)

Couch, W.J., Ellis, R.S., MacLaren, I., Malin, D.F.: A uniformly selected catalogue of distant galaxy clusters. Mon. Not. R. Astron. Soc. **249**, 606 (1991)

Eke, V.R., Cole, S., Frenk, C.S., Patrick Henry, J.: Measuring Omega_0 using cluster evolution. Mon. Not. R. Astron. Soc. **298**, 1145 (1998)

Evrard, A.E., MacFarland, T.J., Couchman, H.M.P., Colberg, J.M., Yoshida, N., White, S.D.M., Jenkins, A., Frenk, C.S., et al.: Galaxy clusters in hubble volume simulations: cosmological constraints from sky survey populations. Astrophys. J. **573**, 7 (2002)

Gladders, M.D., Yee, H.K.C.: A new method for galaxy cluster detection. I. The algorithm. Astron. J. **120**, 2148 (2000)

Hirst, P., Casali, M., Adamson, A., Ives, D., Kerr, T.: The UKIRT wide-field camera (WFCAM): commissioning and performance on the telescope. In: McLean, I.S., Iye, M. (eds.) Ground-based and airborne instrumentation for astronomy, Proceedings of the SPIE, vol. 6269, Orlando, FL, 25–29 May 2006

Jenkins, A., Frenk, C.S., White, S.D.M., Colberg, J.M., Cole, S., Evrard, A.E., Couchman, H.M.P., Yoshida, N.: The mass function of dark matter haloes. Mon. Not. R. Astron. Soc. **321**, 372 (2001)

Kim, J.-W., Edge, A.C., Wake, D.A., Stott, J.P.: Clustering properties of high-redshift red galaxies in SA22 from the UKIDSS Deep eXtragalactic survey. Mon. Not. R. Astron. Soc. **410**, 241–256 (2011). doi:10.1111/j.1365-2966.2010.17439.x

Lonsdale, C.J., Smith, H.E., Rowan-Robinson, M., Surace, J., Shupe, D., Xu, C., Oliver, S., Padgett, D., et al.: SWIRE: the SIRTF wide-area infrared extragalactic survey. Publ. Astron. Soc. Pac. **115**, 897 (2003)

Mullis, C.R., Rosati, P., Lamer, G., Böhringer, H., Schwope, A., Schuecker, P., Fassbender, R.: Discovery of an X-ray-luminous galaxy cluster at z=1.4. Astrophys. J. Lett. **623**, L85 (2005)

Reed, D.S., Bower, R., Frenk, C.S., Gao, L., Jenkins, A., Theuns, T., White, S.D.M.: The first generation of star-forming haloes. Mon. Not. R. Astron. Soc. **363**, 393 (2005)

Romer, A.K., Viana, P.T.P., Liddle, A.R., Mann, R.G.: A serendipitous galaxy cluster survey with XMM: expected catalog properties and scientific applications. Astrophys. J. **547**, 594 (2001)

Rosati, P., Borgani, S., Norman, C.: The evolution of X-ray clusters of galaxies. Annu. Rev. Astron. Astrophys. **40**, 539 (2002)

Stanford, S.A., Romer, A.K., Sabirli, K., Davidson, M., Hilton, M., Viana, P.T.P., Collins, C.A., Kay, S.T., et al.: The XMM cluster survey: a massive galaxy cluster at z = 1.45. Astrophys. J. Lett. **646**, L13 (2006)

Swinbank, A.M., Edge, A.C., Smail, I., Stott, J.P., Bremer, M., Sato, Y., van Breukelen, C., Jarvis, M., Waddington, I., Clewley, L., Bergeron, J., Cotter, G., Dye, S., Geach, J.E., Gonzalez-Solares, E., Hirst, P., Ivison, R.J., Rawlings, S., Simpson, C., Smith, G.P., Verma, A., Yamada, T.: The discovery of a massive supercluster at z = 0.9 in the UKIDSS Deep eXtragalactic survey. Mon. Not. R. Astron. Soc. **379**, 1343 (2007)

Tully, R.B., et al.: Alignment of clusters and galaxies on scales up to 0.1 C. Astrophys. J. **303**, 25 (1986)

Tully, R.B.: The galaxy luminosity function and environmental dependencies. Astron. J. **96**, 73 (1988)

van Breukelen, C., Clewley, L., Bonfield, D.G., Rawlings, S., Jarvis, M.J., Barr, J.M., Foucaud, S., Almaini, O., et al.: Galaxy clusters at 0.6 < z < 1.4 in the UKIDSS ultra deep survey early data release. Mon. Not. R. Astron. Soc. **373**, L26 (2006)

Viana, P.T.P., Liddle, A.R.: The cluster abundance in flat and open cosmologies. Mon. Not. R. Astron. Soc. **281**, 323 (1996)

Chapter 29
UKIDSS UDS Progress and Science Highlights

W.G. Hartley, O. Almaini, S. Foucaud, and UKIDSS UDS Working Group

Abstract In this contribution we provide a comprehensive update of the progress of the UDS survey and the science that has already been achieved through its use. In the context of this meeting (a look-back over the life of UKIRT to date) we begin with a retrospective of the original motivation for the UDS and how it remains well-placed to answer some of the biggest questions in extra-galactic astronomy. Following this is a progress report of the UDS data collection and a summary of the multi-wavelength data covering the field. We then detail a few of the science highlights that have resulted from the survey, concentrating on those which rely most heavily on the infrared data and are of greatest impact. With the survey now less than 3 years from completion, we end with a look forward to the final probable depths and the continuing data acquisition which will maintain the UDS as one of the most important surveys for several years to come.

Introduction

When the science case for UKIDSS was written in 2001 the concordance cosmology had been established for only a few years (Ostriker and Steinhardt 1995). Various observations indicated a Lambda-cold dark matter universe, and N-body simulations of the growth of structure in such a universe suggested that galaxies should form hierarchically (Navarro et al. 1995). It had previously been thought (Tinsley and Gunn 1976) that massive low-redshift elliptical galaxies had formed the bulk of their stars in a single rapid burst very early in their history. Observational evidence in the form of old, elliptical galaxies at high redshift seemed to confirm this hypothesis (e.g. Dunlop et al. 1996), yet it was in apparent contradiction with the hierarchical growth predicted by models. A robust test to differentiate between the new hierarchical model and the old monolithic-style collapse model was clearly required. This test was suggested to be the number density of L* ellipticals at $z > 2$, and it formed the basis for the science case of the Ultra-Deep Survey (UDS). It was felt that if more than a third of present day L* elliptical galaxies were found to already be in place by $z = 2$, then no model of the hierarchical growth of elliptical galaxies could be correct. Prior to the advent of WFCAM on UKIRT such a test had not been possible. Elliptical galaxies are typically very faint or even invisible at rest-frame ultra-violet wavelengths, so to find them above $z = 1$ requires surveys at near-infrared wavelengths. However, previous near-infrared instruments had very small fields-of-view, and so were not suited to the wide-field survey which was required to find sufficient numbers of such objects. In the years between submission of the UDS science case and the start of observations confidence had grown in the cosmological model, and the introduction of feedback mechanisms had eased the tension between the growth of dark matter structure and that of the galaxies within it (e.g. Benson et al. 2003). Instead, attention shifted towards the details of the theory of galaxy formation within the context of the concordance cosmology. The science goals of the UDS remained hugely important however, and the formation of massive elliptical galaxies at high redshift is still one of the greatest difficulties in theoretical galaxy formation models. The UDS will provide tight constraints for the models to adhere to for many years to come.

The central theme of $z > 2$ elliptical galaxies was by no means the only motivation for a deep, wide-field infrared survey. The large-scale clustering properties of a galaxy population are directly related to the mass of the dark matter halos which host them (e.g. Cooray and Sheth 2002). Prior to the UDS it had only been possible to study clustering of fair samples at $z > 1$ on very small scales. Wide-field data sets on the other hand were limited to highly biased samples, such as the Lyman break technique. Through the use of the UDS it is possible to extend clustering studies, and hence halo mass estimates, to galaxies in the redshift range $1 < z < 5$. Further tests of galaxy formation models, such as the evolution with redshift of the Kormendy relation (one of the projections of the fundamental plane of elliptical galaxies), were also expected to be possible. In recent years size evolution in elliptical galaxies has become a highly controversial topic, with the UDS well-placed to be the data set able to resolve the issue. The sub-arcsecond seeing at the top of Mauna Kea is sufficient for robust size measurements of massive elliptical galaxies to be taken and the K-band depth of the UDS allows unbiased selection to $z = 3$. Rare and possibly vital sources for our understanding of the evolution of galaxies, such as sub-mm sources and obscured AGN, are often invisible at optical wavelengths. The link between sub-mm galaxies and the K-selected population was an important piece of the original science case. It continues to be of central importance with the SCUBA II instrument at JCMT recently beginning operations. AGN are thought to be responsible for the termination of some galaxies' star-formation periods (Silk and Rees 1998), but the connection is as yet unproven. The peak of AGN activity is at high redshift ($z \sim 2$), so deep infrared data is vital in characterising robust samples of active galaxies and their links to sub-mm galaxies and the wider galaxy population. In order to reach the depths required for the UDS, data must be taken over many years. Proper motions of the lowest mass stars in the milky way and variability of many other stars are therefore also possible.

Current Progress

The UDS (Lawrence et al. 2007; Almaini et al. 2007) is a single mosaic of four WFCAM (Casali et al. 2001) pointings, covering a 0.88° by 0.88° area. This pattern is the smallest contiguous area possible with WFCAM and thus maximises depth relative to the other UKIDSS surveys. Data began to be taken in the autumn and winter of 2005. In this first semester (released along with the other sub-surveys as Data Release 1, Warren et al. 2007) only J and K-band data were taken, achieving depths of $J = 22.61$ and $K = 21.55$ (5σ, $2''$ diameter aperture magnitudes in the Vega magnitude system). The following year saw the addition of data in the H-band to a similar depth and a small increment in the depth of the J and K-band data. The depths of the data accumulated by the end of 2008, comprising the DR5, were $J = 23.1$, $H = 22.3$ and $K = 22.0$ (Vega). Relative to the revised targets of the 2006 renewal case for UKIDSS, these depths represent only a 9 % completeness of the UDS. At the same rate of data collection, the projected completeness by the end of 2009 (DR7) is only 12 %. The average projected completeness for the other four sub-surveys meanwhile, is approximately 40 %.

There are several reasons for the lack of progress in observing the UDS, but also reasons to be optimistic about its future. Bad weather has played its part in restricting observation of all of the UKIDSS surveys. However, for the UDS poor weather is particularly problematic as it is only visible during a single semester each year. Scheduling of cassegrain instruments and telescope maintenance, which were planned years in advance, happened to coincide with some of the best months for observing the UDS. However, with the maintenance now performed and UKIRT having moved to full-time wide field operation, the restrictions on observing the UDS are far fewer and we anticipate a significant increase in the yearly acquisition of data. In addition, there have been improvements at all stages in the data pipeline, but particularly in the sky-subtraction performed at the Cambridge Astronomical Survey Unit (CASU) and in the handling of large-scale noise. These improvements have allowed us to include more of the K-band data that have been taken and are responsible for much of the K-band depth increment between the DR1 and DR5. The same will follow for the J and H-band data in the future.

For a while it had been suspected that the seeing estimates of fields at high airmass have been a little pessimistic. The large pixel size of WFCAM ($0.4''$/pixel) coupled with the lack of bright stars in the UDS has made this issue particularly severe, which means that even when the field was potentially observable it would not be present in the observing queue. Beginning with the autumn 2009 semester, a new approach has been implemented to tackle this issue. Rather than rely on an estimate extrapolated from the zenith seeing, a nearby standard star field is observed and if the conditions are suitable for the UDS, then it is observed. The enhancement of the mirror cooling system should also have a significant positive impact upon the future data collection of the UDS.[1]

Despite the optimism for the future, the UDS has clearly lagged some way behind the other components of UKIDSS. For this reason the target depths were again revised for the 2009 renewal. These revised targets, assuming that UKIDSS is allowed to run until the end of 2012, are as follows: $J = 24.4$, $H = 23.3$ and $K = 22.8$ (Vega). Achieving these depths will require a total of a further $\sim$1,000 h on source, while over the previous 3 years we have only accumulated $\sim$200 h on source. Given the improvements in both observing strategy and data processing of the poorer quality data, these targets are realistic and will constitute an excellent legacy data set.

The UDS field was chosen to coincide with deep optical data from the Subaru-XMM Deep Survey (SXDS), and many of the science goals require complementary data across a wide range in wavelength. Acquisition of data at other wavelengths is ongoing and the current data that are either in hand or approved are as follows:

- X-ray data from XMM-Newton as part of SXDS: 400 ks over 7 pointings.
- U-band from CFHT megacam to $U = 27.4$ (AB). Approved and taken (now available public from CFHT archive).
- Optical data as part of SXDS: $B = 28.4$, $V = 27.8$, $R = 27.7$, $i' = 27.7$, $z' = 26.7$ (3σ, AB, Furusawa et al. 2008). Additional z'-band data are also being taken (PI: H. Furusawa).
- Spitzer data from a Spitzer legacy survey (SpUDS, PI: J. Dunlop, data in hand) reaching $\sim$24 (AB) at 3.6 and 4.5 μm with further Spitzer warm time data currently being observed; also, data from SWIRE covering 3.6–160 μm (Lonsdale et al. 2003).
- The UDS field is one of the planned fields of the HerMES survey utilising the recently-launched Herschel space observatory.
- Sub-mm data from SCUBA, with SCUBA-II data due to be taken as part of the SCUBA II Cosmology Legacy Survey.
- Radio data at 1.4 GHz from VLA to 12 μJy rms (also now public).

This wealth of data has cemented the position of the UDS as one of the most important fields in extra-galactic astronomy and further data acquisition projects are planned for the future.

The availability of such high-quality data over all wavelengths allows the construction of photometric redshifts from the sources extracted from the UDS.

---

[1] This has indeed been the case in the remainder of semester 2009b.

Fig. 29.1 Photometric redshift against spectroscopic redshift for the sources currently with spectra in the UDS field. The agreement is extremely good, with dispersion in $\Delta z/(1+z)$ only $\sigma = 0.03$

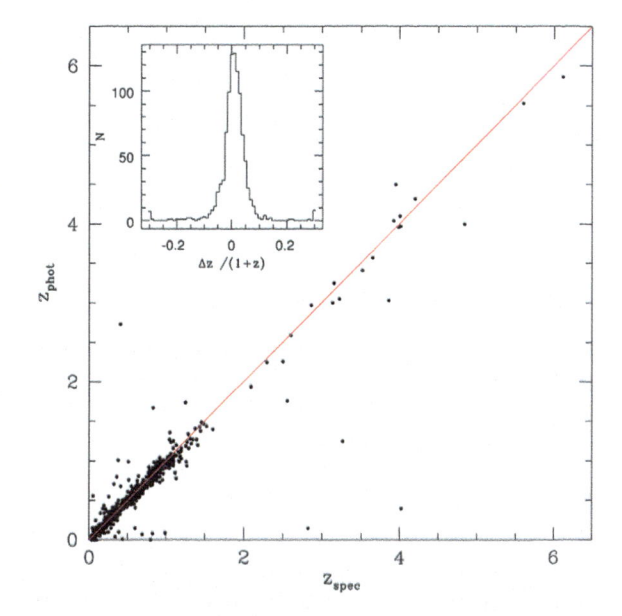

Photometric redshifts are an incredibly powerful tool for studying the distant universe, with no fewer than three independent groups using UDS data to produce them. Figure 29.1 (Cirasuolo et al. 2007, 2010) shows the excellent agreement between the photometric redshifts constructed from the data and the spectroscopic redshifts available in the field. The spread in $\Delta z/(1+z)$ is only $\sigma = 0.03$. The data in the field are so deep that obtaining spectra for all of the sources is beyond even the largest telescopes currently in existence. Photometric redshifts are therefore absolutely vital in achieving the science goals of the survey and will continue to be so for the duration of its impact as a legacy data set. The breadth in wavelength of high quality data available in the field enables robust studies of galaxies across the whole range in redshift.

Science Highlights

As of the 13th of September 2009, the UDS data have been used (at least in part) to produce 26 publications, with a further 5 on the pre-print archive. Of these, the lead author was working in a UK institution in 19 cases and within Europe in a further 7 cases. The remaining five were produced from the wider, world community. This final statistic is particularly impressive, as the DR3 (the first to include H-band data) only became world public in June of the same year. The most highly cited science paper from the UDS, that of Cirasuolo et al. (2007), has already accumulated 56 citations by the time of this meeting. In this section we shall describe a few of the current science highlights from the UDS data.

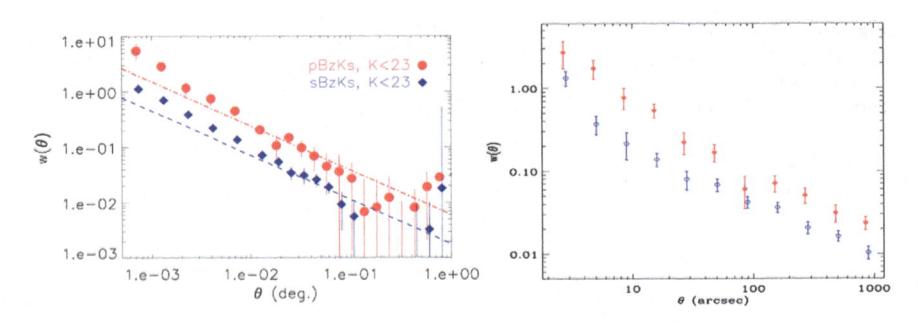

Fig. 29.2 The angular clustering results of Hartley et al. (2008) (*left*) and Williams et al. (2009) (*right*). In each case the passive galaxy sample is found to cluster more strongly than the star-forming samples, indicating that passive galaxies are hosted by more massive dark matter halos

Galaxy Clustering

The clustering properties of a galaxy sample contain a wealth of information regarding both its large-scale and smaller-scale environments. Of particular interest is the amplitude of clustering at large scales (angular separations equivalent to 1 Mpc or greater). The large-scale amplitude is primarily dependent on the typical mass of the dark matter halos which host the population in question. The clustering strength of dark matter halos is a monotonically increasing function of their mass. If two samples are selected at the same redshift, then the more strongly clustered of them will therefore be hosted by more massive halos. This simple argument, using a statistic which is relatively straightforward to compute, is incredibly powerful. By finding the most clustered populations at various epochs, it allows us to broadly link, from one epoch to the next, progenitors of low redshift massive elliptical galaxies. One such link in this chain was studied by Hartley et al. (2008). Using a 2-colour selection method proposed by Daddi et al. (2004), they were able to compute the clustering properties of galaxies selected to be either passive or star-forming over the redshift range $1.4 < z < 2.5$. The result of their measurements is shown in Fig. 29.2. Clearly passive galaxies are more strongly clustered than star-forming galaxies at this epoch. With the addition of photometric redshifts for the sample galaxies, the angular clustering was deprojected to find the correlation scale-length, r0. This correlation length can then be compared directly with the equivalent values for dark matter halos of different masses, predicted by structure formation models (Mo and White 2002). The halo masses implied by the measurements are 10^{13} M_{sun} for the passive sample and 6×10^{11} M_{sun} for the star-forming sample. These typical host masses identify the passive galaxies as being hosted by halos which will become group and cluster mass halos by $z = 0$. Using additional information from their number counts and luminosity function, it was concluded that they were likely to become at least some of the massive, low-redshift ellipticals that we observe.

Following the result of Hartley et al. (2008), Williams et al. (2009) formulated a novel method of selecting the passive members of the red sequence using photomet-

Fig. 29.3 Deprojected correlation lengths of passive and star-forming sub-samples from Hartley et al. (2010). Passive galaxies are more strongly clustered, irrespective of K-band luminosity, to at least $z = 1.5$. By extrapolating the evolution in redshift, they expect that passive and star-forming clustering strengths will become equal by $z \sim 2.5$

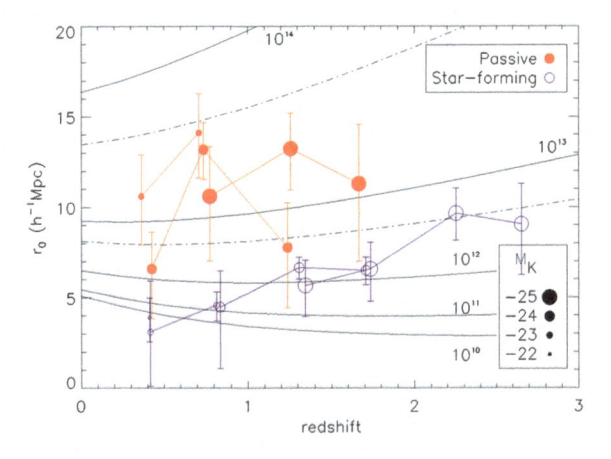

ric redshifts, and rest-frame U–V and V–J colours derived from them. Above $z = 1$ the traditional red sequence, selected from a colour-magnitude diagram, becomes highly contaminated by dusty, star-forming objects. Using just these two colours, Williams et al. were able to isolate galaxies which had spectroscopic redshifts at $z > 1$ and little [OII] emission from the broader population. With their passive and star-forming samples defined, they computed the angular clustering and deprojected correlation lengths for samples over $1 < z < 2$, separated into two luminosity sub-samples. Their angular clustering measurements are shown in the right hand panel of Fig. 29.2. Their results were in general agreement with those of Hartley et al. (2008), finding greater clustering strength for passive galaxies.

Finally, more recently a detailed study of the colour, star-formation rate and luminosity-dependent clustering of galaxies over $0 < z < 3$ has been conducted. Hartley et al. (2010) used synthetic templates of galaxy spectral energy distributions to simultaneously fit the redshifts and star-formation histories of the UDS DR3 galaxies. Using this information they constructed sub-samples in redshift, K-band luminosity, colour and star-formation history. For each sub-sample the deprojected correlation length was computed and hence they obtained the evolution in clustering over the whole $0 < z < 3$ range which is currently accessible from the UDS data. The results of the passive and star-forming galaxy sub-samples were particularly striking. As previously found by Hartley et al. (2008) and Williams et al. (2009) for their much broader samples, passive galaxies were found to be more strongly clustered to at least $z = 1.5$, as shown in Fig. 29.3. However, this difference in clustering was found to be less related to K-band luminosity than to the star-formation rate of the galaxies. Furthermore, if their measurements were to be extrapolated to higher redshift their results indicated that such a difference would disappear. Equal clustering strengths for passive and star-forming galaxies would mean that they are hosted by equal mass halos, and hence be a strong suggestion that the principle epoch for a galaxy to undergo transition from star-forming to passive had been found.

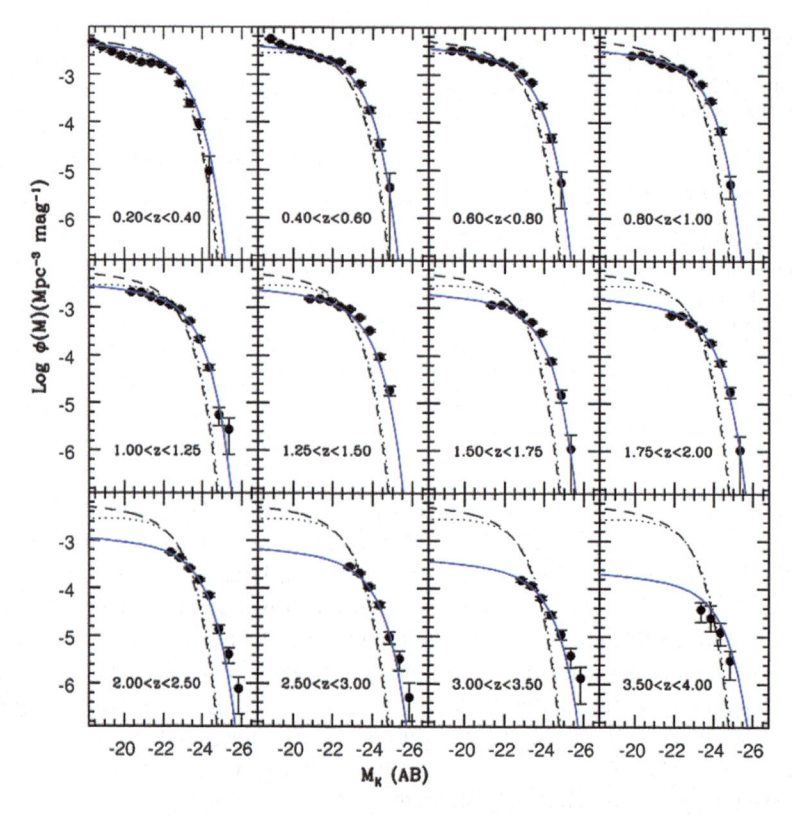

Fig. 29.4 The evolution of the K-band luminosity function from Cirasuolo et al. (2010). The luminosity function is one of the most fundamental statistical quantities of a galaxy population, and the K-band luminosity function in particular. This figure shows how it evolves over much of the universe's history and is an important constraint for any model of galaxy evolution

Evolution of the Luminosity Function

One of the first papers using UDS data to appear was that of Cirasuolo et al. (2007), using the early data release to investigate the build-up of the red sequence and the evolution of the K-band luminosity function. As more data were accumulated and the study of fainter sources made possible, the study of the luminosity function was extended to higher redshift. These measurements, presented in Cirasuolo et al. (2010), are shown in Fig. 29.4 and constitute a genuine benchmark that any model of galaxy formation must match. For many years, the models at the forefront of theoretical research into galaxy formation over cosmic scales were matched and tested against quantities at low-redshift. The lack of an unbiased, observation-derived guide to the global galaxy population across a broad range in redshift was a significant hinderance to those working on so-called semi-analytic models. The results of Cirasuolo et al. (2010) are an example of exactly why the UDS is required. In combination with the many low-redshift statistical properties that semi-analytic modellers use, the evolution of the K-band luminosity function will help ensure that the most important physical processes are highlighted in future models.

Galaxies at z > 5

This topic is presented elsewhere in this volume, but nevertheless deserves to be mentioned as one of the science highlights from the UDS. The study of galaxies at very high redshift has become one of the most important topics in extragalactic astronomy in recent years. The work of McLure et al. (2009) on the massive end of the high-redshift galaxy population was an important milestone in the field. In many ways McLure et al. laid the groundwork for similar studies that will be made using the UltraVISTA survey and the recent results at the lower-mass end using WFC3 data in the Hubble Ultra-Deep Field. With up to 1,000 h more data to be collected in the UDS, it promises to remain a highlight of UDS science in the future.

Future Direction

It is highly likely that UKIRT will remain open until the end of 2012 and therefore that the UKIDSS survey will run to completion (as detailed by the case submitted to the UKIRT board in 2009). For the UDS this means that we should reach the projected depths of $J = 24.4$, $H = 23.3$ and $K = 22.8$ (Vega).[2] In the K-band this projected final figure is almost a magnitude fainter than the current (DR5) data and only 0.2 magnitudes short of the original target. With the current UDS data we are able to probe L* galaxies, the galaxies which dominate the stellar mass of the universe at a given epoch, up to $z = 3$ (Fig. 29.5). In terms of the galaxies that may be studied using the final data set, the $K = 22.8$ limit corresponds to an L* galaxy at $z = 5$ and an L* + 2 magnitude galaxy at $z = 2$. The UDS will therefore represent a census of the galaxy population over $0 < z < 5$, pushing the current science possibilities to higher redshift and fainter magnitudes. We expect to obtain $\sim$1,000 galaxies at $z > 6$, making an accurate determination of the bright end of the UV luminosity function and their large-scale structure at this epoch possible.

---

[2]The improvements at the telescope of a new mirror cooling system and seeing estimation have had an immense impact on the rate of data collection. Since the meeting, the combined exposure time of the UDS has been doubled by the addition of the 2009b semester data. At this rate, the UDS is well on target to reach these projected depths.

Fig. 29.5 Absolute K-band magnitude against redshift for the UDS galaxies. The *curved dashed line* is the evolving value for L* from Cirasuolo et al. (2010). At the current depths we are probing L* galaxies at z = 3, in the final data set we hope to be able to reach L* galaxies at z = 5

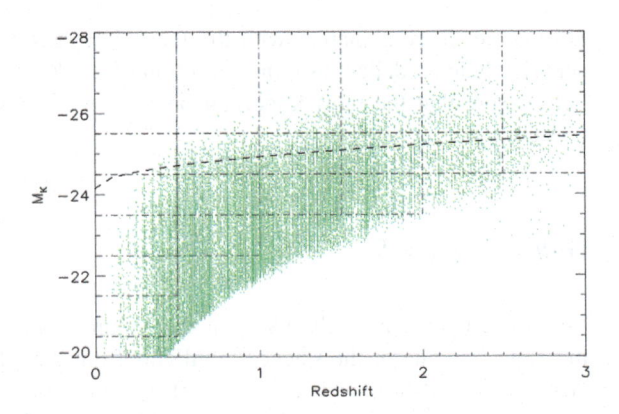

We shall be able to identify massive systems such as groups and clusters to higher redshifts than have previously been possible. Finally, we will be able to study the co-evolution of galaxies and their environments for a wide range in redshift and galaxy stellar mass.

This last project is already well underway. Chuter et al. (2011) have used the UDS DR3 to study the environments of red, blue, passive and star-forming galaxies. They chose as their measure of environment the sky density of galaxies relative to a random distribution within two different sized apertures. These apertures correspond to 500 h^{-1} kpc and 1 h^{-1} Mpc (co-moving) at the photometric redshift of the galaxy in question. Their results are plotted in Fig. 29.6. At low redshift there is a well-known morphology-density relation (Dressler 1980), in which elliptical galaxies are found to dominate higher density regions. This relation is mimicked by colour, with red galaxies located in regions of higher density and blue galaxies dominating the 'field'. Work by Cooper et al. (2007) seemed to suggest that this colour dependence on density would become absent by $z \sim 1.5$. However, as their sample was selected in the B-band they were unable to test this hypothesis. The measurements shown in Fig. 29.6 demonstrate that, although the correlation between colour and density is weaker at z = 1.5, it is still present. At higher redshift there are very few red galaxies and the redshift bin width is consequently very broad. However, the environment of a galaxy appears to be independent of colour at z = 2, a hypothesis that could well be confirmed or otherwise by the full UDS dataset.

Alongside the continued infrared data collection of the UDS, there are further data either in hand or approved, many of which were listed in the summary above. The z'-band data, reaching z' = 26.5 (AB) were a match to the other optical data in the field and at the time of observation it was felt that they would be the deepest that would be required in the field. However, with the addition of the UDS, deeper z'-band data are warranted and have now been approved, with an estimated depth of z' $\sim$ 27.5 (AB, 5σ). Though the field was observed with Spitzer as part of SWIRE, these data were far shallower than the optical and infrared data. A Spitzer Legacy Programme to match the UDS depths, SpUDS, was approved and data reaching magnitudes of $\sim$24 (AB) in 3.6 and 4.5 μm have been collected (raw data are

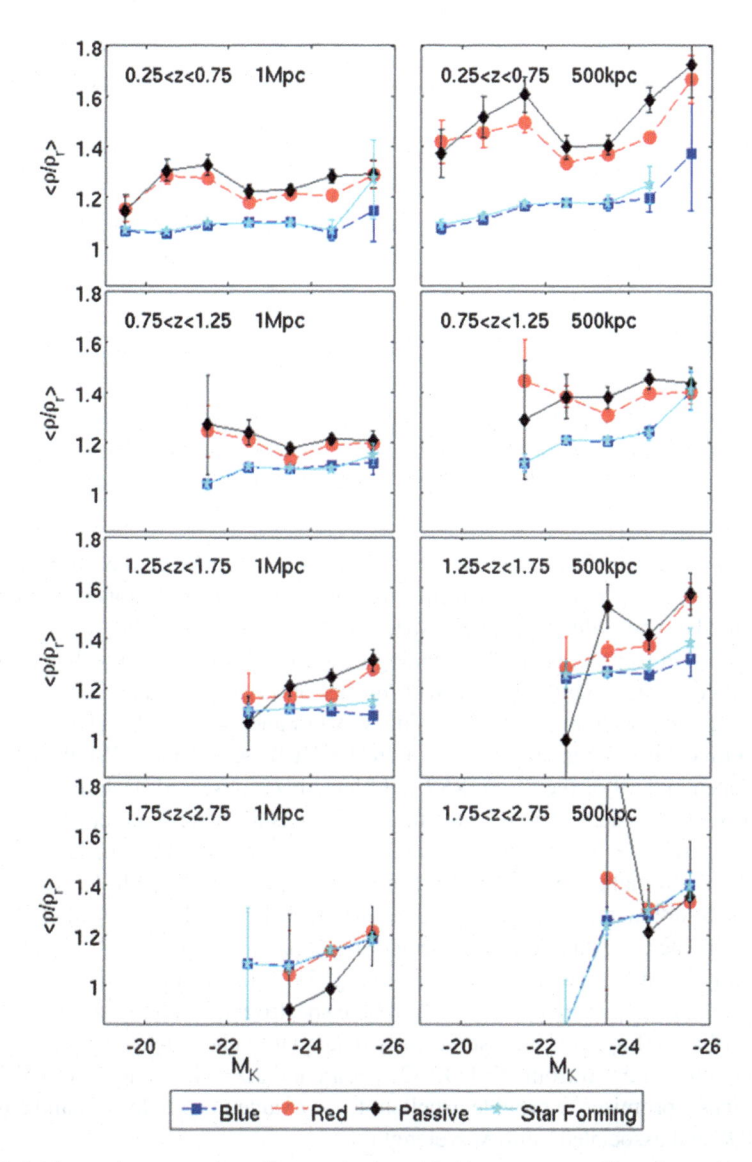

Fig. 29.6 The environment of a galaxy can be quantified in many ways. Here the environmental measure is the sky density of galaxies within an aperture, relative to a random distribution. The apertures chosen are equivalent to 500 h^{-1} kpc and 1 h^{-1} Mpc at the photometric redshift of the galaxy in question. *Red* and passive galaxies are found in denser environments than *blue* or star-forming galaxies to $z = 1.75$

publicly available from the Spitzer archive). In addition, a large amount of Spitzer warm time has been allocated to observe the UDS, making it the deepest data over such an area that will ever be observed with the satellite. Since the meeting, a Multi-cycle Hubble Treasury programme (PI: Faber), with the UDS as one of four fields,

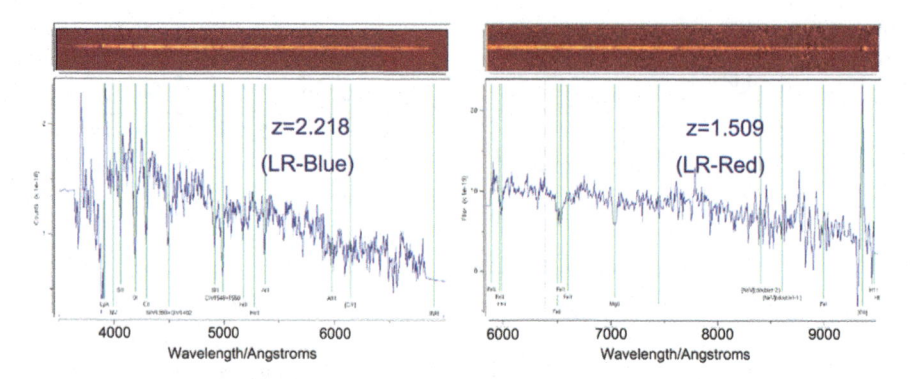

Fig. 29.7 Some example spectra from the recently acquired UDSz spectroscopic survey data. Redshifts are very difficult to obtain at $1 < z < 2$ due to the lack of strong emission lines. However, early indications are that completeness will be reasonably high

has been accepted. This programme will cover a sub-area of the UDS in J and H-bands, allowing morphological information to be used in conjunction with data covering the whole electromagnetic spectrum.

In addition to progress in the imaging of the UDS, the UDS now has excellent spectral coverage from the UDSz, with further observations planned. The UDSz is an ESO Large Programme of 93 h using VIMOS and 142 h with FORS2 to obtain ~4,000 redshifts for galaxies at $z > 1$ in the UDS. All of the VIMOS data have been taken and 19 of the 20 FORS2 field have been observed. Data reduction has progressed well and we are now beginning the redshift determination. Many of the spectra are of objects expected to lie in the 'redshift desert' where the lack of strong emission lines makes accurate redshift determination very difficult. However, early indications are that the final completeness of the data set will be reasonably high, at 70 % or greater. A data release is expected sometime in the summer. An indication of the quality of the data can be gained from Fig. 29.7. This figure shows two VIMOS spectra, the first from the low-resolution blue spectrograph and the second from the red. Further highlights from just the early data quality checking stage are a Lyman-alpha emitter at $z > 6$ from the FORS2 spectra and a $z = 4.8$ quasar from VIMOS. The data set promises to provide excellent legacy value to complement and enhance the UDS and associated multi-wavelength data.

Summary

The UDS, at only 12 % completeness, has already been established as one of the most important fields in the community, and produced some outstanding science results. Data collection, both with UKIRT in the near infrared and across the electromagnetic spectrum, is ongoing and has even accelerated in recent months.

The science exploitation of the survey is also accelerating and is expected to take a significant jump forward with the next data release. Though the future of UKIRT is uncertain beyond 2012, its impact will be felt for many more years. The legacy value of the UDS and associated multi-wavelength data is without question, and many of the results found using the data will have great, lasting scientific impact.

References

Almaini, O., Foucaud, S., Lane, K., Conselice, C.J., McLure, R.J., Cirasuolo, M., Dunlop, J.S., Smail, I., Simpson, C.: ASP Conf. Ser. **379**, 163 (2007)
Benson, A., Bower, R., Frenk, C., et al.: Astrophys. J. **599**, 38 (2003)
Casali, M., Lunney, D., Henry, D., et al.: ASP Conf. Series. **232**, 357 (2001)
Chuter, R.W., Almaini, O., Hartley, W.G., McLure, R.J., Dunlop, J.S., Foucaud, S., Conselice, C.J., Simpson, C., Cirasuolo, M., Bradshaw, E.J.: Mon. Not. R. Astron. Soc. **413**, 1678 (2011)
Cirasuolo, M., McLure, R., Dunlop, J., et al.: Mon. Not. R. Astron. Soc. **380**, 585 (2007)
Cirasuolo, M., McLure, R., Dunlop, J., et al.: Mon. Not. R. Astron. Soc. **401**, 1166 (2010)
Cooper, M., Newman, J., Coil, A., et al.: Mon. Not. R. Astron. Soc. **376**, 1445 (2007)
Cooray, A., Sheth, R.: Phys. Rev. **372**, 1 (2002)
Daddi, E., Cimatti, A., Renzini, A., et al.: Astrophys. J. **617**, 746 (2004)
Dressler, A.: Astrophys. J. **236**, 351 (1980)
Dunlop, J., Peacock, J., Spinrad, H., et al.: Nature **381**, 581 (1996)
Furusawa, H., Kosugi, G., Akiyama, M., et al.: Astrophys. J. Suppl. **176**, 1 (2008)
Hartley, W., Lane, K., Almaini, O., et al.: Mon. Not. R. Astron. Soc. **391**, 1301 (2008)
Hartley, W.G., Almaini, O., Cirasuolo, M., Foucaud, S., Simpson, C., Conselice, C.J., Smail, I., McLure, R.J., Dunlop, J.S., Chuter, R.W., Maddox, S., Lane, K.P., Bradshaw, E.J.: Mon. Not. R. Astron. Soc. **407**, 1212 (2010)
Lawrence, A., Warren, S., Almaini, O., et al.: Mon. Not. R. Astron. Soc. **379**, 1599 (2007)
Lonsdale, C., Smith, H., Rowan-Robinson, M., et al.: Publ. Astron. Soc. Pac. **115**, 897 (2003)
McLure, R., Cirasuolo, M., Dunlop, J., et al.: Mon. Not. R. Astron. Soc. **395**, 2196 (2009)
Mo, H., White, S.: Mon. Not. R. Astron. Soc. **336**, 112 (2002)
Navarro, J., Frenk, C., White, S.: Mon. Not. R. Astron. Soc. **275**, 720 (1995)
Ostriker, J., Steinhardt, P.: Nature **377**, 600 (1995)
Silk, J., Rees, M.: Astron. Astrophys. **331L**, 1 (1998)
Tinsley, B., Gunn, J.: Astrophys. J. **203**, 52 (1976)
Warren, S., Hambly, N., Dye S. et al.: Mon. Not. R. Astron. Soc. **375**, 213 (2007)
Williams, R., Quadri, R., Franx, M., et al.: Astrophys. J. **691**, 1879 (2009)

Chapter 30
Exploring Massive Galaxy Evolution with the UKIDSS Ultra-Deep Survey

R.J. McLure, M. Cirasuolo, J.S. Dunlop, O. Almaini, and S. Foucaud

Abstract We present the results of a study of massive galaxy evolution at $z > 5$ using the deep optical and near-IR imaging available in the UKIDSS ultra-deep survey (UDS). Based on the results of a full SED fitting analysis we are able to examine the evolution of the UV-selected galaxy luminosity function from $z = 5$ to $z = 6$, finding that the characteristic magnitude (M*) dims by a factor of ~ 2 over this redshift interval. Moreover, by exploiting the contiguous area and deep near-IR data available in the UDS, we also present measurements of the clustering properties, dark matter halo masses and typical stellar masses of luminous Lyman-break galaxies at $z \sim 5.5$.

Introduction

Studies of massive galaxies at high redshift have a crucial role to play in testing the current generation of galaxy formation models, which have traditionally struggled to produce sufficient numbers of massive objects at high redshift. Moreover, current evidence from observations of both high-redshift quasars and the cosmic microwave background indicate that the Universe was re-ionized somewhere in the redshift interval $6 < z < 11$ (e.g. Dunkley et al. 2009). Consequently, another principal motivation for studying the properties of galaxies at high redshift is to determine the nature of the objects that re-ionized the Universe.

The Lyman-break technique for photometrically identifying galaxies at high redshift, via a characteristic drop in apparent magnitude at observed wavelengths shortward of the Lyman-alpha emission line, is now very mature. Indeed, largely due to ultra-deep optical observations from the Hubble Space Telescope (HST), it is now possible to routinely identify large samples of $z=4$, $z=5$, and $z=6$ Lyman-break galaxies (LBGs). In fact, due to the remarkable sensitivity of HST, it is now possible to identify objects as faint as 0.1 L* out to a redshift of $z \sim 6$ (e.g. Bouwens et al. 2007). However, due to the small areal coverage of the deep HST observations, the effects of cosmic variance on high-redshift galaxy studies are potentially severe. Furthermore, potential uncertainties due to cosmic variance are especially worrying with respect to massive galaxies, which are simply too rare to appear in HST pencil-beam surveys. It is within this context that deep, wide-field, ground-based observations have a key role to play. In this proceedings we report the results of a study that combines the strengths of wide-field, ground-based observations, with ultra-deep, pencil-beam HST observations to study a sample of LBGs at $z=5$ and $z=6$ spanning a factor 100 in luminosity (McLure et al. 2009).

Selecting Robust Samples of High-Redshift Galaxies

The UKIDSS Ultra-deep Survey (UDS) is unique among 1-sq.deg. survey fields in that it currently has the deepest available imaging from 0.4 to 4.5 μm, combining optical observations from Subaru, near-IR observations from UKIRT and deep mid-IR data from a dedicated Spitzer legacy mission (SpUDS). As a consequence, the UDS field is a powerful resource for studying galaxy evolution in general, and the evolution of massive galaxies at high redshift in particular. In McLure et al. (2009) we exploited the data available in the UDS to investigate the evolution of the bright-end of the galaxy luminosity function between $z=5$ and $z=6$. In order to make maximal use of the high-quality data available, we employed an analysis technique that is somewhat different to what is typically adopted in the literature. Specifically, rather than using the usual approach of colour-colour cuts to isolate potential LBGs within specific redshift windows, we used a full photometric SED-fitting method to estimate the redshift probability distribution for each galaxy in our sample. The main advantages of this approach are that it makes full use of the information available, and allows the impact of potential low-redshift interlopers to be evaluated properly. Moreover, this technique deals naturally with the problem of double photometric redshift minima when estimating the luminosity function.

The High-Redshift Galaxy Luminosity Function

In Fig. 30.1a we show the determination of the $z=5$ and $z=6$ luminosity function from McLure et al. (2009). The data-points at the bright-end are derived from the UKIDSS UDS, whereas the data-points at the faint-end are taken from the study of Bouwens et al. (2007), which is based on ultra-deep, HST pencil-beam surveys. By

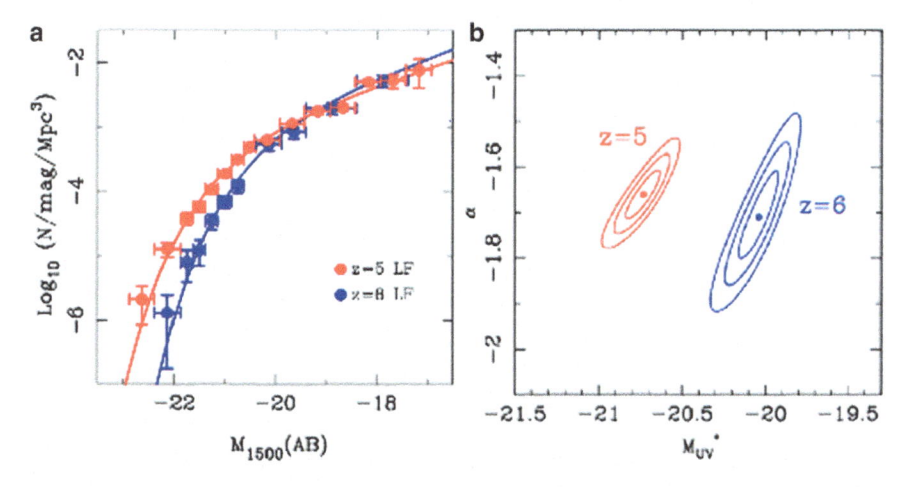

Fig. 30.1 Panel (**a**) shows the estimates of the z = 5 and z = 6 luminosity functions from McLure et al. (2009). The data-points brighter than $M_{1,500} = -20.5$ are from the UDS, while the fainter points are taken from the HST-based study of Bouwens et al. (2007). The *solid curves* show the best-fitting Schechter functions to the combined ground-based +HST data sets. Panel (**b**) shows the 1, 2 & 3σ confidence intervals on the best-fitting faint-end slope and characteristic magnitude at z = 5 and z = 6

combining these two data sets it is possible to perform a maximum likelihood fit to the luminosity function at z = 5 and z = 6 using data spanning a factor of 100 in luminosity (10 L* < L < 0.1 L*). The resulting confidence intervals on the derived faint-end slope (α) and characteristic magnitude (M*) are shown in Fig. 30.1b. We find that the evolution of the luminosity function from z = 6 to z = 5 is well described by a simple change in the characteristic magnitude, with M* dimming by a factor of two between z = 5 and z = 6. In contrast, we find little evidence for any evolution in either the faint-end slope (α) or normalization (φ*). This observed evolution is at least qualitatively consistent with the expectations of hierarchical build-up, with no indication of galaxy downsizing.

Typical Stellar Masses

The final LBG sample analyzed in McLure et al. (2009) consisted of >500 LBGs in the redshift interval 5 < z < 6. By stacking the photometry of this large sample an estimate of the average SED of luminous LBGs at z ∼ 5.5 could be obtained. The SED fit to the stacked photometry (Fig. 30.2a) revealed that L* LBGs at z ∼ 5.5 have a typical stellar mass of 10^{10} $M_\odot$. Moreover, by using the average mass-to-light ratio derived from fitting the stacked photometry we proceeded to convert the LBG luminosity function measurements shown in Fig. 30.1a into estimates of the galaxy stellar mass function. The corresponding estimate of the high-redshift stellar

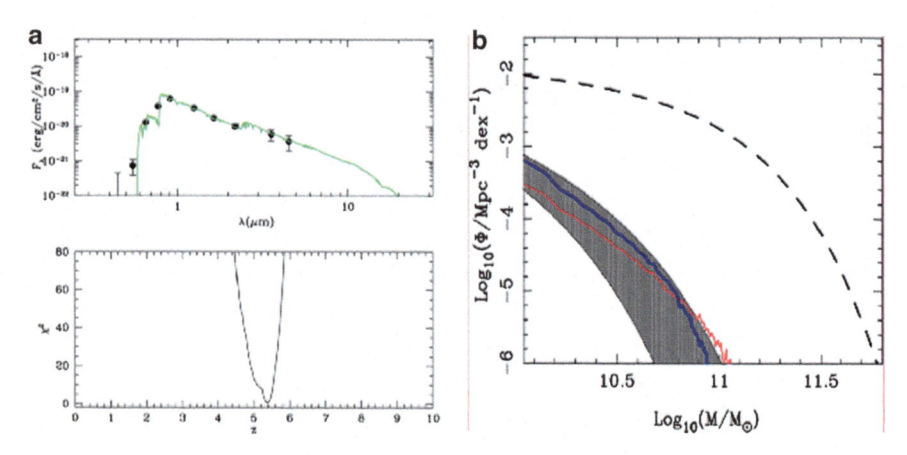

Fig. 30.2 Panel (**a**) shows the best-fitting SED model to the stacked photometry of >500 LBGs in the redshift interval 5 < z < 6, together with a plot of χ^2 versus photometric redshift. The best-fitting mass-to-light ratio from this SED fit makes it possible to convert the luminosity functions shown in Fig. 30.1 into estimates of the stellar mass function at $z \sim 5.5$. Panel (**b**) shows the corresponding stellar mass function estimate, with the *grey shaded* area indicating the estimated uncertainty. The *red* and *blue curves* show the predicted stellar mass function at $z \sim 5.5$ from the Bower et al. (2006) and De Lucia and Blaizot (2007) galaxy evolution models. The *upper dashed curve* shows the stellar mass function at z = 0

mass function at $z \sim 5.5$ is shown in Fig. 30.2b, along with the predicted stellar mass functions from two recent semi-analytic galaxy formation models (Bower et al. 2006; De Lucia and Blaizot 2007). It can be seen that within the large uncertainties the latest galaxy evolution models (both of which incorporate some form of AGN feedback) are able to qualitatively reproduce the observations. Finally, by integrating the galaxy stellar mass function it is possible to estimate the stellar mass density in place at high redshift. The results of this calculation indicate that the stellar mass density in place at z = 5 and z = 6 is $\sim 1 \times 10^7$ $M_\odot$ and $\sim 4 \times 10^6$ $M_\odot$ respectively; in good agreement with other recent estimates (e.g. Yan et al. 2006; Stark et al. 2007).

Clustering Properties

Armed with a sample of >500 LBGs in the redshift range 5 < z < 6, selected over a contiguous area of ~ 0.65 sq.deg., it was possible to measure accurately the angular clustering properties of luminous LBGs at z > 5. The results of this analysis demonstrated that L > L* LBGs at $z \sim 5.5$ are strongly clustered, with a clustering length of $r_0 \sim 8$ Mpc. When this is compared to the latest models of dark matter halo clustering it suggests that luminous LBGs at $z \sim 5.5$ are typically located in dark matter halos with masses of $\sim 10^{11.5}$ $M_\odot$. When combined with our estimate of the typical stellar mass, this indicates that luminous LBGs at $z \sim 5.5$ have stellar:dark matter ratios of ~ 50, in good agreement with model predictions (e.g. Bower et al. 2006).

Conclusions

We have briefly reviewed the main results of a study aimed at exploring the properties of massive LBGs in the redshift range $4.5 < z < 6.5$, based on combining optical and near-IR data in the UKIDSS UDS field with the results of ultra-deep, HST pencil-beam surveys (McLure et al. 2009). The results of this study have revealed that the luminosity function of UV-selected galaxies evolves significantly between $z = 5$ and $z = 6$, with M* dimming by a factor of ~ 2. Moreover, by exploiting the availability of deep near-IR data it was possible to estimate that the typical stellar mass of L* LBGs at $z \sim 5.5$ is 10^{10} $M_\odot$. Finally, using the large contiguous area of the UDS survey an accurate measurement of the clustering length and dark matter halo masses of L* LBGs at $z \sim 5.5$ was obtained. At the time of writing we are analyzing deep optical spectroscopy of a small sub-sample of our $z > 5$ LBG candidates, from which we have already identified >10 luminous Lyman-alpha emitters in the redshift range $6.01 < z < 6.49$ (Curtis-Lake et al. 2012).

References

Bouwens, R.J., et al.: Astrophys. J. **670**, 928 (2007)
Bower, R.G., et al.: Mon. Not. R. Astron. Soc. **370**, 645 (2006)
Curtis-Lake, E., et al.: A remarkably high fraction of strong Lyα emitters amongst luminous redshift $6.0 < z < 6.5$ Lyman-break galaxies in the UKIDSS Ultra-Deep Survey. Mon. Not. R. Astron. Soc. **422**, 1425 (2012)
Dunkley, J., et al.: Astrophys. J. **180**, 306 (2009)
De Lucia, G., Blaizot, J.: Mon. Not. R. Astron. Soc. **375**, 2 (2007)
McLure, R.J., et al.: Mon. Not. R. Astron. Soc. **395**, 2196 (2009)
Stark, D.P., et al.: Astrophys. J. **659**, 84 (2007)
Yan, H., et al.: Astrophys. J. **651**, 24 (2006)

Chapter 31
The UKIRT Planet Finder

Hugh R.A. Jones, John Barnes, Ian Bryson, Andy Adamson, David Henry,
David Montgomery, Derek Ives, Ian Egan, David Lunney, Phil Rees,
John Rayner, Larry Ramsey, Bill Vacca, Chris Tinney, and Mike Liu

Abstract We present a conceptual design for the UKIRT Planet Finder (UPF). It is a fibre-fed high resolving power (R $\sim$ 70,000 at 2.5 pixel sampling) cryogenic echelle spectrograph operating in the near infrared (0.95–1.8 μm) and is designed to provide 1 m/s radial velocity measurements. We identify the various error sources to overcome in order to achieve the required stability. We have constructed models simulating likely candidates and demonstrated the ability to recover exoplanetary radial-velocity (RV) signals in the infrared. UPF should achieve a total RV error of around 1 m/s on a typical M6V star. We use these results as an input to a simulated 5-year survey of nearby M stars, which has the sensitivity to detect of the order of 30 terrestrial mass planets in the habitable zone around those stars. UPF will thus test theoretical planet formation models, which predict an abundance of terrestrial-mass planets around low-mass stars enabling critical tests of planet-formation theories and allowing the identification of nearby planets with conditions potentially suitable for life.

H.R.A. Jones (✉) • J. Barnes
Centre for Astrophysics Research, University of Hertfordshire, Hatfield, UK
e-mail: h.r.a.jones@herts.ac.uk

I. Bryson • D. Henry • D. Montgomery • D. Ives • I. Egan • D. Lunney • P. Rees
Astronomy Technology Centre, Royal Observatory, Edinburgh, UK

A. Adamson
Joint Astronomy Centre, 660 N.A'ohoku Place, University Park, Hilo, HI 96720, USA

J. Rayner • B. Vacca • M. Liu
Institute for Astronomy, University of Hawaii, Hilo, HI, USA

L. Ramsey
Department of Astronomy & Astrophysics, The Pennsylvania State University, University Park, PA 16802, USA

C. Tinney
School of Physics, University of New South Wales, Sydney 2052, Australia

Introduction

The recent announcement of a number of exoplanets with masses below five Earth masses around nearby M dwarfs strengthens the theoretical suggestion that there are large numbers of Earth-mass planets in the habitable zone around M dwarfs.

Based on modelling, laboratory experiments and a conservative scaling of optical results, a UKIRT Planet Finder survey will be sensitive to terrestrial mass planets in the habitable zone around nearby stars. It can thus provide key follow-up targets for other planet search techniques and test theoretical planet formation models, which predict an over-abundance of terrestrial-mass planets around low-mass stars.

The concept of a high precision infrared radial velocity spectrometer has been vigorously supported by a variety of review processes as the most cost effective method to discover Earth-mass habitable-zone exoplanets. Such an instrument is an essential complement to transit missions such as CoRoT, Kepler and UKIRT's own WFCAM Transit Survey that will all require follow-up of objects beyond the limit of optical spectrographs. There is very widespread scientific support for UPF within the UK and wider community. In addition to unprecedented access to low-mass exoplanets, UPF also offers a wide range of other compelling science.

The Potential for Finding Terrestrial-Mass Exoplanets Around Low-Mass Stars

The radial velocity technique has played the dominant role in foundation of this new field. Surveys using this method have discovered almost all the planets known within 200 pc, probe a wide range of properties, and play the critical role in the confirmation of exoplanets discovered by the transit technique. Doppler searches provide minimum planet mass, orbital period, orbital semi-major axis and eccentricity. They also differ from photometric transit and micro-lensing surveys because they specifically target nearby well characterised stars.

Two of the key results coming out of the Doppler surveys so far are that low-mass planets are more common than high-mass ones ($dN/dM \propto M^{-1}$ O'Toole et al. 2009) and that multiple-planet exoplanet systems predominate, particularly for M dwarfs (e.g. 5/8 is the lower limit based on systems discovered to date, Mayor et al. 2009). Improvements in the efficiency and sampling of searches at optical wavelengths promise long-term precisions of ~0.5 m/s and ~5 $M_\oplus$ detections around solar-type stars. While this may be the limit for CCD-based surveys of solar type stars until larger telescopes become available, it is nonetheless feasible to survey lower mass primaries to achieve a corresponding smaller mass limit. Thus the lowest mass Doppler signals have been found around M dwarfs (e.g., GJ581e 2 $M_\oplus$ Mayor et al. 2009) and detections down to $M_\oplus$ might be feasible around mid-type M dwarfs with very large amounts of optical data (e.g., 119 points for GJ581). Therefore the UPF approach is to search around lower-mass primary stars since the RV reflex signal will be larger for lighter primary stars.

31 The UKIRT Planet Finder

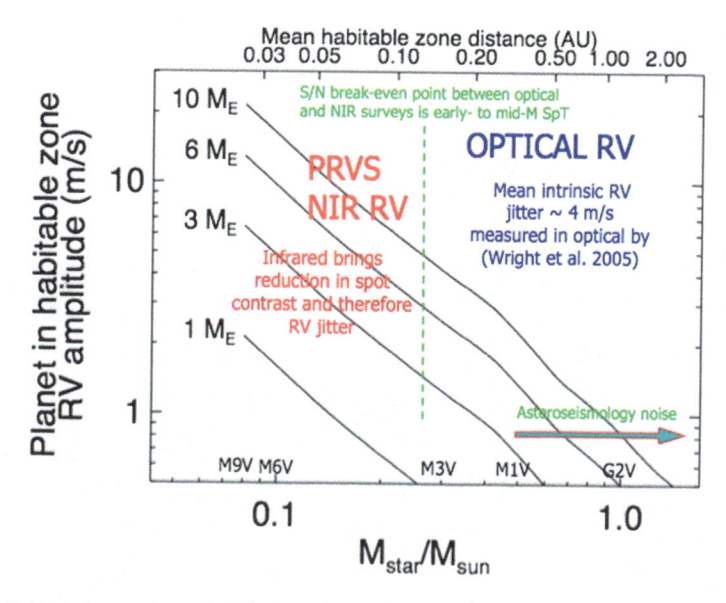

Fig. 31.1 Radial velocity amplitude as function of host mass or mean habitable zone distance comparing UPF with optical radial velocity surveys

While optical RV surveys have been very successful they are restricted to stars more massive than ~0.3 $M_\odot$ (about M4 dwarfs). Very nearby, lower-mass M dwarfs are optically too faint (e.g. the low mass planet with the latest spectral type host star has a 1.9 d period and is around Gl876, an M4 star at 4.7 pc; even this case with maximum signal (nearby and close-orbiting) required regular monitoring with the Keck 10-m from 1997 to 2005 and data on 6 consecutive nights, Rivera et al. 2005). Such work thus requires large investments of time (at high operational cost on large facilities) at M4 and becomes prohibitive in the optical at later spectral types (Fig. 31.1).

The relative lack of flux from M dwarfs in the optical bands has led them to be somewhat overlooked in exoplanet studies and it therefore follows that since more than 72 % of stars within 10 pc are M dwarfs (www.recons.org), it is imperative to have the appropriate infrared instrumentation to study what are probably the most common sites of exoplanetary systems. In fact, despite very strong biases against their detection, a variety of planet search techniques probing very different environments do find evidence for the ubiquity of low-mass planets, e.g. pulsar timing planets around PSR 1257+12 (Konacki and Wolszcan 2003) and the microlensing detection of MOA-2007-BLG-192-Lb (Bennett et al. 2008). Together with recent direct imaging success such as HST astrometry of the circumstellar ring of material around Fomalhaut (Kalas et al. 2008) these results contribute to the clear observational indication that low-mass exoplanets exist in a variety of environments and are indeed ubiquitous. Thus the major new frontier in exoplanet research is to find and characterise the closest (most studiable) examples.

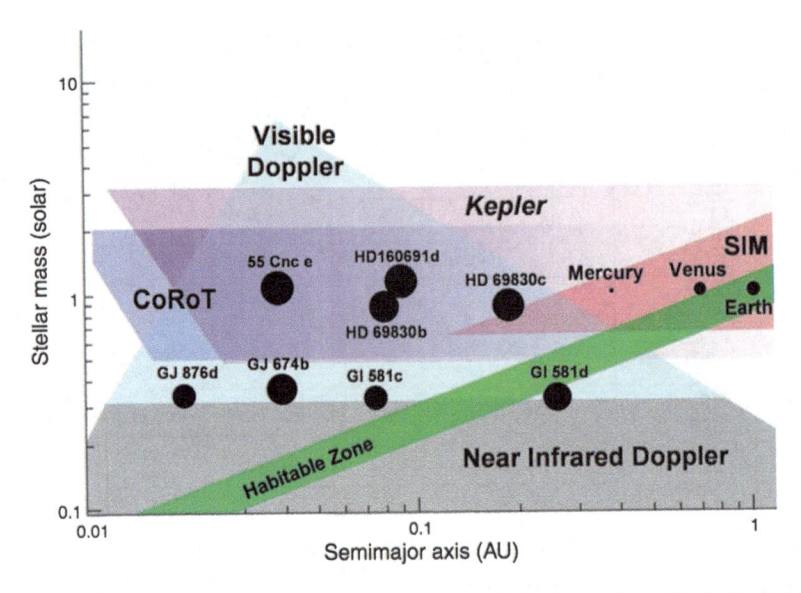

Fig. 31.2 Potential for different techniques, instruments and missions to detect Earth-sized planets in the habitable zones for stars less than a solar mass. The inner solar system and most reported exoplanets with minimum masses less than 15 times that of the Earth are shown as *black circles* (proportional to mass). The *green* zone is the habitable zone in which liquid water is stable on an Earth-like planet (From Gaidos et al. 2007)

In addition to their proximity, extending the exoplanetary census to much lower mass objects will be a powerful and novel avenue for testing planet-formation theories. The favoured core-accretion models predict a large rise in the number of planets below ~ 10 $M_\oplus$, as more massive planets would have rapidly accreted gas and grown to larger masses. Since M dwarfs are fainter than solar-type stars, both the habitable zones and ice lines are located in relatively close proximity to the central stars. In principle this means that data acquired over only a few years can be expected to constrain the migration efficiency and the quantitative features of inner disk regions. The detailed predictions for M dwarfs sensitively depend on a number of factors and are the subject of considerable theoretical modelling (e.g., Currie 2009; Mordasini et al. 2009; Ogihara and Ida 2009). UPF can directly test the current paradigm by having the sensitivity to find these sub-critical rocky cores and to determine the critical core mass, as well as its dependence on host star mass. Similarly, current models predict relatively few 10–100 $M_\oplus$ planets at separations inside 3 AU ("the planet desert"; Ida and Lin 2004). This mass range reflects the planet mass needed to start and end runaway gas accretion, while the outer ~ 3 AU radius is tied to the timescale for gas accretion and the protostellar disk lifetime. Therefore, determining the locus of the planet desert for low-mass stars and brown dwarfs in comparison to solar-mass stars will be a key test of the theory.

Furthermore, M dwarfs are particularly interesting because they offer the possibility to detect terrestrial mass planets in their habitable zone (Fig. 31.2).

The relatively reduced luminosities of M dwarfs means that the habitable zones are much closer to the star. The habitability of these systems is the subject of substantial speculation, e.g., Tarter et al. (2007). Given the exceptionally long main-sequence lifetimes of such low mass stars, any terrestrial planets would have abundant opportunities to develop life.

Scientific Objectives and Deliverables

An infrared RV survey with UPF should survey stars that span a mass range of at least a factor of 5, from <0.08–0.4 $M_\odot$. Four major scientific areas can be addressed by a UPF survey: (1) predictions for giant and terrestrial planet formation, (2) incidence of terrestrial planets in their habitable zones, (3) the identification of the closest planet host stars for direct imaging and (4) characterisation of candidates from transit surveys.

In order to gauge the practicality of conducting a high-precision RV survey in the infrared we intend to construct a notional survey. The input population of ultra-cool dwarfs in the solar neighbourhood is based on the Nstars project (Reid et al. 2004; Cruz et al. 2004; Allen et al. 2006) and the observed IR colours and magnitudes of M and L dwarfs. Our analysis based on the Bouchy et al. (2001) formulation indicates that, for our theoretical models, a S/N of 300 is required to reach a velocity precision of ~1 m/s. As an independent check we have used RV code to extract radial velocities from synthetic spectra and obtained very similar results. On the other hand, the limited real data that we have acquired from HIRES, PHOENIX, NIRSPEC and CGS4 (e.g., Fig. 31.3) suggest that S/N of 125 or less will suffice to reach 1 m/s (noting considerable sensitivity to spectral type and metallicity). In comparison the HARPS time exposure calculator assumes a S/N = 110 for a slow rotating G2V. For simulation purposes a value of S/N = 150 is adopted. As

Fig. 31.3 Keck-HIRES data (*thin line*, S/N > 100, Reiners and Basri 2006) for VB10, the archetypal late-type M dwarf, are shown with synthetic spectra (*thick line*) from Brott and Hauschildt (2005). The poor match between data and observation arises from the lack of high quality molecular line lists appropriate for modelling cool stars

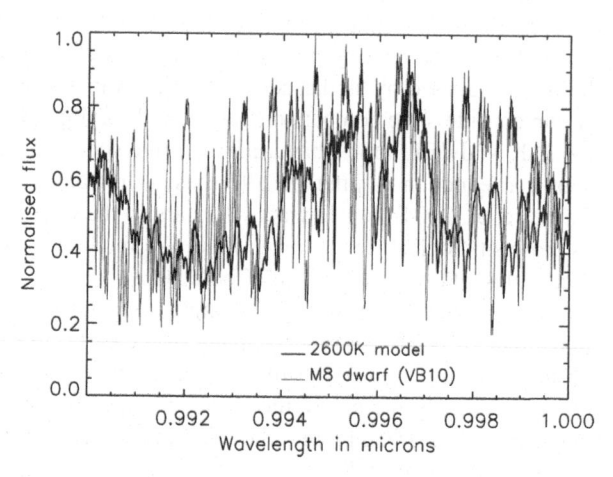

Table 31.1 Illustrates a representative large M dwarf survey

Sp Type	Mass	N (stars)
M2.5 V	0.3	200
M3.0 V	0.24	200
M4.0 V	0.19	200
M5.0 V	0.15	200
M6.0 V	0.12	114
M6.5 V	0.1	37
M8.0 V	0.09	14
M9.0 V	0.08	5
Total		970

It assumes fraction of sky observable = 0.66, fgood an ad hoc parameter which represents the fraction of stars useable for an RV survey (no selection on activity or v sin i), min-max integration per object = 60–4,800.0 s, observing efficiency = 90.0 %, fixed overhead/star/epoch = 60.0 s, avg. # of epochs = 30 (objects with RV signals will be observed more/those without less), max # of targets per LF bin = 200.0, hours of observing per night = 11.0, survey duration = 5.0 year based on 194 nights/year

part of this testing process we also investigated the effect of telluric emission and absorption. Even for extreme cases where telluric features are given motions of 100 m/s, radial velocity information may be recovered. Our simulations are based on 2.5 mm of water and ignoring 30 km/s around telluric features deeper than 2 % (we have not attempted correction, e.g., Bailey et al. 2007). They leave 87 % (Y), 34 % (J) and 58 % (H) of the available UPF wavelength region.

Based on measurements of activity in M stars, 30 % in each luminosity bin are rejected (Rockenfeller et al. (2006). We also reject M stars with v sin i > 10 km/s although our modelling shows that is not necessarily a significant hindrance to measuring precise radial velocities. Based on Wright (2005) there is no evidence for higher radial velocity jitter toward later spectral classes. Indeed we know that RV-scatter caused by stellar activity in M dwarfs is modest, e.g., Reiners (2009). If one considers the comparison between observing a G2 star in the V band and a 2,400 K M dwarf in the J band, for the same spot distribution pattern, the M dwarf shows a reduction in jitter of 1.92 (based on Barnes and Collier Cameron 2001; Barnes et al. 2004). For low-activity stars in the optical a major source of jitter is asteroseismological, for M dwarfs the p–p mode error is expected to reduce by a factor of 4 relative to the Sun (O'Toole 2008).

As expected for a survey spanning a factor of ~ 100 in absolute infrared magnitude, it is easy to observe many of the early-/mid-M types at relatively little cost in telescope time. The required amount of observing time for these bright targets is largely driven by the fixed overhead, not by the integration time, and hence there is a strong premium on minimising the observing overheads (e.g. target acquisition) (Table 31.1).

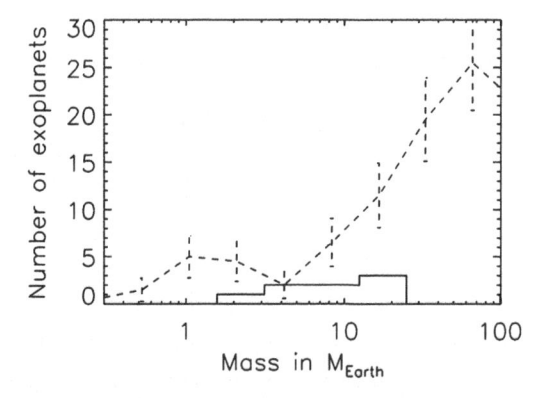

Fig. 31.4 Possible counts (*dashed line*) from a large UPF survey (the known M dwarfs exoplanets from all surveys are plotted as a histogram). The *dashed line* is based on rescaling the Exoplanets Encyclopedia data (exoplanets.eu). All exoplanets with orbital periods of >5 year have been removed. The plot has been rescaled in mass to move from detection around Solar type stars (median value 1.1 $M_\odot$) to detection around 0.2 $M_\odot$ and from 3 to 1 m/s. The expected number has been reduced by a factor of 2 on the basis that UPF will be a single site campaign. Of the UPF exoplanets discovered with <10 $M_\oplus$ we expect that approximately 50 % would be found within their habitable zones (based on the 12/25 known exoplanets with <0.314 $M_{Jupiter}$ being found with 0.1 AU of their host stars). This is likely to be a conservative assumption as simulations suggest that exoplanet mass will scale with host star mass. The plot is based on results from several surveys and is thus drawn from an inhomogeneous sample. The error bars are based on (number)$^{1/2}$ statistics

The key to the characterisation of the most interesting exoplanets is a combination of low RMS error and a large number of epochs. The observational data gathered to date and simulations both stress the importance large numbers of epochs for reliable orbital fitting. For example, Cumming et al. (2008) find that only at 50+ epochs do detection amplitudes become comparable to RMS precision. It can be seen that the most famous and interesting exoplanets have indeed required a lot of epochs to define their orbital characteristics.

The mock survey presented here is intended to be conservative and work is ongoing to design an optimised survey. An allocation of 250 nights per annum for 5 years would allow for the detection of low-mass planets around more than 1,000 of the closest M, L and T dwarfs (Fig. 31.4).

UKIRT – Optimum Location for Near-Infrared Radial Velocity Planet Search

Although UVES is arguably the best radial velocity instrument (e.g., Butler et al. 2004) it does not currently support a long-term radial velocity survey. UKIRT is a natural choice for a number of reasons: (1) infrared optimised telescope, (2) high-speed slew and acquisition, (3) the Mauna Kea location is ideal for minimising the impact of telluric lines – a function of site altitude and precipitable water vapour, (4) a Northern site is preferred since target selection requires knowledge of properties, in particular 65 % of the 800 M dwarfs with known v sin i's (<10 km/s). Furthermore, the nearly lossless transmission of infrared fibres means that UPF is ideal for sharing with other telescopes. Mauna Kea brings particularly attractive potential for linking to Gemini, Subaru and Keck telescopes in order to access fainter primaries.

Other Science

The near-infrared echelle spectrometer CGS4 (on UKIRT) has had a distinguished career (appearing in nearly 1,000 ADS journal entries) being used for an extremely wide range of science. Although UPF has a clearer science focus, it still allows for a wide variety of high-profile science. The stringent instrument requirements for precision radial velocities are also crucial for many other areas; indeed, cases that cannot adequately be addressed using existing facilities such as NIRSPEC (Keck) and CRIRES (VLT) because of their lack of stability, resolution and availability. In addition there are a number of other cases for high resolution infrared echelles, and we refer the reader to 'High resolution Infrared Spectroscopy in Astronomy' (Kaufl et al. 2005), which describes a wide variety of studies of the chemistry, structure, winds and climatology of planetary atmospheres; comets; stellar abundances, pulsations, magnetic fields, disks, in-flows, out-flows, and the chemistry and kinematics of the interstellar medium.

Instrument Design

The instrument is broken into sub-systems, defined so that the interfaces between them are as simple as possible. In many cases the opto-mechanical interface is a fibre that allows a considerable amount of autonomy within the design of each sub-system. The other main interfaces are those to the existing telescope infrastructure. These are made using WFCAM for the fibre pick-off and the Coude room for the spectrograph (Fig. 31.5).

Light from the telescope is re-imaged onto the entrance of an optical fibre located in the Fibre Deployment and Acquisition System (FDAS). This is mounted on top of the WFCAM main cryostat in a replacement field tower section. The object fibre runs from the FDAS through the telescope structure down to the telescope lower floor and into a bench-mounted spectrograph. A second reference fibre runs from a calibration unit located next to the cryostat into the spectrograph. A third calibration fibre feeds light from the calibration unit back up to the FDAS so that calibration light can also be transmitted through the object fibre when selected.

The object and reference fibres are terminated at the cryostat and optically coupled into the spectrograph. There they form a pseudo-slit (by means of an image

31 The UKIRT Planet Finder

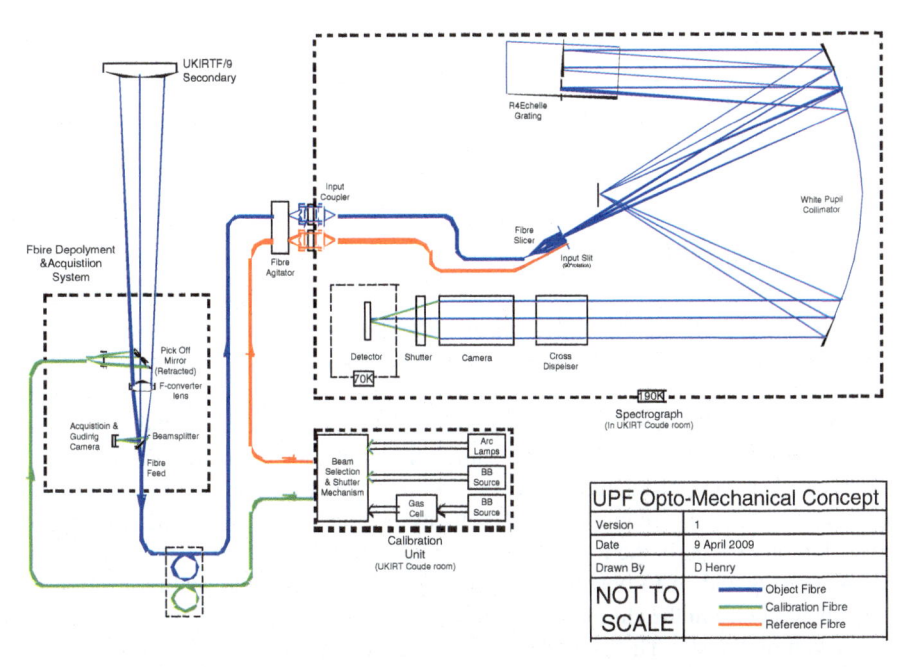

Fig. 31.5 UPF Opto-mechanical Block Diagram. In the FDAS a doublet lens converts the beam to the correct F/ratio for injection into the fibre (f/5.5). Since swift object acquisition is essential for the overall efficiency of the radial velocity survey, and since accurate guiding is needed to meet the required radial velocity precision, the FDAS also includes a dedicated guide camera. This is fed by a beamsplitter, which diverts a small fraction (∼1.5 %) of the beam towards a CCD detector. Light from the calibration fibre can be injected into the main fibre by means of a relay lens and a retractable fold mirror. All components of the FDAS are rigidly mounted together on a common structure, to minimise guiding errors

slicer) at the entrance of the spectrograph. Starlight from the object fibre is dispersed in the spectrograph and forms the object spectrum on which is overlaid a wavelength reference spectrum formed by light from the reference fibre. Radial velocity is measured by measuring the wavelength shift of the object spectrum relative to the simultaneously exposed wavelength reference spectrum.

Two detector systems are used. The spectrograph detector is a 1×2 mosaic of 2048×2048 ∼0.9–1.75 µm devices. For object acquisition and slow guiding a 512×512 CCD camera is used.

Fibre Deployment and Acquisition System

The Fibre Deployment and Acquisition System (FDAS) forms the main opto-mechanical interface between UKIRT and the remainder of the UPF system. It is designed to operate with the f/9 UKIRT secondary mirror. Deployment consists of removing the WFCAM field lens tower (a normal procedure for WFCAM operations) and fitting a second field tower (to which the FDAS is mounted) to the main WFCAM cryostat. When the FDAS is in position observing with WFCAM will not be possible.

Fore-Optics Fibre Assembly

The Assembly (FFA; Fig. 31.6) forms the link between the FDAS and the spectrograph. It comprises the main fibre cable from the FDAS, the optical coupling system for feeding light from the main fibre cable into the spectrograph, the mode scrambler and the fibre slicer. It also includes the fibres that feed from the calibration unit to the spectrograph and from the calibration unit back up to the FDAS.

As well as transporting the light to the spectrograph, the fibre itself plays a critical role in reaching the required radial velocity accuracy by scrambling the spatial distribution of light in the fibre. At a resolving power of 70,000 the slit has a width of 4,300 m/s and so to obtain a velocity precision <1 m/s the spatial position of the photo-centre in the slit must be stabilised to better than 1/4,300 of the projected slit width on the sky. This is achieved by using the spatial scrambling properties of the fibre-feed and by accurate guiding. Single fibres scramble spatial information

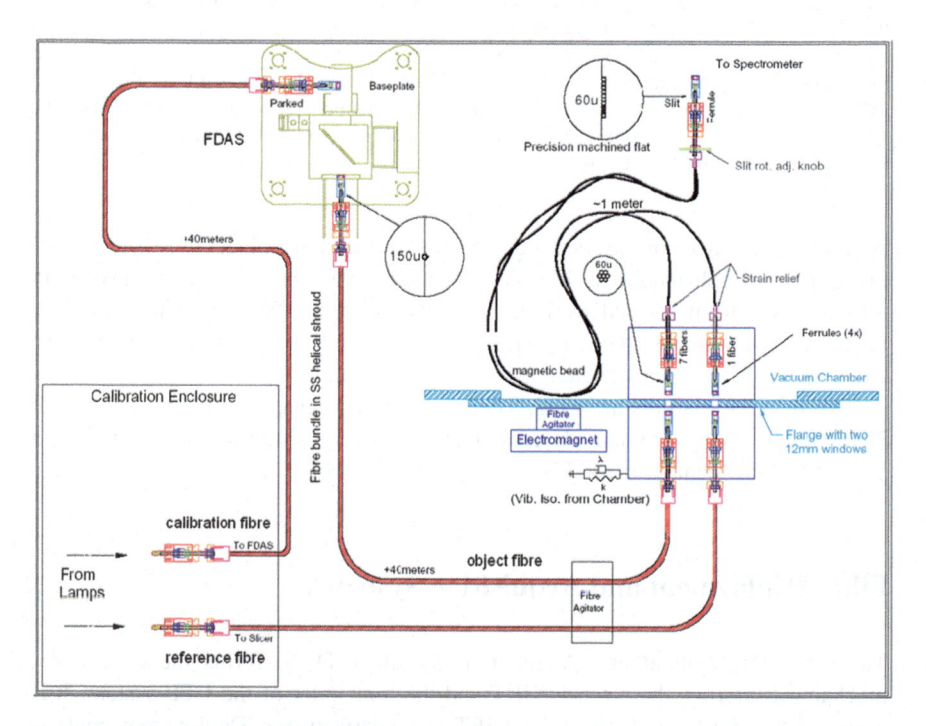

Fig. 31.6 Conceptual schematic of the fore-optics

by a factor of about 20–30. Double-fibre scramblers, which consist of two fibres in series plus transfer optics, scramble spatial information by about 500. However, the scrambling gain is improved at the expense of throughput. Recent experiments (Avila, 2009, private communication) demonstrate scrambling gains in single fibres of 1900 with a 13 % system throughput loss are feasible. We propose to use this technique in UPF.

Great care must be taken in the design of the FFA to minimise the effects of focal ratio degradation (FRD) and modal noise. Experience has shown that FRD losses can be limited to less than 10 % using insertion f/ratios between f/4 and f/6 and by careful construction of the fibre-cables, especially the terminations. For UPF a nominal insertion F-ratio of f/5.5 is assumed. The main fibre-cable is a low-OH fused silica multi-mode fibre with 150 mm core diameter and a numerical aperture of 0.22. The length of the fibre-cable is around 40 m. The two calibration fibre-cables from the calibration unit to the spectrograph and the FDAS are of similar construction. To minimise the effects of modal noise and increase scrambling a mechanical agitator system is used. This vibrates the fibre at a frequency of ~60 Hz with an amplitude of a few hundred microns. Two such agitators are used; one at the entrance to the spectrograph, and one inside the spectrograph close to the fibre slicer.

To achieve the required spatial resolving power it is necessary to use some form of image slicing. For UPF a fibre-slicer will be used. The output end of the main fibre-cable is imaged onto a bundle of 4×60 mm core diameter fibres. These four fibres, together with the calibration fibre and an inactive spacing fibre are arranged into a line and clamped into a ferrule. The output from these fibres forms a pseudo-slit that forms the entrance slit to the spectrograph.

Spectrograph

The UPF spectrograph is a cross-dispersed echelle spectrometer design using a white pupil collimator. The optical design of the spectrograph is similar to other white pupil spectrographs (such as UVES on the VLT, HRS on the HET and, in particular, HARPS on the ESO 3.6 m La Silla telescope). To allow for 10 % focal ratio degradation of the f/5.5 beam entering the fibre at the FDAS, an output f-ratio of f/5 from the fibre-slicer is assumed, and the optical components are sized to allow for an f/5 beam.

The output from the fibre-slicer is fed into a focal reducer lens that changes the f-ratio from f/5 to ~f/13. This beam then passes onto the parabolic collimator mirror (focal length 1,100 mm). This mirror is used in triple pass. Between the 2nd and 3rd passes is an intermediate image plane where a flat fold mirror is located.

The R4 echelle used is from the same grating master as that used in HARPS, UVES and other spectrographs. It has 31.6 lines/mm with an effective blaze angle of 75 deg. It is used in order numbers 35–61. The cross disperser is a 100 line/mm reflective grating used in 1st order with a blaze angle of 4°. It is tilted at 20° to allow the reflected beam to clear the input beam. The camera has a focal length of 450 mm giving an f-ratio of f/5.3. It consists of five lens elements plus a flat detector window. The lenses have all spherical surfaces, and the materials are a mixture of Schott and Ohara standard glass types. The design gives a nominal spectral resolving power of $R \sim 74{,}000$ and sampling of 2.5 pixels per spectral resolution element. Image quality (FWHM) is better than 15 mm across the wavelength range of 1.0–1.7 mm.

Environmental System

The design consists of a vacuum vessel supported on anti-vibration supports of the type used on optical benches. An optical support structure is mounted within the vacuum vessel on an isolating flexure system. The flexure system supports a radiation shield that encloses the optical support structure. It also thermally insulates the optical support structure from the radiation shield. The optical components are mounted within substructure modules, and these in turn are mounted to the optical support structure in a semi-kinematic way. The optical bench and radiation shield are maintained at the operating temperature of 190 K and stabilised to better than 0.05 K by combining a vibration-isolated CTI-1050 closed-cycle cooler with servo controlled resistive heating elements on the optical bench. Liquid Nitrogen plumbing and a dewar is provided to pre-cool the radiation shield and the optical bench. The second stage of the closed-cycle cooler maintains the array mosaic at $\sim$70 K and stabilised to around 0.01 K. A window/feed-through provides the interface for the fibre coupling. There are also breakout panels for the instrument vacuum services, electrical services and detector signal cabling. The total mass of the cryostat is approximately 500 kg. The cryostat bench has temperature control which will maintain the temperature within the requirement at all expected ambient temperatures. However to ensure that there are no gradients induced in the system due to transient temperature changes in the environment, an industrial meat-locker will be constructed around the cryostat to maintain the external temperature to better than ± 1 K.

References

Allen, P.R., Koerner, D.W., Reid, I.N., Trilling, D.E.: Astrophys. J. **625**, 385 (2006)
Bailey, J., Simpson, A., Crisp, D.: Publ. Astron. Soc. Pac. **119**, 228 (2007)
Barnes, J., Collier Cameron, A.: Mon. Not. R. Astron. Soc. **326**, 1057 (2001)
Barnes, J., James, D.J., Cameron, A.C.: Mon. Not. R. Astron. Soc. **352**, 589 (2004)
Bennett, D., et al.: Astrophys. J. **684**, 663 (2008)
Bouchy, F., Pepe, F., Queloz, D.: Astron. Astrophys. **374**, 733 (2001)
Brott, I., Hauschildt, P.H.: In: Turon, C., O'Flaherty, K.S., Perryman, M.A.C. (eds.) The Three Dimensional Universe with GAIA. ESA SPA, vol. 576, p. 525. ESA Publications Division, Noordwijk (2005)

Butler, R.P., et al.: Astrophys. J. **600**, 75 (2004)
Cruz, K.L., Burgasser, A.J., Reid, I.N., Liebert, J.: Astron. J. **126**, 2421 (2004)
Cumming, A., Butler, R.P., Marcy, G.W., Vogt, S., Wright, J.T., Fischer, D.A.: Publ. Astron. Soc. Pac. **120**, 531 (2008)
Currie, T.: Astrophys. J. Lett. **694**, 171 (2009)
Gaidos, E., Haghighipour, N., Agol, E., Latham, D., Raymond, S., Rayner, J.: Science **318**, 212 (2007)
Ida, S., Lin, D.: Astrophys. J. **604**, 388 (2004)
Kalas, P., et al.: Science **322**, 1345 (2008)
Kaufl, H.U., Siebenmorgan, R., Moorwood, A. (eds.): High resolution infrared spectroscopy in astronomy. Springer, Berlin/New York (2005). http://www.springer.com/astronomy/book/978-3-540-25256-6
Konacki, M., Wolszcan, A.: Astrophys. J. Lett. **591**, 147 (2003)
Mayor, M., et al.: (2009)
Mordasini, C., Alibert, Y., Benz, W., Naef, D.: Astron. Astrophys. **501**, 1161 (2009)
O'Toole, S.: Mon. Not. R. Astron. Soc. **386**, 516 (2008)
O'Toole, S.J., Jones, H.R.A., Tinney, C.G., Butler, R.P., Marcy, G.W., Carter, B., Bailey, J., Wittenmyer, R.A.: Astrophys. J. **701**, 1732 (2009)
Ogihara, M., Ida, S.: Astrophys. J. **699**, 824 (2009)
Reid, I.N., et al.: Astron. J. **128**, 463 (2004)
Reiners, A.: Astronomy and Astrophysics. **498**, 853 (2009)
Reiners, A., Basri, G.: Astrophys. J. **644**, 497 (2006)
Rivera, E.J., Lissauer, J.J., Butler, R.P., Marcy, G.W., Vogt, S.: Astrophys. J. **634**, 625 (2005)
Rockenfeller, B., Bailer-Jones, C.A.L., Mundt, R.: Astron. Astrophys. **448**, 1111 (2006)
Tarter, J., et al.: Astrobiology **7**, 30 (2007)
Wright, J.T.: Publ. Astron. Soc. Pac. **117**, 657 (2005)

Printed by Printforce, the Netherlands